Chemistry and Man's Environment

Edward C. Fuller
Beloit College

Chemistry and Man's

Houghton Mifflin Company   Boston

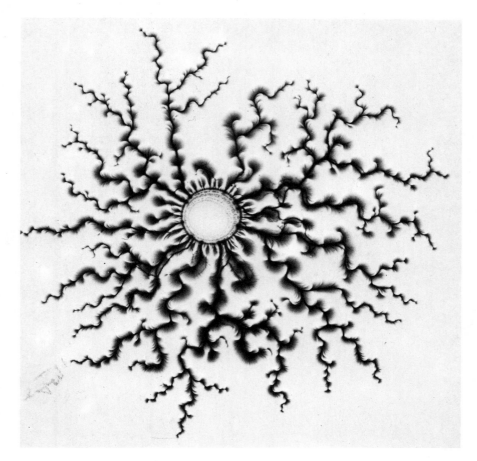

Environment

Atlanta   Dallas   Geneva, Ill.   Hopewell, N.J.   Palo Alto   London

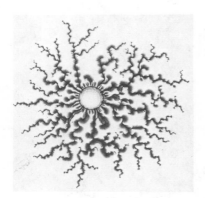

The Lichtenberg figure shows the complex ionization pattern of an electric discharge, similar to the onset of lightning stroke. Courtesy of Prof. A. R. Von Hippel, Massachusetts Institute of Technology.

Printed in the U.S.A.

Library of Congress Catalog Card Number: 73-14019   ISBN: 0-395-17086-9

CONTENTS

Ionization of Acids with More Than One Hydrogen   Summary
New Terms and Concepts   Testing Yourself

PREFACE

This book is for students who want to know more about their environment but who do not expect to spend their lives working as scientists. Because applied science and technology have made such an impact upon our environment in the last few years, the person who wants to understand what's going on has to have some knowledge of basic science. Future businessmen, housewives, lawyers, clergymen, social workers, teachers, politicians—indeed, all citizens—need to understand the potentialities and the limitations of science so that they can vote for the people and the issues which will maximize the good and minimize the bad applications of science to contemporary problems.

Though environmental studies involve many aspects of biology, geology, and physics, solutions to the problems of air, water, and land pollution depend most heavily on the applications of chemistry. If the nonscientist-citizen learns a few basic chemical concepts, he can act more intelligently when faced with issues involving the production of energy, the proper use of natural resources, and the disposal of wastes. The chemical concepts discussed in this book are chosen, then, to form a matrix within which many environmental problems can be understood. A student whose major interest lies outside science need not be acquainted with specific reaction rates, free energy and equilibrium constants, oxidation-reduction potentials, molecular orbitals, etc., in order to grasp the fundamental chemistry of many processes in the environment.

The book is of a length suitable for one semester or two quarters. Further, it may be used with considerable flexibility. For example, some instructors may wish to treat the chapters on chemical concepts in depth, and to spend less time on those concerned with the environment. Others may prefer less emphasis on chemical principles and more on their application to the solution of environmental problems. Previous instruction in chemistry is not essential, although the student who has already studied elementary chemistry will still find much in this book that is new to him.

This book has grown out of my experience in teaching chemistry to students whose major interest lies outside science. My students have contributed constructively to repeated revisions of the text and have done much to make it more useful and more interesting. Discussions with my fellow teachers at meetings of the Division of Chemical Education of the American Chemical Society have helped in many ways. I am especially indebted to Avrom A. Blumberg (DePaul University, Chicago), Alfred B. Garrett (Ohio State University), and William T. Mooney, Jr. (El Camino College), who read the entire manuscript and made scores of valuable suggestions for its improvement.

Christine Correra was a mighty help in the typing of the several revisions of the manuscript which culminated in the final copy. This book would never have come to print but for the editorial help, skillful proofreading, bibliographic assistance, and constant encouragement of my wife, Dorothy. Working together has made writing the book an exciting enterprise rather than a grueling task.

Edward C. Fuller
Beloit, Wisconsin

Chemistry and Man's Environment

Fern fossil in coal. Wise utilization of fossil fuels, such as ethane and
methane shown here, is one of man's greater technological challenges.

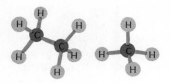

1. Population ⟶ Production ⟶ Pollution

How much pollution did you just inhale? Did it smell good? Did you enjoy the "chlorine cocktail" you gulped from the drinking fountain before you came to class? Did you get any detergent foam with it? How did you enjoy the scenery on your vacation last summer—the smoldering dumps, the bottles, cans, and papers littering the highway, the auto graveyards, the strip mines? Did you find a sign saying "Danger. Polluted Waters. No Fishing, No Swimming, No Boating, No Waterskiing" when you got to Clear Lake?

How did we get into this mess? Can we get out of it? How can we get a breath of fresh air without having to go to the country for it? How can we get a fresh-tasting drink of water without having to buy it in a bottle filled at a mountain spring? How can we get clean streets, highways, and parks? How can we get the coal for our power plants and the iron for our steel furnaces without turning the countryside into a nightmarish moonscape? Is there any hope?

Yes, there is hope. Hope based not on wishful thinking but on scientific fact. Through our creativity we have developed the wonders and delights of modern civilization, but at serious cost to our environment. We are in danger of destroying a delicate relationship with nature that must be maintained if we are to survive. We must use the same creativity and knowledge to restore the balance. This book shows ways in which this can be done. But before we can learn how to get out of our present predicament, we must first learn how we got into it.

1.1 How We Got into Our Present Predicament

environment

We got into the spot we're in because we have been living off nature's bounty for several thousand years without any thought of what we were doing to our *environment*.* In ancient days man hunted and fished and gathered plants for his food and clothing. He found protection in caves and in crude shelters he built. Later he learned to domesticate animals, plow the land, and plant crops. As his skills in agriculture and animal husbandry improved, he was able to live in villages, towns, and cities. With life in cities came health problems which had not existed when men lived in small and isolated groups. Crop failures and epidemic diseases prevented the human population from growing very rapidly in primitive societies. Man used only tiny fractions of such natural resources as forests, metal ores and minerals, animals, and fish; he had very little influence on natural processes such as the falling of rain, the flowing of rivers, the eroding of land, the growth and spread of plant life, and the birth, growth, and death of animals. He was able to dispose of the waste products from his way of life without worrying about their effect on his surroundings. The natural resources remaining dwarfed the amounts he had removed. As long as man's activities were puny compared to the processes of nature, no impact was noted.

technology

About 300 years ago, man began to step up the pace of his investigation of the world, creating the body of knowledge we call science. He began to apply his new knowledge to the invention of things and processes that would make his life easier. This application is *technology*, and its growth, along with that of its partner, science, has mushroomed so much that in this century scientific knowledge has doubled every 10 or 15 years. Scientific discoveries and technological inventions have given us stainless steel and synthetic rubber, LP records and snowmobiles, transistor radios and Boeing 747s, television and DDT, plastic bags and insulin, nylon and napalm.

energy

The industries required to make these products have grown so rapidly that today in the United States our 7 percent of the world's population is using 50 percent of the world's production of raw materials. Americans have consumed more metal ores and mineral fuels in the past 40 years than were consumed by the entire human race up to that time. Our production of *energy* is very high and rising rapidly. The Egyptian Pharaohs who built the great pyramids commanded the labor (energy) of 100,000 slaves. The glories of Athens in its Golden Age were supported by the labor of 125,000 slaves. Life in the United States today is smoothed by energy equivalent to the labor of 40 *billion* slaves. To keep us warm, manufacture our goods, grow and process our foods, transport these necessities and luxuries to our homes, and entertain

*New terms and concepts which appear in color in the margin are defined in an alphabetical list at the end of the chapter in which they are discussed.

"Have you given any thought to what you'll do with your Saturdays when the world's fossil fuels are used up?"

us in our leisure hours, each of us commands the energy equivalent to the work of 200 full-time slaves.

1.2 The Problem of Population

The rate at which we have been producing things to make life easier has grown explosively in this century, as has the number of people to use these things. In the early days of the human race, epidemic disease, famine, and malnutrition kept the population in check. When man learned to combat these plagues through science and technology, the population began to soar (see Table 1.1 and Fig. 1.1). The population of the United States was 100 million in 1918 and 200 million in 1968; by 2000 it will probably be 300 million.

1.3 Production

A hundred years ago three people working on a farm produced enough food and fiber for four people, which meant that only one in four of our population could do any work other than agricultural. Today in the United States

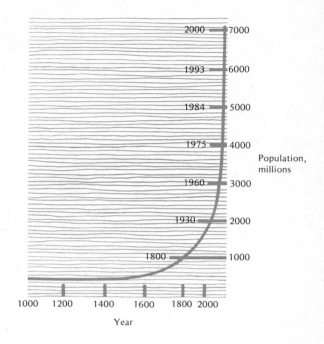

FIGURE 1.1
Population growth in
the last 1000 years.
(From Revelle, Khosla,
and Vinovskis,
*Survival Equation:
Man, Resources, and
His Environment,*
Houghton Mifflin
Company, Boston,
1971, p. 6.)

modern agricultural technology is so efficient that one farm worker can produce enough to meet the needs of twenty people. By studying the details of plant and animal growth we have learned how to produce a dozen ears of corn where only one grew before. By studying the habits and reactions of insects we have learned how to save millions of tons of grains, vegetables, fruits, and meats from these six-legged competitors for our food. By harnessing the gasoline engine to our farm machinery, we have vastly increased the area of land that can be cultivated by one man.

1.4 Pollution

pollutant

Now, suddenly, within the last 10 years, we have been confronted with the results of our growing population and our expanding science and technology. As a consequence of our numbers and our way of life, in 1970 we spewed 142 million tons of *pollutants* into our atmosphere, dumped 300 billion gallons of waste-bearing water into our running streams, threw 200 million tons of garbage and trash into open dumps, and junked 8 million cars. We are running out of fuels, metals, building materials, water, and even land good enough to grow food. The blessing of dramatic increases in food production has been

TABLE 1.1 *World Population Growth*

Time in history	Date	Population
Neolithic man	7500 B.C.	10 million
Beginning of the Christian era	A.D. 1	250 million
Beginning of the scientific era	1650	500 million
	1950	2.5 billion
	1970	3.5 billion
Predicted for	2000	6 to 7 billion

followed by the curse of rivers laden with silt washed from our most productive soils, lakes dying from the rapid growth of algae resulting from overdoses of fertilizer leached from the land, and farm workers and animals sick from exposure to pesticides. Cities shrouded in smog, rivers choked with sewage, and littered landscapes are driving home to us the lesson that no creature can long foul its own nest without a penalty.

1.5 The Relationship between Man and Nature

dynamic
equilibrium

Why did we go wrong? We went wrong because we failed to realize that man must live in balance with his environment. The chemist has a special term for this kind of balance. He calls it *dynamic equilibrium,** and he uses it to describe what happens when two opposing processes are going on at the same time but make no overall change in the surroundings. What is done by one process in a short time is exactly undone by an opposing process in the same time, so that there is no net change in the situation as a whole. For example, if you are walking down an escalator that is moving up, you will stay at the same position between floors if your downward speed exactly equals the upward speed of the escalator. You will then be in a state of dynamic equilibrium with respect to motion between floors. Both your downward motion and the upward motion of the escalator are proceeding, but there is no observable change in your location.

Again, if you empty a tray of ice cubes into some water in an insulated ice bucket, after you keep them there awhile, some of the cubes will join to form irregularly shaped chunks of ice. Evidently, melting and freezing are going

*Chemical equilibrium is discussed in detail in Chap. 14.

on simultaneously; the cubes melt in some places, and the water between them freezes in others. The ice and water are in dynamic equilibrium.

The present crisis in our environment has come about because, in our increasing numbers and with our increasing demands for natural resources, we have taken so much from the earth and have done so much damage to it that the earth is now damaging us. If we are to continue as a species, we must balance our withdrawals by returning materials to the environment in the form of new resources. We must modify our impact so that the balance remains favorable to us.

To maintain a dynamic equilibrium, we don't have to stop using nature's gifts, but we must give back as much as we take. We don't need to stop the processes of civilization, but we must modify them so that they do not interfere with the *natural* processes on which they depend. We must come into dynamic equilibrium with nature. This is the main theme of this book. It will be repeated many times.

1.6 The Environment and Biological Success

If 50 years ago you had suggested to the ordinary man in the street (call him Omits) that man as a species might someday become as dead as the dodo, Omits would probably have laughed at you for even considering such a preposterous notion. If he had studied his high school biology, he would have told you that man had more than met the standards for biological success, which are

1 to be able to live in a wide variety of environments
2 to dominate other species
3 to increase in numbers
4 to adapt to changes in the environment

In terms of these standards, man has been a smashing success. He lives all over the globe, he dominates all other forms of life, and indeed, he has hunted some of them to their extinction. His numbers are growing explosively. But today the danger signals are up. Clearly, new criteria of biological success must be created and met if Homo sapiens is to live up to his name of "man, the wise." The new standard should be that the successful organism is that one which establishes a dynamic equilibrium with its physical and biological environment.

1.7 The Threatening Ecocatastrophe

The existence of a dynamic equilibrium means that the organism draws from the environment only the amounts and kinds of essential materials that can be replaced at a rate equal to their withdrawal. Otherwise, either the withdrawal must be decreased or the rate of replacement increased. For instance, it is estimated that we are now using up fossil fuels (coal, petroleum, and natural gas) at a rate 1 million times the rate of their formation. Since supplies of these resources are finite, they will be exhausted sometime. However far in the future this may happen, Homo sapiens, if indeed he is wise, must start looking *now* for ways of decreasing his drafts on these irreplaceable resources by developing new sources of energy (see Chaps. 9 and 11). Again, as we continually increase the tonnage of fish we take each year from the oceans, we must learn how fast they replenish themselves and what we can do to speed that process lest we destroy them forever.

Today we are using the earth's resources much faster than they can be replaced by nature alone. We must learn to substitute reclaimed materials for new materials. For instance, extraction of metals from natural ores must be heavily supplemented or even replaced by "mining" scrap metal from the mountains of trash that threaten to engulf the technological cultures of the world. We must learn to make more and better paper from reclaimed waste and less and less from new-cut forests.

Overpopulation and overproduction have thrown us out of balance with our environment, and helped to cause the mess we are in. As we try to restore the equilibrium, what of the future? Figure 1.2 shows the rates at which some parts of our economy are expected to grow in the next 30 years, based on their growth in the past 20 years. If this prediction is right, by the year 2000 we will have 4 times as many trucks on our highways as we have today; we will use almost 4 times as much electric power; build 3 times as many houses and 3 times as many cars; use 3 times as many petroleum products, 2.5 times as much paper, and 2.5 times as many cans of beer.

Recently an international team of industrial managers, scientists, and educators used a computer to study the interplay of growth in population, food consumption, use of natural resources, industrial output per capita, and pollution. Their grim conclusions are shown graphically in Fig. 1.3. On the basis of their studies they predicted that "if current physical, economic, and social relationships continue unchanged, nonrenewable natural resources will be exhausted, followed eventually by a soaring death rate and a rapid drop in world population." If, they said, 75 percent of all resources are recycled, if nuclear power is rapidly developed, if pollution is reduced to 25 percent of the 1970 value, if the yield of food is doubled, and if some form of worldwide birth control is effected, then the expected drastic reduction of population

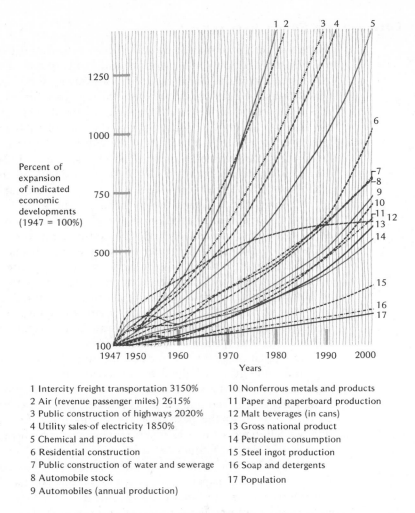

Percent of
expansion
of indicated
economic
developments
(1947 = 100%)

Years

1 Intercity freight transportation 3150%
2 Air (revenue passenger miles) 2615%
3 Public construction of highways 2020%
4 Utility sales of electricity 1850%
5 Chemical and products
6 Residential construction
7 Public construction of water and sewerage
8 Automobile stock
9 Automobiles (annual production)

10 Nonferrous metals and products
11 Paper and paperboard production
12 Malt beverages (in cans)
13 Gross national product
14 Petroleum consumption
15 Steel ingot production
16 Soap and detergents
17 Population

FIGURE 1.2 Actual and anticipated growth rates of certain goods and services during the last half of the twentieth century. (Figure redrawn from David Archbald, "The Population Pollution Syndrome," *The Journal of Environmental Education*, vol. 3, no. 1, Fall 1971.)

will not take place until some time in the second half of the twenty-first century. If we are to postpone or prevent the collapse predicted for 100 years from now, we must make even greater efforts in the years immediately ahead.

1.8 What Can We Do?

Do you want a better life for yourself and the assurance that your children and grandchildren will be able to enjoy their lives? If so, there are some things you must do now. You must understand the problems we all face, you must want to do something about them, you must learn enough about solutions to the problems so that you can make wise decisions as a voter and as a consumer, and you must be willing to change your life-style and redefine your values.

For hundreds of years Western man has worshiped ''progress,'' and we Americans have been especially devout. The poet Stephen Vincent Benét expressed what he regarded as the true spirit of America when he wrote (in the poem ''Western Star'' in 1943)

> We don't know where we're going, but we're on our way!

and

> Americans, who whistle as you go!
> (And, where it is you do not really know,
> You do not really care).

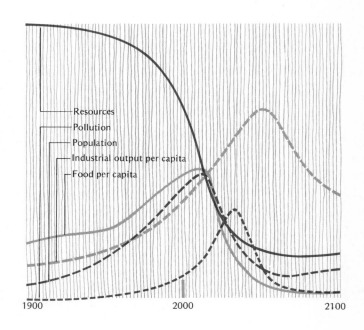

FIGURE 1.3 Production, consumption, pollution, and population in the twentieth and twenty-first centuries. (Figure redrawn from D. H. Meadows, D. L. Meadows, J. Randers, and W. W. Behrens, *The Limits to Growth*, Universe Books, New York, 1972.)

Kenneth Boulding has called our present system a "cowboy economy." We operate like the reckless rider of the giant herd on the illimitable plains, with all the exploitive, romantic, and violent behavior the image conjures up. Boulding contrasts this with the "spaceman economy" that must prevail on this single spaceship earth, where all resources are in limited supply and the recycling of materials is necessary. We must change our ideas about what we need if we are to live a good life. Then we must redirect our massive tools of production to fill these readjusted needs.

1.9 What Do We Need and How Can We Get It?

Clearly, our fundamental needs are breathable air, drinkable water, edible food, and effective clothing and housing to protect us from the weather. We also *want* a world in which we can live with joy, and not just grub out an existence. Technology can provide the necessities. To preserve a good environment will require social and political action.

At the 1972 United Nations Conference on the Human Environment, Lord Zuckerman, former chief scientific adviser to the United Kingdom, said

> The worst enemies of the environment, as well as of mankind, are poverty and ignorance. . . . Pollution is essentially a social and political problem. For the first time we know what it is we are doing as our influence spreads over the globe. It is unscientific to speak as if technological advance is working inexorably toward making a worse world and, in particular, that we are going to be unable to deal with our waste products. I regard the environmental dangers which we face as far more manageable than I do the social and political problems which exist in the world. The physical and measurable problems of the environment are matters with which science and technology are competent to deal. But they cannot deal with value systems which determine which amenities should be preserved. These belong to the political domain where all citizens have a voice.

Before we can use our voices as citizens in the solution of our environmental problems, we must have knowledge. Our immediate job is to learn all we can about how smog is produced and how we can get rid of it; how water gets polluted and how we can prevent it; how food can be raised with the aid of insecticides and fertilizers without contaminating our air and water; how power to run the immense industrial machine required to manufacture the needs of life can be generated without poisoning both air and water with radioactive killers, filling the air with corrosive smoke and gases, and fouling

environmental
science
environmental
technology

our rivers with hot water. In short, we must create a new science—*environmental science*. And we must develop a new technology—*environmental technology*.

In 1971 the National Science Board urged that the federal government immediately launch a large-scale program to support research in environmental science. The board also had this to say about education:

> Of special importance to implementing a national program for environmental science is the existence of an informed citizenry, both as a source of future scientists and as the necessary basis for national understanding and motivation of the entire program.

This book is designed to help produce an "informed citizenry." It has two objectives:

1 to present the chemical knowledge required for understanding our environmental problems
2 to inspire the reader to want to help solve these problems

1.10 What Is Environmental Science?

According to the National Science Board,

system

ecology

> Environmental science is conceived . . . as the study of all the *systems* [italics added] of air, land, water, energy, and life that surround man. It includes . . . meteorology, geophysics, oceanography, . . . *ecology* [italics added], . . . physics, chemistry, biology, mathematics, and engineering.

1.11 What Are Science and Technology?

The body of knowledge that man has accumulated about the universe is called science. The objective of science is to interpret and understand the universe. When man first began to understand happenings (like fire and lightning) and things (like rocks, water, plants, and animals), he began to apply his knowledge to make new fabrics, foods, medicines, alloys, fuels, etc. This applied science is called technology.

When primitive man first had time left over from finding food and shelter, he wondered about the world around him. He saw many kinds of plants and animals; he saw rocks, earth, valleys and mountains, springs, rivers, lakes, and oceans. He learned to cope with rains, snows, floods, fierce winds, fires,

The Gateway to the West in St. Louis: a symbol of progress in the midst of urban decay.

earthquakes, and volcanic eruptions. For thousands of years people developed the skills and accumulated the knowledge that enabled them to survive, and they passed this wisdom on by word of mouth. With the invention of writing about 7000 years ago, man became able to store up this knowledge and hand it on more effectively. The invention of printing in Germany in the middle of the fifteenth century set off in the Western world a revolution in the acquisition and transmission of knowledge which still profoundly affects our daily lives.

When I can record my observations of the world and compare them with the printed record of what others have seen, there is the beginning of natural history—the accumulation and classification of knowledge about nature.

When I can test my theories about natural phenomena against those others have published, there is the beginning of natural philosophy—the systematic interpretation of natural history. When most investigators agree on the facts they have observed and on theories to interpret them, there is, in these facts and theories, the beginning of *natural science*. Modern natural science, then, is the sum total of human wisdom about nature that has been generally accepted by those who are familiar with specific phenomena or problems. No scientist knows all of science. His acceptance of science as a whole is based on his faith in his fellow scientists.

natural science

Understanding and interpreting the universe is a huge job. The scientist simplifies this job by classification. The classification of living things as plants or animals, the classification of animals as birds, reptiles, fish, mammals, etc., makes it easier to organize knowledge about nature. Out of the natural history and natural philosophy of the eighteenth century came the more limited natural sciences of the nineteenth: biology, chemistry, geology, physics. When no one man could master a whole science, subdisciplines and specialties like acoustics (in physics), organic chemistry (in chemistry), vertebrate zoology (in biology), and paleontology (in geology) blossomed in increasing numbers. Later came interdisciplinary subjects like biochemistry and geophysics. Today there is a reversal in the trend toward compartmentalization in the sciences. Environmental science, space science, moon science, and oceanography are developing out of the cooperative efforts of scientists from different traditional sciences (Fig. 1.4).

1.12 What Is Chemistry?

chemistry

Chemistry is bread baking, wine fermenting, the gasoline in your car engine burning. Chemistry is the science concerned with the composition of materials, the changes they undergo, and the energy associated with these changes. By the use of an electric current we can break down water into hydrogen and oxygen. We can burn hydrogen in oxygen to produce water and heat. And so we conclude that water is composed of hydrogen and oxygen, and that energy is liberated when water is formed and consumed when water is decomposed. Chemistry is also concerned with the structures of materials, the arrangement of the parts that make up complex patterns. By using methods described later in this book, we can determine that water has a structure like this: $H-O\diagdown_H$, where H stands for a hydrogen atom and O for an oxygen atom, and the bond lines indicate that these atoms are held together in this pattern. We shall consider the evidence for the existence of atoms in Chap. 3.

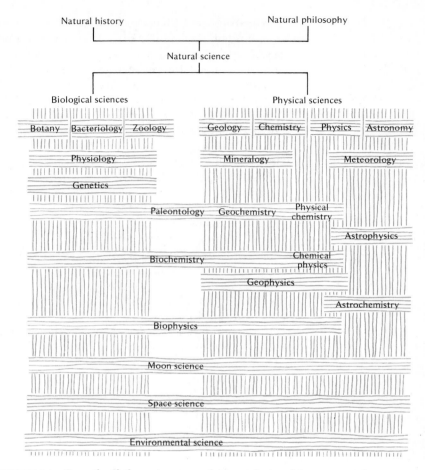

FIGURE 1.4 Growth of the sciences and interrelationships among them

physical properties

chemical properties

To recognize changes which take place in a material, we need to describe it before and after the change. We call descriptions such as color, texture, melting point, boiling point, etc., the *physical properties* of the materials. We call the behavior of the material when it changes into a different substance its *chemical properties*. For instance, iron is a gray, shiny solid that melts at a high temperature; these are some of its physical properties. When we expose iron to moist air, it becomes covered with a powdery red material we call rust. The interaction of iron with moist air is a chemical property.

Chemistry, then, is the sum of our knowledge of the physical and chemical properties of materials, their structures, and the processes by which they are changed into different materials.

1.13 How Chemistry Fits into Environmental Science

Understanding cross-disciplinary sciences like environmental science requires that we build on the basic concepts of the fundamental sciences—biology, chemistry, geology, and physics. This book develops some concepts in chemistry that are basic to an understanding of our physical environment. It will help you answer general questions: What is the world around us made of? How is it put together? How do some things change into others? How can we stop these changes if they are undesirable, or speed them up if we wish? Can man bring about changes that do not occur in nature? This book will also help you answer more particular questions: What are the pollutants in our air and water? How can we get rid of them? Can't we solve many of our pollution problems by just not using insecticides and fertilizers? Can't we clean up the mess by "going back to nature" and throwing out our technological polluters? What can we do with trash? How can we protect our natural resources without abandoning modern technological culture? Can we turn pollutants into assets? Is worldwide starvation inevitable? How might it be prevented? What can the citizen who is not a scientist contribute to the improvement of the environment?

The chemistry involved in making use of solid wastes, preventing air and water pollution, increasing the generation of electric power without more pollution, and improving food production is discussed in some detail in Chaps. 4, 7, 11, 13, and 21.

Science and technology have grown together, and now together they must take a new turn, away from unlimited expansion and heedless consumption of what they grow on, to tidying up the mess they've made and to careful nurture of scarce resources.

New Terms and Concepts

CHEMICAL PROPERTIES: The behavior of a substance when it changes into other substances.

CHEMISTRY: The science concerned with the composition of materials and the changes in composition that they undergo.

DYNAMIC EQUILIBRIUM: If one process taking place in a system offsets another so that there is no net change, the processes are continuous but in dynamic equilibrium. (Remember the escalator.)

ECOLOGY: The science of the relationships between organisms and their environments.

ENERGY: The capacity for doing work.

ENVIRONMENT: The total of conditions surrounding an organism or group of organisms; the complex of social and cultural conditions affecting the nature of an individual or community.

ENVIRONMENTAL SCIENCE: The study of the systems of air, land, water, energy, and life that surround man.

ENVIRONMENTAL TECHNOLOGY: The application of environmental science to improve the quality of the environment.

NATURAL SCIENCE: The sum total of human wisdom with respect to nature that has been generally accepted by those who are familiar with specific phenomena or problems. A science that deals with matter, energy, and their interrelations or with objectively measurable phenomena.

PHYSICAL PROPERTIES: The characteristics of a substance, such as its color, texture, melting point, boiling point, conductance of heat or electricity, etc.

POLLUTANT: Any undesirable substance that is present in air, water, food, or solid waste; the presence of undesirable heat is considered thermal pollution.

SYSTEM: A group of interrelated units operating together as a whole.

TECHNOLOGY: The application of science to industrial or commercial objectives.

Testing Yourself

1.1 How has technology in the United States made life better for Omits?

1.2 In what ways has technology made life worse for Omits?

1.3 "People today are no happier than their forefathers, but they do not suffer as much." Do you agree?

1.4 "Man's greatest failure is his biological success." Discuss.

1.5 Cockroaches have existed longer than almost any other kind of organism now living. Are cockroaches more successful than man?

1.6 On the basis of the curves shown in Figs. 1.1 and 1.2, what would you say is the most critical change we need to bring about in our society between now and the year 2000?

1.7 Do Boulding's concepts of the cowboy economy and the spaceman economy help us understand our present predicament? If so, in what ways must we modify our cowboy economy if we are to achieve a spaceman economy by the year 2000?

1.8 What is the relation between environmental science and the natural sciences?

1.9 In what ways is environmental science related to the social sciences?

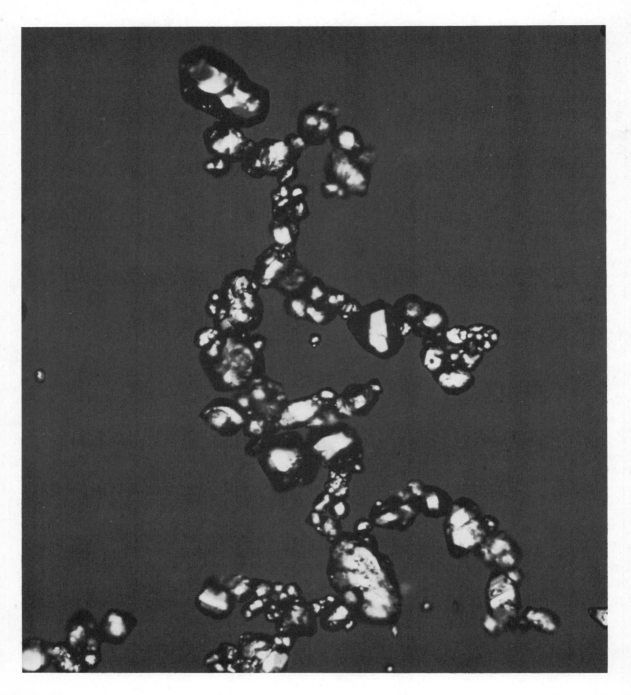

Compounds of the element sulfur, whose photomicrograph
and molecular structure are shown here, are major pollutants.

2. How the Chemist Simplifies the Problem of Understanding the World

What is the world made of? Rocks and lakes and trees and people look very different. Are they made of entirely different substances? Is there a different kind of substance for every different thing? Is there any kind of substance in you that is also in a rock or a lake? How is a rock like a piece of ice? How is it different? Are all rocks alike? Are they made of the same material? If they are not, what is the difference between the kinds of substance in different kinds of rocks? What is snow? When snow melts, we get water. Why? When water stands in open air, it disappears. Where does it go?

These are the kinds of question man has been trying to answer ever since he has had time to wonder and to think. Because we are surrounded by so many different things in nature, we have to fit them into some scheme of organization if we are to make sense of them. This chapter describes how chemists organize our knowledge of the substances in the world around us.

2.1 Matter and Energy

The job of interpreting the world is a big one, but it can be made easier by a method that chemists learned early to use: *simplification by classification*. The chemist observes that the universe is made of two great entities, matter

matter

and energy. He defines *matter* as that which occupies space, has weight, and can be perceived by one or more of our five senses. Rocks, trees, houses, the ocean, and the gas that bubbles up from the bottom of a marshy pond are composed of various kinds of matter. You can see and feel rocks, trees, houses, and the ocean. You can taste some kinds of rocks and the ocean. You can smell the gas from a marshy pond.

The scientist defines energy as the ability to do work. Forms of energy such as heat, light, and electricity can do work when they are harnessed by machines such as steam engines, solar heat engines, and electric motors. When you do physical work, you transform some of the chemical energy in your body into mechanical motion by means of various chemical processes in your muscles.

2.2   The States of Matter

solid
volume

liquid

gas

states of matter

You may conveniently classify matter further as solid, liquid, or gas. A *solid* has both a definite *volume* (occupies a definite space) and a definite shape. A *liquid* has a definite volume but takes the shape of its container. If the volume of the liquid is less than the volume of the container, the liquid fills only the bottom part of the container. A *gas* has neither a definite volume nor a definite shape, and it completely fills its container; if it is not kept in a closed container, it will escape. We call these categories the *states of matter:* the solid state, the liquid state, the gaseous state.

2.3   Homogeneous and Heterogeneous Matter

homogeneous
heterogeneous

A solid may be *homogeneous* (have the same composition throughout, like a piece of glass) or *heterogeneous* (show differences in composition in different places, like a chunk of concrete that contains pieces of rock held together by cement). A liquid such as raw milk is also heterogeneous, because droplets of butterfat (cream) will separate from the milk and rise to the surface. We call muddy water a heterogeneous liquid because we can see the dirt settle out when the liquid is allowed to stand undisturbed. Tea, however, is a homogeneous liquid, because its color and taste are the same throughout. Clean air is a homogeneous gas, but dusty air is heterogeneous. If we pass dusty air through a porous cloth (as in a vacuum cleaner bag) we separate the solid dust from the gaseous air. If we pass reddish-brown nitrogen dioxide gas into a bottle of air, it mixes uniformly with the air to produce a homogeneous gas. Mixtures of gases with no solid or liquid particles in them are always homogeneous.

If we stir some sugar into a cup of coffee, it disappears; we have a homo-

	geneous liquid that has the color and taste of coffee and the sweet taste of
mixture	sugar. This is a *mixture*—a material composed of more than one substance
components	in which each substance shows the properties it has when by itself. The *components* of the mixture (coffee and sugar) have not been changed by mixing.
dissolve	The sugar has *dissolved* in the coffee to form a *solution*, a homogeneous mixture. The solutions most commonly encountered in the environment are
solution	ture. The solutions most commonly encountered in the environment are
solvent	homogeneous mixtures containing mostly water. We call the water the *solvent* in the solution. In syrup, the water is the solvent and sugar is the *solute*.
solute	*vent* in the solution. In syrup, the water is the solvent and sugar is the *solute*.
soluble	We say that sugar is *soluble* in water.

2.4 Pure Substances

No matter what store we buy it in, sugar is sugar—a white, granular, sweet stuff. If all the samples of a given material have the same physical and chemical properties, we call it a *pure substance*. Purity in this sense has nothing to do with suitability for use as a food. Both sugar and the substance commonly called "arsenic" are white, granular solids, but the resemblance ends there. Pure sugar is a good food, pure arsenic is a poison familiar in murder mysteries. (Remember *Arsenic and Old Lace?* On the other hand, we would not call pancake syrup a pure substance, for though it is quite suitable as food, different brands contain different percentages of sugar and water. Likewise "pure" milk is not a "pure" substance, for different shipments contain different percentages of butterfat.

pure substance

How do you know when you have a pure substance? All samples of a given pure substance have the same properties. All samples of pure water are colorless, odorless, tasteless, and freeze at 0°C (Celsius). (Temperature scales are discussed in Sec. 5.2.) A watery-looking liquid that has color or odor is not pure water. A colorless, odorless liquid that freezes at some temperature other than 0°C is not pure water. Seawater is colorless and odorless but has a salty taste and freezes at a temperature below 0°C. It is not pure water.

2.5 Separating Pure Substances from Mixtures

steam

Can you get pure water from seawater? If you put some seawater in an open dish and boil it, *steam* is given off. If you boil it dry, a salty white solid is left in the dish. Apparently seawater is a solution of salty solids in water. If you hold a cold object in the steam from the boiling seawater, droplets of liquid collect on it. This liquid is colorless, odorless, and tasteless. Is it water? To answer this question, you need more of the liquid so you can see whether it freezes at 0°C. To get more liquid, you can set up the apparatus shown in Fig. 2.1.

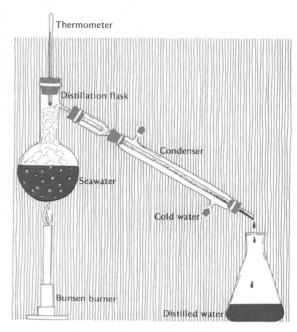

FIGURE 2.1 A still for purifying water

Heat the seawater in the round-bottomed flask (called the boiler) to boil the water and drive steam into the tube slanting down to the right. This tube is surrounded by cold water. When the steam passes through the tube, it is cooled and forms droplets on the inner walls. This process is called *condensation*, and the part of the apparatus in which it takes place is the *condenser*. The entire process of boiling and condensation is *distillation*, and the apparatus is called a *still*. You put heat energy into the liquid water in the boiler and produce gaseous steam. Then you take heat from the steam in the condenser and produce liquid water. The condensation of steam in the condenser is the reverse of the *evaporation* of water in the boiler.

The *freezing point* of the distilled water is 0°C. This proves that you have separated pure water from seawater. The process of distillation is widely used on ocean-going vessels to provide drinking water from the sea for the passengers and crew. Distillation is also used to separate gasoline, kerosene, and fuel oils from crude petroleum.

Can you get drinking water from a muddy river without distillation? Yes, you can separate mud from water by *filtration* (Fig. 2.2). A conical funnel supports a circular piece of filter paper folded into the shape of a cone. The

condensation
condenser
distillation
still

evaporation
freezing point

filtration

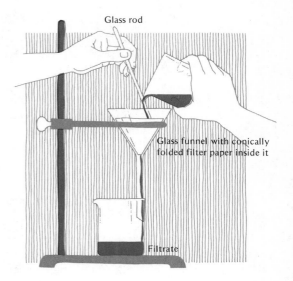

FIGURE 2.2 Apparatus for filtration

Glass rod

Glass funnel with conically folded filter paper inside it

Filtrate

filter paper, which is porous like blotting paper, lets water through its pores but holds back the mud. (When you pour some of the muddy water from the container in your right hand, you can avoid spilling it by guiding it into the inside of the paper cone with a rod held in your left hand.) The liquid passing through the paper is called the *filtrate*. Although it is clear, it may still contain solutes that have also passed through the pores of the filter paper. It may also contain bacteria, singled-celled organisms so small that they, too, have passed through the pores. To make filtered water safe for drinking, boil it for 20 min to kill any bacteria that may be present.

filtrate

If we treat a substance by all the procedures we can think of to remove impurities, and then find that the purified material still has unchanged properties (such as melting point), we can assume that we have a pure substance.

2.6 Mixtures and Compounds

A simple experiment in the laboratory will show what can happen when two pure substances are mixed. If we put a little pile of powdered particles of gray iron on a piece of paper and bring a magnet close to them, the magnet picks them up. If we add a few of the particles to some liquid carbon disulfide, they do not dissolve. If on another piece of paper we put some powdered particles of yellow sulfur, these particles are not attracted by the magnet and they do dissolve in liquid carbon disulfide.

Now we mix the two piles of powder thoroughly. The mixture has a grayish-yellow color, a composite of the original colors of the iron and the sulfur. When we stir up some of the mixture with the magnet, the iron particles stick to it and can be separated from the sulfur. When we shake up some of the mixture with carbon disulfide, the sulfur dissolves and leaves the iron behind. If we filter the mixture of solid iron and liquid carbon disulfide solution, the liquid goes through the filter paper and the iron does not. If we let the filtrate evaporate, the dissolved sulfur remains as a yellow solid.

We thus have several observations.

1 The color of the mixture of iron filings and sulfur is a composite of the colors of the substances before mixing.
2 The iron can be recovered by a magnet.
3 The sulfur can be recovered by dissolving it in carbon disulfide.

From these observations we conclude that the mixing of iron and sulfur has not caused any interaction between these two pure substances.

If we put some of the iron-sulfur mixture in a test tube and heat it, a bright-red glow spreads through the mix, the test tube gets red-hot, and the entire mass inside the tube solidifies into one lump. If after it cools we grind this lump, the powder is black, it is not attracted by a magnet, and it is not soluble in carbon disulfide. Because the properties of the new substance are so different from those of the original iron and sulfur, we conclude that these two have interacted chemically to produce a new substance, which we call *iron sulfide*. We say that the iron and sulfur are *constituents* that have *combined* by *chemical reaction* to form the *compound* iron sulfide. In more general terms we say that the *reactants* (iron and sulfur) have undergone a chemical reaction to form the *product* (iron sulfide).

constituents
combination
chemical reaction
compound
reactants
product

2.7 Chemical Elements

If we warm some silvery, shiny liquid mercury in a dish open to the air in a well-ventilated fume hood* for a few hours, a powdery red substance collects on the surface of the mercury. If we remove this powder, put it in a test tube, and heat it strongly in the hood, it changes to liquid mercury. If we hold a splinter of wood with a glowing spark in the test tube, the spark ignites into a flame. Since this ignition of a spark is a characteristic of oxygen gas, it seems reasonable to conclude that the red powder contained mercury and oxygen. We call this compound *mercury oxide*. This compound must have been

*Mercury and its compounds are very poisonous (see Sec. 13.9), and they must be handled with extreme care.

formed by the combination of the warmed mercury with oxygen from the air. We summarize these observations by saying that

1 mercury and oxygen combine at a slightly elevated temperature to form the compound mercury oxide

decomposition

2 mercury oxide heated to a higher temperature *decomposes* into mercury and oxygen

metal
oxide

We find that many *metals* combine with oxygen to form *oxides*. When heated, a few of these decompose to form the metal and oxygen. Many of them must be supplied with energy such as light or electricity by more complicated processes to yield the metal. But, no matter how much energy we supply to mercury, we cannot decompose it into other substances. The same is true of all other metals. Likewise, no matter how drastically we treat oxygen, we cannot decompose it into other substances. The same is true of sulfur,

nonmetal

iodine, carbon, and many other *nonmetallic* substances.

chemical element

Those pure substances that we are unable to decompose by any means we call *chemical elements*. We define a compound, then, as a pure substance that may be decomposed into two or more substances. And we define an element as a pure substance that cannot be so decomposed. Figure 2.3 summarizes the chemist's "simplification by classification."

The writings of ancient peoples show that they knew the elements gold, silver, copper, iron, lead, tin, mercury, sulfur, and carbon. Between 600 and 1700 the elements arsenic, antimony, bismuth, and phosphorus were discovered, and in the eighteenth century, zinc, cobalt, nickel, manganese, hydrogen, nitrogen, and oxygen. By the middle of the nineteenth century about 60 elements were known. All 90 elements now known to occur in natural substances were isolated and characterized by 1940. About this time the first manmade elements were created. With the exception of neptunium and plu-

FIGURE 2.3 An example of simplification by classification

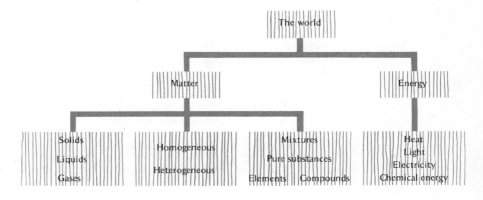

tonium, manmade elements are very unstable and exist for only short periods of time after they are created. Although they may have been present on earth millions of years ago, their short lifetimes resulted in their disappearance from nature. They are technetium, astatine, neptunium, plutonium, americium, curium, berkelium, californium, einsteinium, fermium, mendelevium, nobelium, lawrencium, rutherfordium, and hahnium. You may recognize some of these names as derived from the names of people who made important contributions to modern chemistry, and from the home of the University of California, where many of the manmade elements were first prepared.

For convenience we generally use symbols rather than names for the elements when we are writing about them. The symbol consists of a capital letter that is usually the initial of the name of the element, often followed by a lowercase letter contained in the name. Cl stands for chlorine, Al for aluminum, etc. We use single letters for some elements: H for hydrogen, O for oxygen, etc. The symbols that don't seem to match the English names of the elements were taken from their Latin names. These are listed in Table 2.1.

When you see the symbol for an element, it should make you think of that substance, and after you have learned the symbols and used them for a while, you will find yourself immediately thinking "carbon" when you see "C." The names, symbols, and some properties of some common elements found in nature or commonly prepared by man are given in Table 2.2. Most of the elements are metals, some are nonmetallic solids, several are gases, and two are liquids at room temperature and ordinary pressures.

TABLE 2.1 *The Elements Whose Symbols Are Derived from Latin Names*

Element	Latin name	Symbol
Antimony	Stibium	Sb
Copper	Cuprum	Cu
Gold	Aurum	Au
Iron	Ferrum	Fe
Lead	Plumbum	Pb
Mercury	Hydrargyrum	Hg
Potassium	Kalium	K
Silver	Argentum	Ag
Sodium	Natrium	Na
Tin	Stannum	Sn

All the elements are listed alphabetically in Appendix A together with their symbols.

TABLE 2.2 *Some Common Elements*

Element	Symbol	Description
Solids		
Metals		
Aluminum	Al	A light, silvery metal much used in construction
Copper	Cu	A reddish metal very useful for conducting electricity
Gold	Au	A yellowish precious metal much used in jewelry and to some extent in industry because it resists corrosion
Iron	Fe	A grayish metal widely used for construction
Magnesium	Mg	A light, silvery metal used to build lightweight structures like airplanes
Platinum	Pt	A very heavy silvery metal used in jewelry and in industry because it resists corrosion
Silver	Ag	A precious metal used in jewelry and to some extent in industry
Tin	Sn	A grayish metal used in "tin cans," which are made of iron and coated with tin to slow corrosion
Zinc	Zn	A grayish metal used to "galvanize" iron, that is, to coat iron with zinc
Nonmetals		
Carbon	C	A black solid present in large amounts in coal and soot from burning fuels
Sulfur	S	A pale-yellow solid that burns readily in air
Iodine	I	Present as I_2 in a grayish-violet solid; gives a brown solution in alcohol (tincture of iodine)*
Gases		
Chlorine	Cl	Present as Cl_2 in a very poisonous greenish-yellow gas*
Hydrogen	H	Present as H_2 in a colorless, odorless, highly *combustible*† gas*
Nitrogen	N	Present as N_2 in a colorless, odorless, noncombustible gas*
Oxygen	O	Present as O_2 in a colorless, odorless gas that supports *combustion*†; many elements and compounds burn vigorously in oxygen*
Liquids		
Bromine	Br	Present as Br_2 in a heavy reddish-brown liquid; very dangerous because it burns living tissue fiercely*
Mercury	Hg	A heavy liquid with a silvery metallic luster; an excellent conductor of electricity

*The doubling of atoms in some elements is discussed in Sec. 6.8.
†These terms are defined at the end of the chapter under New Terms and Concepts.

2.8 Atoms and Molecules

As early as the time when Greek culture flourished in the Aegean area more than 2000 years ago, thoughtful men were concerned with the ultimate structure of matter. They reasoned that if one were to take a piece of any substance and cut it into smaller and smaller bits, a final "uncuttable" portion would be reached. This tiniest bit they called "atomos"—the "uncut" (indivisible). The Greek atomists Leucippus and Democritus postulated many different kinds of atoms, whose motions with respect to one another would account for changes of all sorts. Since many kinds of change are not perceived as motion, they conceived these atoms to be invisible. They attributed all differences of quality to differences in the sizes and shapes of atoms. These were thought to collide and adhere to one another, to separate under impact, and to recombine in other configurations. Combinations of atoms might yield infinitely various aggregates. The Greeks thought of the world as being in perpetual internal motion, undergoing countless changes by the combination and separation of the ultimate atoms.

This promising theory remained only a philosophical exercise, because no attempts were made to devise experiments to test it. Aristotle rejected it because it failed to describe the phenomena of life. His followers preferred to think of the elements as earth, air, fire, and water. Because of the great influence of Aristotle, the theories of Leucippus and Democritus sank into oblivion, but they were preserved for posterity in the great poem *De Rerum Natura* (*On the Nature of Things*), written by the Roman Lucretius in the first century B.C.

atomic theory

weight

From the time of the ancient Greeks through the eighteenth century so many different hypotheses were presented to account for the transformations of matter from one kind into another that no single interpretation was generally accepted. The *atomic theory* as we now conceive it (see Chap. 3) was not developed until large amounts of data had been collected to show the *weights* of substances produced or consumed in chemical transformations. The somewhat vague and qualitative concepts of the early thinkers were finally explained clearly and quantitatively by the nineteenth-century atomists.

molecule

By assuming that atoms are the fundamental particles of the elements and that these atoms combine with one another in various ways to form the multitudes of pure compounds then known, the nineteenth-century scientists explained most convincingly both the qualitative and the quantitative aspects of chemical changes. The smallest grouping of atoms present in a compound was called a *molecule* of that substance. The numbers of atoms and molecules in elements and compounds are enormous. The few grains of salt that you sprinkle on your hamburger at lunch contain about 1,000,000,000,000,000,000,000 atoms!

Summary

We can simplify our task of understanding our environment by remembering that the universe is composed of two entities, matter and energy. Matter is anything which has weight; energy is the capacity for doing work. Matter exists in the solid, liquid, and gaseous states. It may be homogeneous or heterogeneous, a pure substance or a mixture. And a pure substance may be a chemical element or a compound composed of more than one element.

Every pure substance has a set of properties, such as melting point, boiling point, solubility, and chemical reactivity, which are different from those of every other pure substance. When two or more pure substances are put together in a mixture, each substance retains its own properties, but the properties of the mixture are a composite of the properties of its components.

When two or more substances interact to form a chemical compound, the properties of the compound are entirely different from those of its constituents.

The existence of atoms has been assumed since the time of the ancient Greeks, but general acceptance of the atomic theory of matter was not achieved until the middle of the nineteenth century.

New Terms and Concepts

ATOMIC THEORY: The theory that all substances are composed of tiny, "uncuttable" particles called atoms.

CHEMICAL ELEMENT: Any one of a limited number of distinct varieties of matter which, singly or in combination, compose substances of all kinds.

CHEMICAL REACTION: The decomposition of one substance into two or more other substances or the interaction of two or more substances to form one or more new substances.

COMBINATION: The union of two or more substances to form a new substance.

COMBUSTIBLE: Burnable.

COMBUSTION: The burning of a substance with oxygen to produce heat.

COMPONENTS: The substances present in a mixture.

COMPOUND: A pure substance composed of atoms of more than one element in definite proportions.

CONDENSATION: The process in which a gas becomes a liquid with the liberation of heat.

CONDENSER: A device in which a gas is cooled and condensed to a liquid.

CONSTITUENTS: The elements that are combined into a chemical compound.

DECOMPOSITION: The breakdown of one substance into two or more other substances.

DISSOLVE: To pass into solution.

DISTILLATION: The process of vaporizing a liquid, collecting the vapor, and condensing it to a liquid.

EVAPORATION: The passage of a substance from the liquid form to the gaseous form.

FILTRATE: The substance passing through the filtering medium when a suspension of a solid in a liquid is filtered.

FILTRATION: The process of removing a solid from its suspension in a gas or a liquid by passing the suspension through a material permitting the passage of the gas or liquid but not the solid.

FREEZING POINT: The temperature at which a liquid changes to a solid with the liberation of heat.

GAS: A sample of matter that has no definite shape or volume but occupies the entire volume of its container.

HETEROGENEOUS: Consisting of parts that are unlike.

HOMOGENEOUS: Uniform in structure and composition throughout.

LIQUID: A specimen of matter that has no definite shape but occupies a definite volume.

MATTER: That which occupies space, has weight, and can be perceived by one or more of the five senses.

METAL: A chemical element that is shiny, malleable, ductile, and a good conductor of heat and electricity.

MIXTURE: A complex of two or more ingredients that do not bear a fixed proportion to one another and that, however thoroughly commingled, are conceived as retaining a separate existence.

MOLECULE: A unit of matter; the smallest portion of an element or compound that retains chemical identity with the substance in mass.

NONMETAL: A chemical element that does not have metallic properties.

OXIDE: A compound composed of oxygen and another element.

PRODUCT: The material produced by a chemical reaction.

PURE SUBSTANCE: A material that has a fixed composition and a definite set of properties.

REACTANTS: The starting materials for a chemical reaction.

SOLID: A specimen of matter with a definite size and shape.

SOLUBLE: Capable of being dissolved.

SOLUTE: The dissolved substance in a solution.

SOLUTION: A homogeneous mixture.

SOLVENT: A liquid that dissolves another substance.

STATES OF MATTER: Matter may exist in three forms called the solid, liquid, and gaseous states.

STEAM: Water in the gaseous state above the boiling point of liquid water.

STILL: A device for separating a liquid from a solution by boiling the solution and condensing the vapor.

VOLUME: Quantity of space occupied.

WEIGHT: The attraction of the earth for another object.

Testing Yourself

2.1 Name some common substances that are (*a*) solids, (*b*) liquids, (*c*) gases.

2.2 Name some common substances that are (*a*) homogeneous solids, (*b*) heterogeneous solids, (*c*) homogeneous liquids, (*d*) heterogeneous liquids.

2.3 Name some common substances that are (*a*) solutions, (*b*) solutes in solutions, (*c*) solvents.

2.4 What do we mean by the "properties" of a substance?

2.5 How can we tell when we have a pure substance?

2.6 Name some common substances that are useful because they have the following properties: (*a*) they burn readily; (*b*) they do not burn; (*c*) they conduct heat very well; (*d*) they conduct heat very poorly.

2.7 How can we determine whether the following substances are mixtures or compounds? (*a*) maple syrup; (*b*) wine; (*c*) salt; (*d*) city air.

2.8 What is the essential difference between elements and compounds?

2.9 What do we mean by the term "atom"? "Molecule"?

Sodium nitrate is a compound of the elements sodium, nitrogen, and oxygen.

3. Symbols: The Language of Chemistry

So far we have taken a bird's-eye view of the damage being done to our environment by our numbers and our way of life. We have seen that we must create a dynamic equilibrium between what we take from the environment and what we must restore to it. And we have begun to get some idea of what the chemist can do to help us understand and solve our problems. It is evident that our "cowboy economy," based on maximum exploitation, must be replaced by a "spaceman economy," based on minimum consumption of natural resources and due concern for their replacement. A dynamic equilibrium with the environment must be placed high on our list of human needs. Achieving it is a social process, but it must be supported by scientific knowledge in order to encourage political action.

Chemistry is at the heart of this scientific knowledge, for many pollutants are produced by chemical processes, and other chemical processes must be used to minimize or cancel their effects. In Chap. 2 we began our study of chemistry and the environment by seeing how chemists classify matter by kinds—that is, qualitatively. But qualitative descriptions won't answer all our questions, and we soon must deal with questions of "how much." The environmentalist asks, How much polluted water is a paper mill discharging into a river? How much will it cost to purify the discharge? How much sulfur dioxide does a factory smokestack emit in one day? And the chemist knows how much of one substance combines with another and how much energy will be produced or consumed by a certain reaction. In this chapter we shall see how the chemist measures and describes chemical phenomena—and gain some insight into the fundamental truth that in the physical world quantitative considerations always give us deeper insights than we can achieve by studying only qualitative relationships.

3.1 Measuring Length, Area, and Volume

When you buy a roll of fish line or a spool of thread, you want to know what it's made of (cotton, linen, silk) and *how much length* you're getting for your money (100 ft, 100 yd). Feet and yards are quantities—units of length, the distance between two points. When you buy a piece of carpet you want to know what it's made of (wool, nylon, acrylic) and *how much area* you're getting (square feet or square yards). Area is another quantity—a measure of the amount of surface enclosed by a boundary. When you buy gasoline, you want to know what quality you're getting (regular or premium) and also *how much volume* (gallons). You pay for the gas you cook with by volume —so much per cubic foot. Gallons and cubic feet are units of volume, the amount of space occupied by something.

In ancient and medieval times, crude measurements were made by comparing objects to a finger, handspan, foot, or stride. The ancient cubit was the distance from the elbow to the finger tip; the inch was the length of the endjoint of the thumb. By the time of the French Revolution (1789), so many different units of length were in use in European countries that trade was hampered and scientific measurements made in one nation were almost useless in another. One of the declared objectives of the Revolution was to establish the supremacy of reason, and it was agreed to establish rational standards of measurement based on the nature of the earth and its most abundant substance, water.

The new unit of length was the meter, 1/10,000,000 of the distance from the earth's equator to the North Pole. In 1791 the French bureau of standards made a platinum bar with two scratches on it separated by this distance, called

meter
centimeter

one *meter* and abbreviated 1 m. This bar is kept at the bureau in Sèvres near Paris. The meter is divided into hundredths (*centimeters*, abbreviated cm) and thousandths (millimeters, mm). Long distances are measured in thousands of meters (kilometers, km).

Areas are measured in square meters (m^2), square centimeters (cm^2), or square millimeters (mm^2). Volumes are measured in cubic meters (m^3),

liter

cubic centimeters (cm^3), or cubic millimeters (mm^3). A convenient unit of volume is the *liter*, which is 1000 cm^3.

So useful was the meter that in 1875 an international treaty, the Metric Convention, was signed by representatives of 17 nations, establishing the meter as a worldwide standard. Replicas of the original meter bar are kept in the bureaus of standards in many countries. Ours is in the National Bureau of Standards in Washington, D.C. The modernized metric system, used by most countries, is called the International System of Units, abbreviated SI units. The United States and Great Britain are in the process of changing from the British system to SI units. The names and symbols for some of the SI units and their equivalents in British units are tabulated in Appendix D.

Recently scientists have made more precise measurements of length than is possible with a meter stick. Instruments now measure the length of light waves (Sec. 16.4) with fantastic precision. We no longer define the meter as the distance between two scratches on a metal bar. The SI unit of length is 1,650,763.73 wavelengths in vacuum of the orange-red light in the spectrum (Sec. 16.5) of the chemical element krypton.

3.2 Measuring Time

We measure time as the duration of an event. One solar day is the time between the passage of the sun over a given spot on earth at noon (its highest position) on one day and noon on the next. We divide this day into 24 hr, the hour into 60 min, the minute into 60 sec. Thus there are $24 \times 60 \times 60$ or 86,400 sec in 1 day. When we use a clock that measures time precisely, we find that the time from noon to noon varies slightly from year to year and place to place. So we define 1 sec as 1/86,400 of an average solar day at the equator. From data kept over long periods it turns out that the year 1900 is a good one for comparison. So we further refine our definition of a second as 1/31,556,925.9747 of the year at the equator in 1900.

Modern scientists need a unit of time more precise than fractions of the variable solar day. Again, the most precise measurements can be made in terms of waves. In SI units, 1 sec is the duration of 9,192,631,770 vibrations (waves) produced by atoms of the chemical element cesium when they are shot through the space between the ends of electromagnets.

3.3 Measuring Quantity of Matter: The Concept of Mass

Liquids and gases are readily measured by volume, but it is not convenient to measure solids in this way. Even in ancient times amounts of solids were measured by balancing them on a beam supported in the middle (Fig. 3.1).

If we hang one stone in one loop and another in the other loop, and the beam is still horizontal, we sense intuitively that there is the same amount of matter in each stone, even though they may be composed of different materials and may have different sizes and shapes. It is convenient, then, to take some one piece of stone and call it one unit of quantity of matter. If two other stones balance one another and together just balance the unit of stone, each of them would contain one-half unit of matter. A series of stones can be smoothed off to serve as other fractions or multiples of the unit. The Egyptian unit of matter was the *beqa*. Pieces of limestone like those shown in Fig. 3.1 were smoothed off to serve as multiples or fractions of one beqa. This balance was probably used to measure gold. With a pair of balanced pans hanging from the loops it could be used to compare amounts of gold, whether they were in the form of dust, foil, nuggets, or bars.

What is it that we measure on an equal-arm balance? The term itself makes us think of holding one object in the right hand and another in the left. If the right arm has to exert more effort than the left, we intuitively sense that there is more matter in the right hand. If the two arms are "balanced," i.e., seem to exert the same effort, we think of the objects as containing the same amount of matter. The earth attracts pieces of matter—exerts a pull on them —that we must overcome when we hold them up in our hands. We call this pull *gravitational attraction* and take it to be a measure of the *quantity* of matter in a given object. We invent the term *gravitational mass* to express quantity of matter. If two bodies are attracted to the earth with the same pull, we say they contain the same quantity of matter, or gravitational mass. For convenience, we often say just "mass."

To express the mass of an object as a number, we must set up a standard unit of mass and make the balancing masses simple multiples or fractions of this standard. The French bureau of standards constructed a block of platinum as heavy as 1000 cm³ (1 liter) of water at 4°C, and decreed that this amount of mass was to be called 1 *kilogram*, abbreviated 1 kg. A thousandth of this mass is called 1 *gram* (1 g), a unit chemists find more convenient than the kilogram. The original standard of mass is preserved at the French bureau

gravitational
attraction
gravitational
mass

kilogram
gram

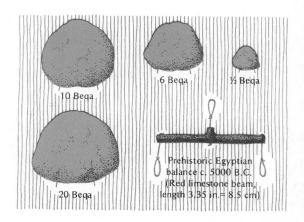

FIGURE 3.1 A balance and weight used in Egypt 7000 years ago. (Figure redrawn courtesy of the Science Museum, London, England.)

10 Beqa

6 Beqa   ½ Beqa

20 Beqa

Prehistoric Egyptian
balance c. 5000 B.C.
(Red limestone beam,
length 3.35 in.= 8.5 cm)

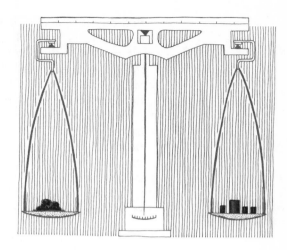

FIGURE 3.2 Schematic diagram of a precise equal-arm balance

of standards, and precise replicas are kept in many countries.

To measure the mass (amount of matter) in a given object, we place it on one pan of an equal-arm balance and put objects having known masses (usually called weights) on the other until the balance beam is horizontal. This process should be called "massing" the object, but the term "weighing" is more commonly used. In this book when we say weight, we really mean mass. We need not be concerned with the difference between them.

3.4 Mass and Chemical Reactions

When heated in air, many metals become coated with a powdery substance that has quite different properties from the metal itself. The coating on mercury is red, that on iron is black; on zinc, aluminum, tin, magnesium, and many other metals, it is white. Because we can recover oxygen from these coatings we conclude that they contain oxygen, and so we call them oxides of the metals. If we weigh the metals before heating them in air and again after some coating has formed, the mass of the metal plus its coating is always greater than that of the metal alone. Is this increase due solely to the addition of oxygen to the metal to form the oxide, or has the mass of the metal changed? We cannot answer this question by this experiment, because we cannot weigh the oxygen that has combined with the metal.

To answer the question, we can heat the metal with oxygen in a closed container. If we weigh an ordinary photographic flashbulb very precisely and then flash it, the aluminum wool in it burns in the air inside the bulb and forms white aluminum oxide. After the reaction we weigh the cool bulb and find that its mass has not changed by any measurable amount. We conclude that the mass of the aluminum oxide formed inside the flashbulb is equal to the masses of aluminum and of oxygen used up. Chemists have performed thousands of experiments in which the reactants were weighed before a chemical reaction and the products were weighed after the reaction. In all cases, the chemical reaction produced no measurable change in total mass.

natural law

When a number of scientists have carried out a large number of experiments which all lead to the same conclusion, the behavior of matter under the prescribed conditions is called a *natural law*. A natural law is not an edict that "Nature" compels matter to "obey." It is a description of the behavior of matter. Since all experiments on the masses of reactants and products in chemical reactions lead to the conclusion that the total mass is unchanged, we formulate the *law of the conservation of matter: In a chemical reaction, matter is neither created nor destroyed.** The formulation of such laws would not be possible without precise measurement.

law of the
conservation
of matter

3.5 Relationships between Masses of Reactants and Products

When a jet of hydrogen gas is ignited in pure oxygen gas, it burns. If we hold a cold object above the flame, water condenses on the cold surface as it did when we boiled seawater (Sec. 2.5). We conclude that the two elements combine to form the compound thus:

$$\text{Hydrogen} + \text{oxygen} \longrightarrow \text{water}$$

The arrow stands for the word "produce."

This experiment gives us the qualitative information that hydrogen and oxygen combine to form water. To answer the question "How much?" we need to do some experiments using a thick-walled closed container.

1 We mix 1.0 g hydrogen with 1.0 g oxygen in the container and ignite the mixture. It explodes, and we find water inside the container when we cool it to room temperature. We also find some hydrogen remaining in the container.
2 We mix 1.0 g hydrogen with 10.0 g oxygen in the container and ignite the mixture. It explodes and we find water and some unused oxygen.
3 We mix 1.0 g hydrogen with 8.0 oxygen in the container. After the explosion we find 9.0 g water but no hydrogen or oxygen.
4 We mix 2.0 g hydrogen with 16.0 g oxygen in the container. After the reaction we find 18.0 g water but no hydrogen or oxygen.

From these experiments we note the following.

*In an uncommon kind of reaction called a nuclear reaction (Chap. 9), there may be a gain or a loss of mass.

1 When the ratio of grams of hydrogen to grams of oxygen is 1:1, as in experiment 1, some hydrogen is left over.

2 When the ratio is 1:10, as in experiment 2, some oxygen is left over.

3 When the ratio is 1:8, as in experiment 3, there is no gas left over, and 9.0 g water is formed.

4 When twice as much hydrogen and oxygen are used, and the ratio in grams is still 1:8, there is no gas left over, and 18.0 g water is formed.

Apparently there is a fixed ratio between the masses of hydrogen and of oxygen that combine to form water. The hydrogen is $\frac{1}{9}$ or 11.1 percent of the mass of water, and the oxygen is $\frac{8}{9}$ or 88.9 percent. Further experiments with different masses of hydrogen and oxygen lead to the same conclusion.

As chemistry was developing in the late eighteenth century, many chemists studied the composition of many materials. Some performed syntheses, putting pure substances together to produce a new pure substance, like water from hydrogen and oxygen. Others performed qualitative and quantitative analyses to find what elements are in a given pure compound and in what percentages. By 1808 the data from laboratories all over Europe led to the following conclusion: *A pure compound always contains the same elements in the same percentages by mass.* This is the *law of constant composition.* Pure table salt is always 39.3% sodium and 60.7% chlorine. Pure cane sugar always contains 42.1% carbon, 6.4% hydrogen, and 51.5% oxygen. And so on.

law of constant
composition

3.6 Can Two Elements Unite to Form More Than One Compound?

Gasoline is composed of compounds of carbon and hydrogen. If we burn it in open air, we get carbon dioxide and water. If we burn it in a limited supply of air (as in the cylinder of an engine), we get carbon dioxide, water, and some carbon monoxide. Carbon dioxide will not burn in air; carbon monoxide will. We exhale carbon dioxide from our lungs, but we are poisoned by breathing carbon monoxide. So the same elements, carbon and oxygen, combine in different proportions to form two quite different compounds: carbon *dioxide* and carbon *monoxide.*

The qualitative differences between these two gases were known in the eighteenth century, but not until the early nineteenth century were they analyzed quantitatively. Carbon dioxide contains 72.73% oxygen and 27.27% carbon; carbon monoxide contains 57.12% oxygen and 42.88% carbon. From these figures we can calculate the number of grams of oxygen per 1 g carbon in each compound.

For carbon dioxide: 100 g of the compound contains 72.73 g oxygen and 27.27 g carbon.

$$\frac{72.73 \text{ g oxygen}}{27.27 \text{ g carbon}} = 2.66 \text{ g oxygen/g carbon}$$

For carbon monoxide: 100 g of the compound contains 57.12 g oxygen and 42.88 g carbon.

$$\frac{57.12 \text{ g oxygen}}{42.88 \text{ g carbon}} = 1.33 \text{ g oxygen/g carbon}$$

Twice as much oxygen is combined with 1 g carbon in carbon dioxide as in carbon monoxide.

Many other elements form more than one compound with a second element, but the mass of the first element combined with 1 g of the second is always a simple multiple of some smallest number. For example, there are five different oxides of nitrogen. The number of grams of nitrogen combined with 1 g oxygen in these compounds is 0.571, 1.142, 1.713, 2.284, 2.855. Each of these is a simple multiple of the smallest.

$$\frac{0.571}{0.571} = 1 \qquad \frac{1.142}{0.571} = 2 \qquad \frac{1.713}{0.571} = 3$$

$$\frac{2.284}{0.571} = 4 \qquad \frac{2.855}{0.571} = 5$$

law of multiple proportions

Data like these led to the *law of multiple proportions*: *If there exists a series of two or more compounds containing the same elements, there is a ratio of small whole numbers between the different masses of one element that combine with any given mass of another element.* As we shall see in Chap. 7, the various oxides of carbon and nitrogen are important in air pollution.

3.7 A Mental Model to Explain Chemical Reactions

How can we explain the changes in the properties and composition of matter when elements combine to form compounds or compounds interact to form new substances? How can we account for the law of constant composition? In his *New System of Chemical Philosophy* (1808–1810) John Dalton created a mental model of matter and its behavior, often called Dalton's atomic theory, which answered a good many questions about what went on in chemical reactions. Following is the "model" Dalton's theory provides:

OBSERVATION Matter is neither created nor destroyed in a chemical reaction.

DALTON'S INTERPRETATION All matter is composed of chemical elements. All chemical elements are composed of extremely small, invisible *atoms* that are neither created nor destroyed in a chemical reaction.

John Dalton, 1766–1844

EXAMPLE Iron sulfide is composed of the elements iron and sulfur. Iron is composed of iron atoms; sulfur is composed of sulfur atoms. When iron and sulfur react, atoms of iron become attached to atoms of sulfur, forming the compound iron sulfide.

OBSERVATION Regardless of the sources from which different samples of a given chemical element are prepared, they have the same properties.

DALTON'S INTERPRETATION All atoms of a given element have the same size, shape, and mass.

EXAMPLE All the atoms of iron are alike, so no matter where it is found in nature or from what compounds it is derived, a piece of iron is always a grayish metal that is ductile (can be drawn into a wire), is a good conductor of heat and electricity, reacts with sulfur, etc.

OBSERVATION Each element has a set of properties that is different as a set from those of any other element.

DALTON'S INTERPRETATION The atoms of a given element are different from the atoms of any other element.

EXAMPLE   Atoms of iron are different from atoms of sulfur, so pieces of iron are different from pieces of sulfur. The differences between iron and sulfur in bulk quantities arise from differences between iron and sulfur atoms. Every element has its own particular set of properties because its atoms are different from any other atoms.

OBSERVATION   The percentage composition of a given compound does not vary from one sample to another regardless of the size or the source of the sample.

DALTON'S INTERPRETATION   A compound is formed by the union of a simple whole number of each of the kinds of atoms in it. The mass of a molecule of the compound is the sum of the masses of the atoms in it. All atoms of a given element have the same mass. Therefore the percentage of the total mass of the molecule contributed by a given atom is always the same. And the percentage of the total mass of any sample (large collection of molecules) contributed by a given element in a compound is always the same.

EXAMPLE   If one atom of element X unites with one atom of element Y to form one molecule of compound XY, and if an atom of X has a mass of 5 units and an atom of Y has a mass of 10 units, a molecule of XY will have a mass of 15 units. The composition of the molecule will be $\frac{5}{15}$ or $\frac{1}{3}$ or 33.3% X and $\frac{10}{15}$ or $\frac{2}{3}$ or 66.7% Y. One molecule or 1 billion billion molecules will have the same percentage composition; so any sample of the compound will contain 33.3% X and 66.7% Y.

OBSERVATION   When two elements form more than one compound, there is a ratio of small whole numbers between the masses of one element that combine with any given mass of the other element.

DALTON'S INTERPRETATION   Different whole numbers of atoms may be involved in the formation of different compounds by the union of two elements.

EXAMPLE   Carbon dioxide contains twice as many grams of oxygen per gram of carbon as is found for carbon monoxide. Therefore, 1 molecule of carbon dioxide contains twice as many atoms of oxygen as 1 molecule of carbon monoxide. If carbon monoxide is formed by the union of 1 atom of carbon with 1 of oxygen, then carbon dioxide is formed by the union of 1 atom of carbon with 2 atoms of oxygen.

3.8 Formulas: Chemical Shorthand

The letter symbols chemists use to stand for the names of chemical elements are combined with numerals that represent the number of atoms in com-

formula

pounds. These combinations are called *formulas*. Here are formulas for the compounds discussed so far, together with their "longhand" equivalents.

Water, H_2O — 2 atoms of hydrogen combine with 1 atom of oxygen to form 1 molecule of water. (Absence of a subscript means "1.")

Carbon monoxide, CO — 1 atom of carbon combines with 1 atom of oxygen to form 1 molecule of carbon monoxide.

Carbon dioxide, CO_2 — 1 atom of carbon and 2 atoms of oxygen form 1 molecule of carbon dioxide.

Carbon disulfide, CS_2 — 1 atom of carbon and 2 atoms of sulfur form 1 molecule of carbon disulfide.

Sugar, $C_{12}H_{22}O_{11}$ — 12 atoms of carbon, 22 atoms of hydrogen, and 11 atoms of oxygen combine into 1 molecule of sugar.

As we shall see in Sec. 6.8, a number of the gaseous elements exist as diatomic molecules (2 atoms per molecule). They are hydrogen H_2; oxygen, O_2; nitrogen, N_2; fluorine, F_2; and chlorine, Cl_2.

3.9 Chemical Equations: Shorthand for Chemical Reactions

One step beyond the formula is the chemical equation, which tells what happens in a chemical reaction. Consider the statement "Hydrogen and oxygen combine to form water," or

$$H_2 + O_2 \longrightarrow H_2O$$

which describes what happens in terms of atoms and molecules, but does not take into account the conservation of matter. There are 2 atoms in the oxygen molecule, but only one of them is used in the water molecule. Our equation would be a better description if we wrote it to show that matter is conserved, that there is the same number of atoms of each element before and after reaction.

If we multiply the water molecule by 2, we have the same number of oxygen atoms before and after the reaction:

$$H_2 + O_2 \longrightarrow 2H_2O$$

But we now show 4 atoms of hydrogen in the water molecule and only 2 in the hydrogen molecule. If we also put a coefficient of 2 in front of the hydrogen molecule, we will have the same number of hydrogen atoms before and after reaction:

$$2H_2 + O_2 \longrightarrow 2H_2O$$

Consider another example. If we connect a gas burner to a source of natural gas (which is mostly methane, CH_4) and light it, a flame is produced by the burning of the gas with the oxygen of the air. If we hold a cold dish above the flame, water condenses on it as it did when we burned hydrogen. If we withdraw some of the gas produced by the flame, we find that it is odorless. When the gas is drawn through clear limewater, the limewater becomes clouded with a finely divided solid. Carbon dioxide is the only odorless gas that causes this cloudiness when passed through limewater, so we conclude that the burning of methane by oxygen in the air produced water and carbon dioxide. This conclusion can be stated in words: Methane combines with oxygen to produce water and carbon dioxide.

We can state this fact much more concisely by using chemical symbols and formulas:

$$CH_4 + O_2 \longrightarrow H_2O + CO_2$$

This symbolic statement describes our observations correctly in terms of the atoms and molecules involved, but it does not take into account the conservation of matter. Again we must modify our symbolic statement to show that there are the same numbers of carbon, hydrogen, and oxygen atoms before and after reaction.

If we place a coefficient of 2 before the formula for water, we will have the same number of hydrogen atoms before and after reaction; the four H atoms in one molecule of methane will produce two molecules of water, each molecule containing two H atoms, thus accounting for four H atoms. The statement then becomes:

$$CH_4 + O_2 \longrightarrow 2H_2O + CO_2$$

Now our symbolic statement shows the conservation of H and C atoms. The C atom in $1CH_4$ is the C atom in $1CO_2$. The four H atoms in CH_4 are the four H atoms in $2H_2O$. But there are only two atoms of oxygen in $1O_2$ and four atoms of oxygen in $2H_2O$ and $1CO_2$. We show the conservation of oxygen atoms by placing the coefficient 2 before O_2. The statement then becomes

$$CH_4 + 2O_2 \longrightarrow 2H_2O + CO_2 \tag{3.1}$$

Now the numbers of C, H, and O atoms are the same before and after the reaction.

When our symbolic statement shows the conservation of all atoms, we have a balanced chemical equation, that is, a symbolic statement about a chemical reaction: the formulas of all reactants are followed by an arrow and the formulas of all products; the total number of atoms of each element in the reactants must equal the total number of atoms of that element in the products. Equation (3.1) is the balanced chemical equation for the combustion of methane. Constructing it is called "writing a balanced equation" or "balancing an equation."

A balanced chemical equation is always derived from actual experiments in the laboratory that show that the reactants are *observed* to give the products to the right of the arrow. The chemical equation grows out of the experiment; we cannot just write down some formulas, balance an equation, and expect it to be a real statement of what actually happens in nature.

Here are several examples of statements about chemical reactions discussed in this chapter.

EXAMPLE 3.1 When metallic mercury, a silvery liquid, is heated in air, a red solid forms on its surface. Analysis shows that the coating is mercury oxide, HgO. What is the balanced chemical equation for this reaction?

ANSWER

In words: Mercury + oxygen $\longrightarrow$ mercury oxide
In symbols: Hg + O_2 $\longrightarrow$ HgO

Two oxygen atoms are shown on the left of the arrow, only one on the right. Multiply HgO by 2 to balance the oxygen atoms:

$$Hg + O_2 \longrightarrow 2HgO$$

Oxygen atoms are balanced but mercury atoms are not. Multiply Hg by 2 to balance mercury atoms:

$$2Hg + O_2 \longrightarrow 2HgO \tag{3.2}$$

This is a balanced chemical equation.

EXAMPLE 3.2 Iron and sulfur heated together form a black substance, iron sulfide. Analysis shows it to be a compound, FeS. What is the balanced chemical equation?

ANSWER

In words: Iron + sulfur $\longrightarrow$ iron sulfide
In symbols: Fe + S $\longrightarrow$ FeS (3.3)

This is a balanced equation.

EXAMPLE 3.3 When octane, an important component of gasoline, is burned in the open air, it reacts with the oxygen of the air and yields water and carbon dioxide. The formula for octane is C_8H_{18}, that for water is H_2O, and that for carbon dioxide is CO_2. What is the balanced chemical equation for this reaction?

ANSWER

In words: Octane + oxygen ⟶ water + carbon dioxide
In symbols: C_8H_{18} + O_2 ⟶ H_2O + CO_2

Eight carbon atoms are shown on the left, only one on the right. Multiply CO_2 by 8 to balance the carbon atoms:

$$C_8H_{18} + O_2 \longrightarrow H_2O + 8CO_2$$

Multiply H_2O by 9 to balance the hydrogen atoms:

$$C_8H_{18} + O_2 \longrightarrow 9H_2O + 8CO_2$$

There are 9 oxygen atoms in $9H_2O$ and 16 in $8CO_2$, or a total of 25. We need 12.5 molecules of oxygen to get 25 oxygen atoms on the left:

$$C_8H_{18} + 12.5O_2 \longrightarrow 9H_2O + 8CO_2$$

But fractional coefficients are awkward, so we avoid them. If we multiply each coefficient by 2, we get whole numbers for all:

$$2C_8H_{18} + 25O_2 \longrightarrow 18H_2O + 16CO_2 \tag{3.4}$$

We now have 16 C, 36 H, and 50 O on each side of the equation, though the 50 O atoms on the left are in one substance and those on the right in two (18 plus 32).

EXAMPLE 3.4   When octane is burned in an automobile engine, the oxygen supply is limited, and so some water and carbon monoxide are produced. The formula for carbon monoxide is CO. What is the balanced chemical equation for this reaction?

ANSWER

In words: Octane + oxygen ⟶ water + carbon monoxide
In symbols: C_8H_{18} + O_2 ⟶ H_2O + CO

Eight C atoms are shown on the left, only one on the right. Multiply CO by 8 to balance C:

$$C_8H_{18} + O_2 \longrightarrow H_2O + 8CO$$

Multiply H_2O by 9 to balance H:

$$C_8H_{18} + O_2 \longrightarrow 9H_2O + 8CO$$

There are 9 oxygen atoms in $9H_2O$ and 8 in 8CO, a total of 17 on the right. We need 8.5 oxygen molecules to get 17 oxygen atoms on the left:

$$C_8H_{18} + 8.5O_2 \longrightarrow 9H_2O + 8CO$$
$$2C_8H_{18} + 17O_2 \longrightarrow 18H_2O + 16CO \tag{3.5}$$

This is the simplest balanced equation for this reaction.

3.10 The Masses of Atoms: Atomic Weights

atomic mass unit

Atoms are much too light to be weighed individually; even the tiniest speck of an element contains billions of billions of atoms. But the masses of individual atoms and molecules can be determined on a complex instrument called a mass spectrometer. An atom is so light that its mass expressed in grams would be a very awkward fraction of a gram. For convenience we invent a new unit called the *atomic mass unit* (amu), with a value of 1.66 $\times$ 10^{-24} g;* this is 1.66 millionths of a millionth of a millionth of a millionth of a gram! Atomic masses run from 1 to about 250 amu, numbers of a size convenient to work with.

isotope

When a sample of a pure element is studied in the mass spectrometer, we find that not all the atoms have the same mass. We give the name *isotopes* to the atoms of an element that have slightly different masses. There are three isotopes of hydrogen, having atomic masses of 1, 2, and 3 amu, respectively. There are two isotopes of chlorine, having atomic masses of 35 and 37 amu. Most of the elements occurring in nature have two or more isotopes. Tin has the largest number, 10. The mass spectrometer determines not only the different masses of isotopes but also what percentage of each is present in the element as it occurs in nature. These numbers can be used to calculate the average mass for all the atoms of an element. This average is called the *atomic weight of the element*. During the the 4 to 5 billion years of the earth's existence, the isotopes of the elements have been pretty well mixed up, so that a sample of almost any element (with a few exceptions) taken from one place in the earth has the same atomic weight as that of any other. The atomic weights of the elements are recorded in Appendix A. As we shall see in Chap. 9, some isotopes of some elements are unstable; they decompose by a process we call radioactivity. For these elements, the atomic weight of the stable isotope is recorded in Appendix A.

atomic weight of an element

3.11 Do Chemical Equations Help Solve Environmental Problems?

As we shall see in Chap. 11, air pollution by electric power plants burning coal or oil is a severe problem. These fuels generally contain significant amounts of sulfur, and when they burn, the sulfur reacts as follows:

$$S + O_2 \longrightarrow SO_2 \tag{3.6}$$

*Expressions of numbers as powers of 10 are explained in Appendix B.

Sulfur dioxide, SO_2, is a gas that is very irritating to the respiratory system and may cause irreversible damage to the lungs. When SO_2 leaves the smoke-stack of the power plant, it reacts with oxygen and water vapor in the air to form sulfuric acid:

$$2SO_2 + O_2 + 2H_2O \longrightarrow 2H_2SO_4 \qquad (3.7)$$

The sulfuric acid gets washed to the earth in rain. It is very corrosive and does severe damage to living plants and animals and to many kinds of stone used in statuary and buildings.

To determine the amount of air pollution from a given fuel, we must know the *weight* of sulfur in the fuel and what *weight* of SO_2 and H_2SO_4 will be produced from this weight of sulfur. That is, we need to know the relation between the *weights* of the *reactants* and the weights of the *products* in the chemical reactions. Only a knowledge of sulfur chemistry can help solve some air pollution problems.

3.12   Weight Relations in Chemical Reactions

Equation (3.6) states that

1 atom of sulfur reacts with 1 molecule of oxygen

to produce

1 molecule of sulfur dioxide

Equation (3.7) states that

2 molecules of SO_2 react with 1 molecule of O_2 and 2 molecules of H_2O

to produce

2 molecules of H_2SO_4

The weight of 1 molecule of SO_2 must be the sum of the weights of 1 atom of S and 2 atoms of O. Using the table of atomic weights in Appendix A (and rounding off the values to the nearest 0.1 unit to simplify our calculations), we see the following relationships.

Substance	Weight, amu
1 atom of S	1 × 32.1 or 32.1
2 atoms of oxygen	2 × 16.0 or 32.0
1 molecule of SO_2	64.1
1 molecule of O_2	2 × 16.0 or 32.0

Substance	Weight, amu
2 atoms of H	2×1.0 or 2.0
1 atom of oxygen	1×16.0 or 16.0
1 molecule of H_2O	18.0
2 atoms of H	2×1.0 or 2.0
1 atom of S	1×32.1 or 32.1
4 atoms of oxygen	4×16.0 or 64.0
1 molecule of H_2SO_4	98.1

We can add this information to Eq. (3.6) thus:

$$S \quad + \quad O_2 \quad \longrightarrow \quad SO_2 \qquad\qquad (3.6)$$

1 atom + 1 molecule $\longrightarrow$ 1 molecule

$$\underbrace{32.1 \text{ amu} + 32.0 \text{ amu}} \longrightarrow 64.1 \text{ amu}$$
$$64.1 \text{ amu} \qquad\quad \longrightarrow 64.1 \text{ amu}$$

We note that this last statement checks with the law of conservation of matter.

We can also add this information to Eq. (3.7) thus:

$$2SO_2 \quad + \quad O_2 \quad + \quad 2H_2O \quad \longrightarrow \quad 2H_2SO_4 \quad (3.7)$$

2 molecules + 1 molecule + 2 molecules $\longrightarrow$ 2 molecules

2(64.1 amu) + 1(32.0 amu) + 2(18.0 amu) $\longrightarrow$ 2(98.1 amu)

$$\underbrace{128.2 \text{ amu} + 32.0 \text{ amu} + 36.0 \text{ amu}} \longrightarrow 196.2 \text{ amu}$$
$$196.2 \text{ amu} \qquad\qquad\qquad \longrightarrow 196.2 \text{ amu}$$

Again note that this last statement checks with the law of conservation of matter.

Unfortunately, we do not have balances that can weigh such small masses as 128.2 amu, and so we think about Eq. (3.7) as follows:

IF

128.2 amu of SO_2 reacts with 32.0 amu of O_2 and 36.0 amu of H_2O to form 196.2 amu of H_2SO_4

THEN

1 million $\times$ 128.2 amu of SO_2 reacts with 1 million $\times$ 32.0 amu of O_2 and 1 million $\times$ 36.0 amu of H_2O to form 1 million $\times$ 196.2 amu of H_2SO_4

AND

1 billion $\times$ 128.2 amu of SO_2 reacts with 1 billion $\times$ 32.0 amu of O_2 and 1 billion $\times$ 36.0 amu of H_2O to form 1 billion $\times$ 196.2 amu of H_2SO_4

AND

Any number × 128.2 amu of SO_2 reacts with *the same number* × 32.0 amu of O_2 and *the same number* × 36.0 amu of H_2O to form *the same number* × 196.2 amu of H_2SO_4

AND

N × 128.2 amu of SO_2 reacts with N × 32.0 amu of O_2 and N × 36.0 amu of H_2O to form N × 196.2 amu of H_2SO_4

When we set up a quantitative study of a chemical reaction in the laboratory, it is convenient to weigh from one to a couple of hundred grams of the substances involved. To make this possible we use a very large value for N so that when it multiplies the very small value of 1 amu in grams (1 amu = 1.66 × 10^{-24}), we'll get convenient amounts of substances to be weighed.

The number we use as the value for N is

602,000,000,000,000,000,000,000

Avogadro's number
mole

which is more conveniently written as $6.02 × 10^{23}$. This number is frequently referred to as *Avogadro's number* (Sec. 6.8): it is also called 1 *mole*. The word "mole" comes from the Latin "moles," which means a heap or a pile. In the chemist's vocabulary *a mole is a number*, like a dozen or a score. One mole of sulfur means $6.02 × 10^{23}$ atoms of sulfur; 1 mole of sulfur dioxide means $6.02 × 10^{23}$ molecules of sulfur dioxide. One mole of sulfur dioxide contains $6.02 × 10^{23}$ atoms of sulfur and $2 × 6.02 × 10^{23}$ atoms of oxygen.

We use the value $6.02 × 10^{23}$ for 1 mole because it converts the weights of *atoms* and *molecules* expressed in amu into weights of *substances* expressed in grams. Thus we have

1 atom of S weighs 32.1 amu
1 mole of S atoms ($6.02 × 10^{23}$ S atoms) weighs 32.1 g

1 molecule of SO_2 weighs 64.1 amu
1 mole of SO_2 ($6.02 × 10^{23}$ molecules of SO_2) weighs 64.1 g

1 molecule of H_2SO_4 weighs 98.1 amu
1 mole of H_2SO_4 ($6.02 × 10^{23}$ molecules of H_2SO_4) weighs 98.1 g

This information can be added to Eq. (3.7) thus:

$$2SO_2 \; + \; O_2 \; + \; 2H_2O \longrightarrow 2H_2SO_4 \qquad (3.7)$$
$$2(64.1) \; + \; 32.0 \; + \; 2(18.0) \longrightarrow 2(98.1)$$
$$\text{grams} \qquad \text{grams} \qquad \text{grams} \qquad \text{grams}$$

molecular weight
of a compound

This last statement also checks with the law of conservation of matter. The weight of 1 mole of a compound is often called the *molecular weight of a compound*.

3.13 Solving Environmental Problems with the Aid of Chemical Equations

EXAMPLE 3.5 How many grams of sulfur dioxide are produced when 100 g coal containing 3% sulfur is burned?

ANSWER 100 g of this coal contains 3 g sulfur. The sulfur burns thus:

$$S \quad + O_2 \longrightarrow SO_2$$

1 atom	$\longrightarrow$ 1 molecule
1 mole	$\longrightarrow$ 1 mole
32.1 g	$\longrightarrow$ 64.1 g
3 g	$\longrightarrow$ x g

The last two statements may be translated thus:

If 32.1 g S produces 64.1 g SO_2
Then 3 g S will produce x g SO_2

This may be restated as a proportion:

32.1 g is to 64.1 g as 3 g is to x g

or

32.1 g : 64.1 g = 3 g : x g

To solve a proportion, we equate the product of the extremes with the product of the means:

$$32.1x = (64.1)(3)$$

Solving for x: $\qquad x = \dfrac{(64.1)(3)}{32.1}$ or 6 g

When 100 g of this particular brand of coal is burned, 6 g sulfur dioxide will appear in the stack gases.

EXAMPLE 3.6 How many grams of sulfuric acid are produced when 100 g coal containing 3% sulfur is burned and the sulfur dioxide produced in the stack gases reacts with moist air?

ANSWER The 3 g sulfur in 100 g coal is transformed into sulfuric acid in two steps:

$$S + O_2 \longrightarrow SO_2 \quad \text{and} \quad 2SO_2 + O_2 + 2H_2O \longrightarrow 2H_2SO_4$$

We have already found that 3 g sulfur in the coal produces 6 g sulfur dioxide in the stack gas, so we need to use only the second equation to solve this problem.

$$2SO_2 \;+\; O_2 + 2H_2O \longrightarrow 2H_2SO_4$$

$$\begin{array}{ll}
\text{2 moles} & \longrightarrow \text{2 moles} \\
\text{2(64.1) g} & \longrightarrow \text{2(98.1) g} \\
\text{128.2 g} & \longrightarrow \text{196.2 g} \\
\text{6 g} & \longrightarrow \text{x g}
\end{array}$$

Then

$$128.2 \text{ g} : 196.2 \text{ g} = 6 \text{ g} : x \text{ g}$$
$$128.2x = (196.2)(6)$$
$$x = \frac{(196.2)(6)}{128.2} \text{ or } 9.18 \text{ g}$$

When we burn 100 g coal containing 3% sulfur, we pollute the atmosphere with 9.18 g sulfuric acid.

EXAMPLE 3.7 A medium-sized electric power generating plant burns 1000 tons/day of coal. If the coal contains 3% sulfur, how many tons of sulfuric acid are dumped into the atmosphere from such a plant each day?

ANSWER From Example 3.2 we learned that burning *100 g* of coal containing 3% sulfur produced *9.18 g* of sulfuric acid. Therefore, burning *100 tons* of coal containing 3% sulfur will produce *9.18 tons* of sulfuric acid. The power plant burning 1000 tons/day of this coal pollutes the atmosphere with 10 times as much sulfuric acid, or 91.8 tons/day sulfuric acid.

Summary

To understand and deal with our environment, we must be able to measure it quantitatively as well as make qualitative distinctions about it, since questions about "how much" are often as important as questions about what kind. In their measurements scientists now use the International System of Units (SI units)—a modern outgrowth of the metric system created by the French at the end of the eighteenth century—in which the principal units are fractions and multiples of the meter for length, area, and volume; the kilogram for mass or weight; and the second for time.

After extensive experiment and measurement, chemists have evolved three natural laws concerning mass (weight) relationships: the law of the conservation of matter, the law of constant composition, and the law of multiple proportions. These laws describing the behavior of matter were the basis for a mental model, the atomic theory, proposed by John Dalton in his *New System of Chemical Philosophy* (1808–1810). The atomic theory holds that all matter is composed of chemical elements; that each element is made up of a unique combination of invisible particles called atoms; that all atoms of a given element have the same size, shape, and mass; and that atoms of different elements form molecules with masses equal to the masses of the elements within them.

The weights of atoms have been measured, and are given in grams per mole in the table of atomic weights.

Chemists abbreviate the names of elements and use these symbols in formulas, which describe the chemical content of substances, and equations, which describe reactions. All chemical thinking involves both qualitative and quantitative considerations, and neither is fully meaningful without the other. Thus the chemist and the environmentalist need to know not only what pollutants are being released into the air but in what quantities, if they are to correct the situation.

New Terms and Concepts

ATOMIC MASS UNIT: 1 amu is 1.66×10^{-24} g.

ATOMIC WEIGHT OF AN ELEMENT: The weight (mass) of 1 mole of atoms of the element.

AVOGADRO'S NUMBER: 6.02×10^{23}.

CENTIMETER: $\frac{1}{100}$ of a meter.

FORMULA: Symbol for the kinds and relative amounts of chemical elements in a compound.

GRAM: $\frac{1}{1000}$ part of the standard of mass, the kilogram.

GRAVITATIONAL ATTRACTION: All matter exerts a force of attraction on all other matter; this attraction is called gravitational.

GRAVITATIONAL MASS: The mass of a sample of material which is determined by balancing it against known reference masses on a balance.

ISOTOPE: Any of two or more species of atoms of a chemical element; isotopes differ in mass per atom.

KILOGRAM: The standard of mass in the metric system.

LAW OF THE CONSERVATION OF MATTER: In a chemical reaction, matter is neither created nor destroyed.

LAW OF CONSTANT COMPOSITION: All samples of a particular pure compound contain the same percentage of constituent elements.

LAW OF MULTIPLE PROPORTIONS: If two elements form more than one compound, the mass of element A combined with a given mass of element B in the first compound is a simple fraction or simple whole number multiple of the mass of element A combined with the same given mass of element B in the second compound.

LITER: The volume occupied by one kilogram of water at 4°C.

METER: The standard of length in the metric system.

MOLE: 6.02×10^{23} particles. The word is often used as a short way of saying "the mass of 1 mole."

MOLECULAR WEIGHT OF A COMPOUND: The mass of 1 mole of the molecules of the compound.

NATURAL LAW: A description of the behavior of matter.

Testing Yourself

3.1 What is the law of constant composition? The law of multiple proportions? What kinds of experiments led to the formulation of these laws?

3.2 What information does a chemical formula convey? Write the formulas for the following common compounds: carbon monoxide, carbon dioxide, sulfur dioxide, sulfur trioxide, carbon tetrachloride.

3.3 Write balanced chemical equations for the following statements about chemical reactions.
   a. Copper sulfide (Cu_2S) burns in air at a high temperature to form copper oxide (CuO) and sulfur dioxide.
   b. Ethyl alcohol (C_2H_5OH) burns to form carbon dioxide and water.
   c. Hydrogen peroxide (H_2O_2) is a colorless liquid that readily decomposes into water and oxygen.

3.4 What is a mole? Sugar has the formula $C_{12}H_{22}O_{11}$. How many atoms are present in 1 molecule of sugar? How many atoms are present in 1 mole of sugar? How many atoms of carbon are present in 1 mole of sugar?

3.5 What is the atomic weight of copper? How many copper atoms are contained in this weight of copper?

3.6 What is the molecular weight of (a) hydrogen, (b) oxygen, and (c) water?

3.7 How many moles of carbon dioxide are present in 44 g of the gas? In 100 g of the gas?

3.8 How many moles of water are present in 1 liter of pure liquid water?

3.9 If you want to measure out 1 mole of liquid water, what volume should you take?

3.10 What is the percentage of oxygen present in water (H_2O)?

3.11 Write a balanced chemical equation for the burning of glucose ($C_6H_{12}O_6$) in oxygen to produce carbon dioxide and water. When you eat glucose, you convert it to carbon dioxide and water in your body. How many grams of carbon dioxide do you have to excrete in your breath when you eat a spoonful of honey that contains about 25 g glucose?

The basic substance in recycled paper, shown in the
electron micrograph, is the cellulose unit shown on page 57.

4. What Can We Do about Solid Wastes?

Now that you know a bit about elements and compounds, atoms and molecules, formulas and chemical reactions, we can consider some of the simpler chemical problems associated with environmental pollution. It is convenient to classify pollutants in three main kinds: solid wastes that litter our landscapes, gases and smokes that pollute our air, and liquids and solids that are discharged into our water supplies.

In this chapter we shall consider the problem of solid waste disposal. In Chap. 7, when you know more about the nature of gases and how they behave in the earth's atmosphere, we shall study air pollution. And in Chap. 13, when you have a broader understanding of the nature of solutions, the kinds of chemical compounds responsible for pollution, and how these compounds affect us, we shall discuss water pollution and how we can keep pollutants out of our water supplies.

Solid wastes include (1) domestic wastes, such as garbage, paper, glass, and metal cans; (2) paper from business offices, stores, packaging plants, publishers, and printers; (3) rubble from demolished buildings, including brick, stone, concrete, plaster, wood, asphalt and tile roofing, asbestos insulation, ceramic tile, and other rubble of many kinds; (4) old automobiles, trucks, tractors, and buses; (5) old rubber tires; (6) sludge from sewage-treating plants; and (7) industrial wastes, which include solids and sludges left over from various manufacturing processes.

Relatively little is known about effective procedures for the disposal of industrial wastes, and what is known is extremely complex. Consequently we shall not say much about that topic in this book. We shall, however, deal in Chap. 13 with the disposal of sludge from sewage. And in this chapter, we shall discuss the first five of the seven items listed above.

4.1 Our Problem

Seventy-one billion cans, 38 billion bottles, 200 million rubber tires, 8 million junked cars—that's the annual harvest of four kinds of solid waste in the United States today. All told, we discard 360 million tons/yr of trash. Our mines, farms, forests, factories, businesses, and households are turning out more than 100 lb/day of solid wastes for each of us. Solid debris piles up in our city streets and clutters our countryside. In the past 25 years we have been on a consumers' spree. We have been converting the wealth of our natural resources to consumer goods and using them up at an ever-increasing pace. In the production, distribution, and marketing of consumer products we have astounded the rest of the world. We pick and choose from a prodigious array of widely advertised goods; then we throw away the paper, glass, plastic, and metal containers they came in. Production-consumption has been a one-way street (Fig. 4.1).

Only now are we beginning to see how the *law of the conservation of matter* affects our daily lives. This law might well be stated thus: "Stuff doesn't go away; it just stays there." We must learn how to make a circle of the one-way street, reusing glass and plastics, paper and metals, lumber and building materials (Fig. 4.2).

Our failure to complete the cycle accounts for our consigning valuable and recoverable resources to dumps and fouling the air and water by open burning. The Bureau of Solid Waste Management (a unit of the U.S. Environmental Protection Agency) reports that 90 percent of all waste is dumped without treatment, and of this 80 percent is burned in open dumps. Approximately 94 percent of the dumps in the United States are not just poor but unacceptable from the standpoint of pollution control.

FIGURE 4.1 The one-way street of the "cowboy economy"

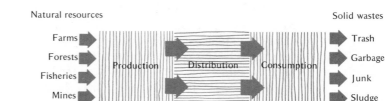

FIGURE 4.2 The conservation cycle of the "spaceman economy"

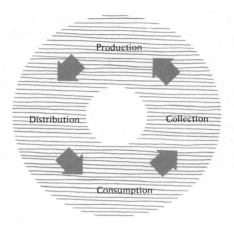

Seaboard cities have simply hauled their trash and garbage out to sea and dumped it. New York City did this for years until 1946, and created an undersea mountain range of refuse a few miles off the shores of Long Island. Complaints of citizens outraged by the resulting pollution of beaches finally put a stop to this practice.

In 1968 we dumped in the seas along our coasts 5 million tons of industrial waste, 5 million tons of sewage sludge, 26,000 tons of garbage and refuse, 500,000 tons of construction and demolition debris, and 15,000 tons of outdated military explosives and chemicals. The 5 million tons of industrial wastes included 2.5 million tons of waste acids, 500,000 tons of oil refinery wastes, 300,000 tons of pesticide wastes, and 150,000 tons of paper mill wastes. In addition we dumped into the seas 50 million tons of dredged material from the deepening of harbors.

At present there is scant data on the effects of these pollutants on the sea and its inhabitants. But realization of the folly of continuing to dump wastes into the seas has given rise to various international restraints. Twelve European countries (Belgium, Denmark, Finland, France, West Germany, Iceland, the Netherlands, Norway, Portugal, Spain, Sweden, and the United Kingdom) have agreed to ban dumping of industrial wastes in the North Atlantic. Unfortunately, the Baltic and Mediterranean Seas are not included in the ban. Pollution is steadily increasing in both these seas, and oceanographer Jacques Piccard predicts that they will be "dead" seas* in 25 years if the dumping is not abated. At the U.N. Conference on the Human Environment in Stockholm in 1972, it was agreed that all dumping of human and industrial wastes in the oceans should be curtailed, but an international treaty to this effect

*A *dead sea* is one in which the balance among various plants and animals has been so upset by pollution that desirable fish can no longer live in it.

is yet to be drawn up. In the summer of 1970, the U.S. Army sank the *S. S. LeBaron Russell Briggs* (Fig. 4.3) in the Atlantic Ocean with tons of nerve gas aboard. The outrage aroused by this incident caused the United States to assure the world that it would never again dump weapons of chemical warfare into the sea.

4.2 Disposal in Sanitary Landfills

sanitary landfill

The simplest, cheapest, and easiest alternative to throwing refuse on an open dump or into the sea is the *sanitary landfill*. This is a disposal area where solid waste is deposited, spread in layers, compacted, and covered daily with earth. This process buries solid wastes under an earth cover to seal off odors and starve rats and other carriers of disease. On flatland a deep trench is dug with a large enough area for the operation of dump trucks and bulldozers. The earth cover each day is obtained by extending the excavation of the trench (Fig. 4.4).

Landfills 100 ft deep have been constructed. They subside about 4 ft a year for the first five years as the refuse decomposes and settles. After this, subsidence is slight. In areas which have undesirable gullies and ravines, or abandoned strip mines or quarries, sanitary landfills can level up terrain for

FIGURE 4.3 *S. S. LeBaron Russell Briggs*, bearing a cargo of nerve gas, sinks into the Atlantic.

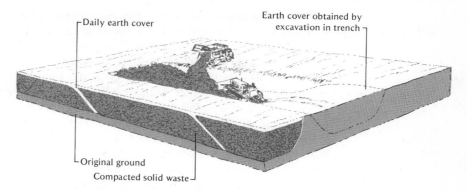

Daily earth cover

Earth cover obtained by excavation in trench

Original ground

Compacted solid waste

FIGURE 4.4 The technique of sanitary landfill

recreational facilities such as golf courses, playgrounds, and picnic groves. To produce hills on flatlands, refuse may be built into a sizable mound and covered with earth each day. One such construction is "Mount Trashmore" southwest of Chicago, where a developer expects to create a ski slope and winter sports area in this otherwise flat countryside.

The primary problem with sanitary landfills is how to control blowing litter and dust during the daily dumping and covering. Close-mesh fences can reduce blowing, and dampening the earth used for fill will minimize dust. Secondary problems arise from the movement of gases and groundwater through the fill. Care must be taken to provide escape channels for gas through loose soil, and to minimize the flow of water into adjacent water-bearing strata of underground rock, sand, or gravel, or into surface streams. About 20 percent of our untreated refuse is now handled in sanitary landfills. Though sanitary landfills are highly preferable to open dumps, they do not without some sorting process allow recycling of precious natural resources, which we are using up at an alarming rate.

In Japan, where there is much less land per person than in the United States, refuse is compacted before it is used as landfill. Osaka, Japan's most highly industrialized city, has a plant for compressing 600 tons/day of unsorted industrial and household refuse. The material is placed in presses and reduced to one-fifth its original volume, forming 2-ton blocks about 4 ft square. These blocks are packed in wire netting, coated with asphalt, and then used as landfill. Small compactors for home use are coming onto the United States market.

4.3 Composting

When plant residues (such as dead leaves, stalks, and flowers), animal manure, and dead animals lie on moist ground, they quickly become food for a

variety of organisms (living creatures) including bacteria, fungi, worms, insect
metabolism   larvae, and molds. The *metabolism*—the sum of the chemical processes in the
building up and destruction of living organisms and their components—of
these creatures produces gases that pass into the air, new chemical substances
that are used as food by other organisms, and a dark-colored, formless, and
humus   structureless mass called *humus*, which is an important ingredient in the
composition of all rich soils.

nutrient   Humus serves important purposes in the life of soil, all of which contribute
to its fertility. It retains essential plant *nutrients* such as potassium, nitrogen,
and phosphorus. It holds moisture easily and so prevents soils from drying out
rapidly. It also keeps soil porous and permits the ready flow of air needed for
root growth. The pores also help in the percolation of water downward to provide good drainage.

composting   The processes involved in the formation of natural humus can be greatly
speeded by *composting* dead plant and animal material. Leaves, straw, grass
clippings, garbage, paper, animal manures, plant and animal wastes from food
processing plants—any dead plant or animal material—can be composted.
The material is shredded and formed into long piles 4 or 5 ft high and 8 to
aerate   12 ft wide. It is moistened and turned periodically to keep it well *aerated*
microorganism   (supplied with air). Under these conditions *microorganisms* (living creatures
native to the soil, too small to be seen by the naked eye but visible with a
microscope) grow rapidly. The heat liberated by their metabolic chemical
reactions causes a rise in temperature which rapidly destroys disease-producing bacteria. The growth of the soil microorganisms produces no objectionable
odors if the compost is kept well aerated. The yield of compost is 40 to 60
percent of the volume of the refuse treated. It is dried and sold to farmers
and gardeners for use as a soil conditioner. Composted animal manures are
rich in plant nutrients and are used as fertilizers for high-value crops like
flowers, strawberries, and vegetables.

Before mixed refuse can be composted, metal trash must be removed from
it. Glass and plastic do not interfere with composting if they are broken into
small bits by the shredding process. Microorganisms attack the sharp corners
of fractured glass and round off cutting edges. Though the cost of composting
municipal waste is 8 to 10 times that of disposal by sanitary landfill, composting is widely used in Europe, where land is too scarce to be used for sanitary
landfill. As we run out of empty lands near major cities, it is likely that composting will increase in our country.

4.4   Fermentation to Produce Fuel Gas

fermentation   *Fermentation* is the name given to the chemical processes which take place
in a substance due to the metabolism of microorganisms. The fermentation

of sweet and starchy liquids by yeast to produce alcoholic beverages has been used by man through recorded history. The leavening of bread (the production of CO_2 gas bubbles to lighten the texture) by the action of yeast in dough has also been used for countless centuries.

aerobic
anaerobic

When dead plant and animal materials are acted upon by microorganisms *anaerobically* (in the absence of air) instead of *aerobically* (in the presence of air, as in composting), combustible gas is produced. The principal component of this gas is methane, CH_4, which is also the major component of natural gas. When burned with plenty of air, methane reacts with oxygen to produce CO_2, water, and heat. If methane is burned in a limited supply of air, it produces carbon monoxide, CO. This poisonous gas can be released into the home if methane is burned in an improperly vented heater.

With plenty of air:

$$CH_4(g)^* + 2O_2(g) \longrightarrow CO_2(g) + 2H_2O(g) + \text{heat}$$

With limited air:

$$2CH_4(g) + 3O_2(g) \longrightarrow 2CO(g) + 4H_2O(g) + \text{heat}$$

The chemical reactions producing methane by fermentation are complicated and not well understood.

If metals and rocky material, such as concrete, bricks, and rubble from demolition, are first removed from municipal waste, the remainder may be used to produce methane. The plant and animal material is shredded and placed in closed tanks which are insulated to minimize loss of heat. The ensuing fermentation raises the temperature to the range of 120 to 150°F, where gas is produced 3 times faster than at room temperature. It is estimated that anaerobic fermentation of the 2.5 billion tons of agricultural wastes, and the 400 million tons of fermentable urban wastes produced each year in our country might supply as much as 25 percent of the nation's consumption of methane.

Moreover, when no more gas is evolved, the remaining end product is humus.

4.5 Incineration

About 10 percent of our solid wastes is burned in air incinerators. Burning at usual furnace temperatures reduces waste to about a quarter of its original volume and a third of its weight. If the waste is burned at 1500°F, 95 percent of the weight of the combustible material it contains is burned off. The noncombustible glass and metals accumulate in the ashes. Unfortunately, 75 per-

*The symbol (g) following a formula means that the substance is a gas.

cent of the incinerators operating in 1970 were deemed inadequate by the Bureau of Solid Waste Management because they lacked adequate control of air pollution. Many improvements need to be made before incineration provides an acceptable solution to our solid-waste problems.

One of the most attractive features of incinerators is the possibility of using the heat they generate to make steam for industrial heating or to produce electricity. Chicago's new northwest incinerator burns 1600 tons/day of municipal waste and provides 100,000 lb/hr of low-pressure steam to local industries. The cost of incineration is between $5 and $8 per ton of refuse. The sales value of by-product steam is about $5.50 per ton of refuse. This brings the net cost of incineration down near the cost of disposal in sanitary landfill (Table 4.1).

It is possible to burn relatively small amounts of highly objectionable wastes—such as dead animals and manures from meat-raising or meat-processing plants, pesticide and drug manufacturers' refuse, and waste industrial chemicals—in incinerators fired by gas or fuel oil. But while burning may solve a problem of local disposal for a particular industry, its use of precious natural gas or fuel oil prevents its being applied on a large scale. Some indus-

TABLE 4.1 *Costs of Solid Waste Disposal*

Method	Dollars/ton
Promiscuous dumping and littering	0.00
Cleaning up litter	4.00 to 40.00
Open dumps, usually with burning	0.50 to 2.00
Sanitary landfill	1.00 to 3.50
Additional cost of gas and odor control (estimated)	0.10 to 1.00
Additional cost of water pollution control by encapsulation* (estimated)	0.20 to 1.00
Additional cost of landscaping	0.05 to 0.50
Incineration by current technology	8.00 to 14.00
Additional cost of air pollution control	1.00 to 2.00
Additional cost of heat recovery	2.00 to 4.00
Composting	8.00 to 30.00
Sea disposal of bulk material	1.00 to 10.00
Sea disposal of baled, barreled, or otherwise contained material	7.00 to 50.00

Encapsulation is the use of waterproof membranes around deposits of solid wastes to prevent the seepage of groundwater through the deposit.

tries are developing schemes for the complete treatment of all their wastes—gaseous, liquid, and solid—by incineration. They would then recover and use waste heat in manufacturing processes or in steam-driven electric generators. Some incinerators have been designed to run at 2000 to 3000°F. These will accept unsorted solid wastes including such items as refrigerators, bicycles, auto tires, and sewage sludge, and produce a residue with only 3 percent of the weight of the feed. Molten metals and glass are also produced in this furnace; these by-products can be withdrawn from time to time and used as raw materials for various manufacturing processes (see Sec. 4.7).

4.6 Pyrolysis

pyrolysis

When solid wastes consisting largely of animal refuse, garbage, and combustible trash are heated for about 6 hr at about 900°C in a container that excludes air, they produce a complicated mixture of gases, liquids, and solids that can be used as fuel. This method of heating without air is called *pyrolysis*.

In a second method of pyrolysis, carbon monoxide (CO), steam (H_2O), and sodium carbonate (Na_2CO_3) are injected into solid wastes in the pyrolysis chamber at high pressure and at 380°C. In 20 min, a ton of dry solid waste produces about 120 gal of a heavy oil that is an excellent fuel for steam boilers. As we have seen, the burning of fuels containing sulfur is a source of air pollution (Chap. 3). Oil from pyrolysis of trash contains only 0.35% sulfur—a figure much lower than the percentages in fuel oils prepared from many widely used petroleums. The second method of pyrolysis promises well for the handling of animal manures from feedlots, now a great source of water pollution. Feedlots may contain 1000 to 50,000 animals at a time, and the disposal of their manure is a formidable problem. The solid waste produced by our farm animals is 20 times that of the human population. If all agricultural solid wastes produced each year in this country were converted to oil by the second method of pyrolysis, they would provide almost half of our present annual consumption of fuel oil.

4.7 Recycling of Materials in Solid Wastes

Depletion of high-grade copper and lead ores, those with a high percentage of metal, has necessitated the increasing use of lower grades that require much more extensive and costly processing. Consequently, the prices of copper and lead have risen so rapidly in the last two decades that almost half our current copper and lead production is from salvaged scrap. The recovery of aluminum, zinc, and nickel is also profitable, and a fifth of the annual production of these metals is from salvage. A fourth of our steel production is from scrap.

The iron in tin-coated cans—"tin cans"—is useful in releasing copper from

Proper disposal of solid wastes, such as soft-drink cans, can help clean up our recreational areas.

solutions prepared by treating copper ore with sulfuric acid. When shredded "tin cans" are added to this acid solution, the iron dissolves and the dissolved copper comes out as a metal powder on the surface of the shreds. This powder is then washed off, collected, and melted into copper ingots. The acid solution containing the iron is thrown away, creating water pollution problems as noted in Chap. 13. Over 100,000 tons of cans are used in this process in the southwestern United States each year. This is only 2 or 3 percent of the total national tonnage of cans discarded, and it cannot be greatly increased because it costs too much to ship tin cans from other parts of the country. Consequently, significant increases in the use of discarded tin cans must involve returning them to iron and steel mills.

Recycling of plastics is complicated by the many different compounds in different products. This makes it likely that plastic containers in trash will have to be returned for reuse in their original function. Other forms of plastics

biodegradable

in trash will not be returned for refabrication. They may be chopped up and used for some new purpose such as building material or soil conditioner. Acid- and corrosion-resistant "plastic" pipes can be fabricated from crushed glass and shredded plastic. Some manufacturers of plastics are now seeking ways of making their products *biodegradable*, that is, decomposable by natural processes to yield harmless or even useful components to the soil.

The recycling of glass is now a well-organized business. Glass manufacturers and processors grind up glass for use as aggregate (the rocky component) in special types of concrete, for filler in bitumen paving, in the making of bricks and tiles, and for remelting to fabricate glass forms in which color and transparency are not important. Metal needs to be removed because it causes black specks in remelted glass. Glass industry spokesmen have indicated that scrap might supply half the material used to make glass. Because the cost of trash glass at the factory is somewhat higher than that of the natural raw materials for glassmaking (sand, limestone, and soda), its recycling must somehow be paid for. But scrap is easier to melt than the raw materials, and so it has some advantages. It would be far more economical if glass containers were returned whole for reuse. To accomplish this will require a major effort in consumer education.

Rubber from discarded tires can be used as a filler in dressings for heavily used blacktop driveways and parking lots. It improves durability, resiliency, and resistance to abrasion. Research on this process indicates that about 25 percent of the rubber from the 200 million tires discarded annually could be used this way. Rubber stripped from old tires and shredded into thin strands can be made into an artificial turf for use on highway median strips and in playgrounds. A mixture of shredded rubber tires and plastic scraps can absorb oil spills to produce a conglomerate useful as an extender for asphalt in roofing, paving, and floor surfacing. Another promising process is pyrolysis, which produces carbon black (an oily soot consisting largely of finely divided solid carbon) and many other valuable chemical raw materials. Carbon black is used extensively in the fabrication of rubber tires; researchers with Cities Service–Goodyear estimate that all the tires scrapped each year could be used as a source of the carbon black that tire manufacturers need. By changing the rate of heating, the temperature, and the pressure in pyrolysis, the Firestone Tire Company produces different materials, including combustible gas, oils for use as fuel or solvents, and solid residues suitable for smokeless fuel.

cellulose
lignin

About half of the municipal trash is paper, and the inherent value of the *cellulose* fibers in paper make it potentially a very profitable material for recycling. The best paper is made by chemically dissolving the *lignin*, the "glue" holding the long, strong cellulose fibers together in a wood particle. The fibers are then bleached to a pure white, whipped up with water, and matted together to form paper. It is cheaper to grind wood to a pulp in water and then mat this into paper. Grinding shortens the cellulose fibers in the pulp so that the paper is not so strong, and the lignin makes the pulp hard to bleach. Paper made from such "groundwood" is of inferior quality and is used mostly for newsprint.

When paper is recycled, it is repulped by beating in water. This shortens the cellulose fibers so they form a weaker paper when matted into a sheet. Because it is necessary to remove all dyes, pigments, inks, waxes, plastics, glues, and other adhesives added to the paper to make it once more suitable for its original use, it is very costly to purify repulped fibers sufficiently to make them into strong, white paper. Food cartons, towels, tissues, paper cups, and plates are now generally made from repulped paper. Corrugated cardboard boxes and newsprint can be repulped to form new boxes, newsprint, wallboard, and egg cartons. Mixed office waste is used to make roofing felt, shoe boxes, and tablet backs. Newspapers are usable without repulping; they can be formed under very high pressures into building bricks or sheets of building and insulating board. About 20 percent of the paper now being made is recycled.

4.8 The Economics of Solid Waste Disposal

The direct costs of getting rid of municipal solid wastes are summarized in Table 4.1. These costs do not include collection or transportation. From this comparison it is easy to see why 90 percent of all municipal wastes are dumped without treatment. Dumping is far cheaper than any other alternative. When we become sufficiently upset by the pollution of land, air, and water from open dumps, we will be willing to pay higher taxes to cover the increased costs of sanitary landfill. When we can find no more land to fill, we will have to pay still higher taxes to build and run fermentation plants, incinerators, and furnaces for pyrolysis. Pyrolysis looks especially promising because of the value of the by-products. Composting has not been able to compete in cost with disposal by furnace. If the separation of metals, glass, and building debris from trash becomes widespread, composting may become competitive in the treatment of the remaining material, especially in regions where crops of high value are grown near the source of compostable debris.

As the technology of incinerating or pyrolyzing solid wastes is improved, it is likely that the recovery of by-products such as steam, metals, glass, and usable ash will lower the net cost of disposal almost to the cost of landfill. The Aluminum Association has prepared detailed plans for a plant costing $15.9 million to process 500 tons/day of municipal waste from a city of 200,000. This operation could produce salable products worth about $1.5 million a year and would provide a net annual revenue of about $133,000.

TABLE 4.2 *Estimated Capital Expenditures for Solid Waste Treatment*

Type of equipment	Dollars/yr (millions)	
	1970	1980
Onsite incinerators (commercial, industrial, institutional, government, apartment)	165	650
Compactors	25	100
Vehicles (packers, sweepers, trucks)	511	2200
Municipal incinerators	200	500
New processing (separation, composting, pyrolysis)	—	150
Miscellaneous	75	350
Total	975	3950

In the city of Franklin, Ohio, municipal wastes are beaten to a pulp with water. Heavy materials are ejected and iron removed by an electromagnet. Paper fiber is recovered for reuse. Glass is extracted and separated into various colors by an optical sorting device. Aluminum is reclaimed from the residue.

Chem Systems, a waste disposal consulting firm in New York City, estimates that spending for solid waste handling equipment will increase markedly in the 1970s (Table 4.2).

4.9 Governmental Action to Improve Waste Disposal

The Solid Waste Disposal Act of 1965 marked the first significant interest by the federal government in management of solid wastes. The act provided for assistance to state and local governments and to others involved in solid waste problems, by financial grants to demonstrate new technology, technical assistance through research and training, and encouragement of proper planning for state and local solid waste management programs. The Resource Recovery Act of 1970 amended the act of 1965 to provide a new emphasis on the recovery of valuable waste materials.

The United States Environmental Protection Agency (EPA) was established December 2, 1970, to bring together for the first time in a single agency the major environmental control programs of the federal government. The EPA is charged with organizing an integrated, coordinated attack on problems of air and water pollution, solid wastes, pesticides, radiation, and noise. At present the EPA

1 performs research to find improved methods in all aspects of solid waste management and provides technical assistance to speed the application of new knowledge

2 makes financial grants for the construction and operation of plants for demonstrating new technology (the plant at Franklin, Ohio, is supported by the EPA)
3 is developing a plan for a system of national disposal sites for storage and disposal of hazardous wastes
4 provides training to develop scientific and technological personnel to design, operate, and maintain complex regional systems
5 provides financial assistance to state and local governments and interstate agencies for preparing plans for managing solid wastes

By 1971, a total of 50 state or interstate agencies had used assistance from the EPA in developing plans for solid waste disposal. Various states have set up their own agencies for protecting the environment. New Jersey was the first to make the handling of solid waste a public utility, like the production and distribution of electricity and gas, with franchises available for different areas.

Taxation is sometimes used to encourage environmental improvement. In an effort to reduce the litter from nonreturnable plastic containers, New York City assessed a sales tax of 3 cents a bottle and 2 cents a tube. The Society of the Plastics Industry is seeking a court test of the constitutionality of this tax, contending that it discriminates against plastic containers and discourages recycling.

4.10 Education for Environmental Improvement

The Resource Recovery Act of 1970 authorized $463 million for an all-out recycling approach to the solid waste problem. Individual support of recycling has been strong. In Madison, Wisconsin, and Louisville, Kentucky, where collection programs were well organized, 50 to 70 percent of the newspapers distributed were collected for recycling. In some areas as much as 45 percent of the aluminum cans distributed have been recycled. In spite of these encouraging reactions by the public, governments and private business enterprises have responded poorly. Of the $172 million appropriated by Congress for solid waste studies in 1972, only $29 million was used by the executive branch of the government. Industries are reluctant to spend money on research and development for disposal processes when government is so hesitant. Of the 50 largest cities in the United States, 25 have landfill space sufficient for less than 5 years; 5 more have space for less than 10 years. But with the taxpayer bitterly complaining about the high cost of government, it appears that nothing short of major crises of waste accumulation in our large cities will force the expenditure needed to build and operate furnaces and fermenters to replace dumps. Clearly, a strong program for educating people in solid waste problems and their solutions must be launched. Only an aroused citizenry

The community of Omaha, Nebraska, used glass containers collected by its members to make glassphalt, a material used for paving and repairing city streets.

can force government action now to avert impending crises. Subsidy by government is necessary to provide the new knowledge and technology that must be developed in the next few years.

As long ago as 1966 we spent about $25 billion a year for packaging, and we discarded 90 percent of this material. Today the packaging industry is considerably larger and is promoting increased use of materials for wrapping, sacking, canning, bottling, and otherwise protecting and selling consumer goods. Advertising campaigns extol the convenience of throwaway containers even as our natural resources diminish and our mountains of trash grow higher and higher. The need for education in environmental problems is dramatically exemplified in the controversy over returnable versus throwaway metal or glass containers for beer and soft drinks, for which about half of all food containers are used. We pay 30 percent more for soft drinks in glass throwaways than we would in returnable containers. In addition, as taxpayers we pay for picking up litter, hauling it away, and disposing of it by landfill or some more costly process. Yet we continue to buy billions of throwaways because we don't know these facts. A study of converting the beverage container system to returnables in the state of Illinois showed that this would yield a net increase of 6500 jobs and save consumers $71 million annually.

Not only the packaging industry but wholesale and retail businesses also prefer the throwaway. In 1960 about 40 percent of the roofed area of a supermarket was devoted to nonselling storage; in 1970 only 10 percent was so used. The packaging and distributing industries find that throwaways increase their profits. Consumers must be educated to the real costs of throwaway convenience if we are to eliminate this rapid mode of transferring our natural resources to our dumps.

Summary

The disposal of solid wastes from homes, businesses, and industries on open dumps and in the oceans must stop. Sanitary landfills are better than open dumps, but we must develop systems for collecting and treating solid wastes that will enable us to recycle glass, plastics, paper, and metals. Until efficient recycling processes are developed, we should incinerate solid wastes and use the heat generated in the process. As soon as practicable, incineration should be replaced by pyrolysis to recover fuel oil from dead plant and animal material. Where there are large concentrations of plant and animal wastes (as at vegetable and fruit canneries, meat processing plants, and feedlots), anaerobic fermentation should be used to generate fuel gas.

The processing of solid wastes to recover reusable materials and to use other materials without polluting our air and water supplies will be very expensive. But we must be prepared to pay a high price to maintain an environment of high quality. Governmental agencies are now working toward environmental improvements by better management of solid wastes. But these activities will fade out unless they are supported by citizens who understand the need for them.

New Terms and Concepts

AERATE: To expose to the circulation of air.

AEROBIC: Requiring molecular oxygen or air to live.

ANAEROBIC: Able to live in the absence of molecular oxygen or air.

BIODEGRADABLE: Capable of being transformed by biological processes from one form into another.

CELLULOSE: A complex carbohydrate having the formula $(C_6H_{10}O_5)_x$.

COMPOSTING: Transformation of organic litter into humus.

FERMENTATION: A chemical process that takes place in a system due to the metabolism of microorganisms.

HUMUS: A brown or black mixture of chemical substances consisting of partially or wholly decayed vegetable matter. An important natural ingredient in many good soils.

LIGNIN: A substance similar to cellulose, with which it makes up the essential part of woody tissue.

METABOLISM: The chemical changes in a living cell by which energy is provided for vital processes and activities, new material is assimilated for growth or repair of the cell, and waste products are eliminated.

MICROORGANISM: An animal or plant of microscopic size (can be seen only with the aid of a microscope).

NUTRIENT: A substance supplying food.

PYROLYSIS: Chemical decomposition produced by heat in the absence of air.

SANITARY LANDFILL: A dump of solid wastes covered with earth and designed to minimize pollution of groundwater.

Testing Yourself

4.1 Should the recycling of solid wastes be a responsibility of private enterprise or of some governmental agency? What arguments support your conclusion?

4.2 What are the advantages of a sanitary landfill as a way of disposing of solid wastes? What precautions must be taken in building such a landfill in order to keep it really sanitary?

4.3 What are the advantages and disadvantages of composting as an alternative to sanitary landfill?

4.4 What problems do you see in adopting the process of fermentation of plant and animal material in solid wastes to produce fuel gas for domestic or industrial use? Can you think of any situations where such a gas generator might be economically as well as ecologically valuable?

4.5 What are the advantages and disadvantages of incineration as a method for disposing of solid wastes?

4.6 What advantages does pyrolysis have over incineration? What disadvantages?

4.7 What kinds of solid wastes are most profitably recycled? Least profitably? What can we do to encourage recycling?

4.8 If Omits says to you, "Why should I go to the bother of buying stuff in returnable bottles and toting the empties back to the store?" how would you answer him?

4.9 Do you think we should encourage the development of biodegradable plastics for containers, or require the recycling of the present almost indestructible containers for reuse?

4.10 What do you think should be the role of federal, state, county, and municipal governments in meeting the problems of solid waste disposal?

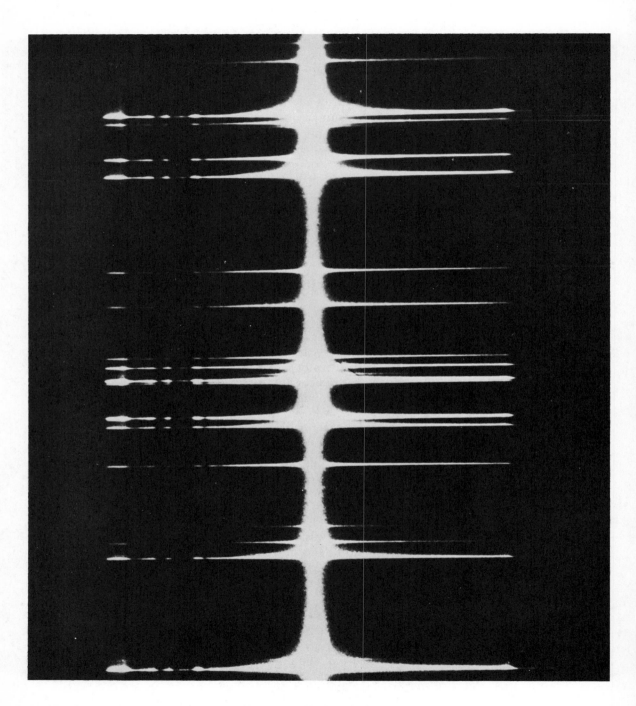

The crystal structure of solid xenon is shown on page 75. A spectrogram
of xenon gas exhibits effects of increased pressure and temperature.

5. The Effects of Heat and Pressure on Gases, Liquids, and Solids

You don't have to be a meteorologist to know that some days are hot, some cold; some sunny, some cloudy; that rains fall, the skies clear, and the ground dries up; that the temperature drops and it snows; that a warm wind blows and the snow melts, or a cold wind blows on a sunny day and the snow disappears without melting; that lakes freeze in winter and thaw in spring; that frost decorates the trees on a cold, damp night and vanishes in the morning sun. All these phenomena involve changes in temperature, and many of them are accompanied by changes in the state of water: from liquid to solid or solid to liquid; from liquid to gas or gas to liquid; from solid to gas or gas to solid.

Generally when we heat an object, it gets warm. What is warmth? Can we measure it? Is it the same as heat? Can we measure heat? If we take an ice cube from a pitcher of ice water and heat it, it doesn't get warm, it melts. Why? Where does the heat go? Why doesn't the ice cube get warmer? When we heat water to 100°C it boils away. Where does it go?

In this chapter we shall look for answers to questions such as these. For changes in *temperature* and changes of state in our environment are often related to changes in energy.

Changes in energy are also often related to changes in *pressure*, when we are dealing with gases. When the mixture of air and gasoline in the cylinder

temperature

pressure

of an engine is ignited by the sparkplug, the explosion (very rapid burning of gasoline) creates a very high pressure. This pressure drives the piston downward in the cylinder, and the motion of the piston turns the crankshaft of the engine. This mechanical energy (energy of motion) is transmitted to the wheels, which move the car. The gasoline engine is essentially a device for changing the chemical energy of the mixture of air and gasoline into mechanical energy. The chemical energy of the air-gasoline mixture comes from the fact that changing the arrangement of the atoms in the mixture to another arrangement of atoms in the products of combustion releases large amounts of heat.

In this chapter we shall see how changes in pressure affect the volumes of gases, liquids, and solids. These changes are of great importance in helping us to understand what goes on in the environment, especially in the air.

5.1 The Difference between Temperature and Heat

Temperature is a measure of the degree of hotness of a substance, but it does not tell us how much heat is present. If we fill a 2-qt bottle with hot water, it stays warm longer than a 1-qt bottle filled with water at the *same temperature*. The total heat in the 2-qt bottle is greater than that in the 1-qt bottle. The *amount of heat* that is absorbed or given off by a substance when its temperature is raised or lowered or it undergoes some other changes is measured as described in Sec. 5.4.

5.2 Measuring Temperature

A simple device for measuring temperature is a mercury thermometer. A bulb containing about a cubic centimeter of mercury is attached to a tube of glass with a very small but uniform inside diameter. When we immerse the bulb in a dish of warm water, the mercury expands as it warms and rises to a certain point in the tube (Fig. 5.1a). As the water cools, the mercury decreases in volume and the thread of mercury in the tube shortens. If we put the bulb in a mixture of ice and water (Fig. 5.1b), the end of the thread of mercury stays in one position as long as both ice and water are present. This position can be used as a fixed point on the thermometer, and a scratch can be made on the glass tube to indicate this temperature.

If we place the bulb in boiling water (Fig. 5.1c), the mercury rises through the tube and stays in one position as long as the water continues to boil. We make a scratch on the glass at the top of the thread of mercury to mark this temperature.

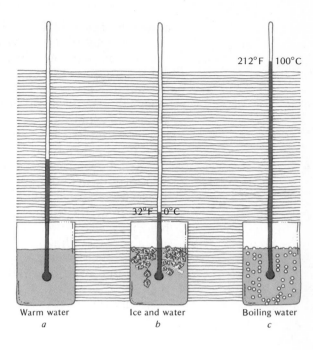

FIGURE 5.1
Lengths of a
column of
mercury in a glass
thermometer
under different
conditions

212°F   100°C

32°F   0°C

Warm water
a

Ice and water
b

Boiling water
c

Fahrenheit scale
of temperature
Celsius scale
of temperature

We now have scratches at two fixed points. On the *Fahrenheit temperature scale* the boiling point of water is marked 212° and the freezing point 32°. On the *Celsius scale* (sometimes called the centigrade scale) established by Anders Celsius, the boiling point of water is called 100°* and its freezing point is called 0°. We distinguish between the two scales by adding F or C to the temperature reading. On a Fahrenheit thermometer there are scratches to show 180 even divisions between 32 and 212°, each to mark 1°F. On a Celsius thermometer there are scratches to show 100 even divisions between 0 and 100°, each to mark one degree Celsius. All scientists and the people of the non-English-speaking world use the Celsius scale; some English-speaking people use the Fahrenheit scale. Various temperatures are compared in °F and °C in Appendix C.

5.3   How Heat Affects the States of Matter

You know that snow melts in the springtime when the thermometer stays above 32°F for awhile, and that ice forms on a skating pond when the temper-

*Both Fahrenheit and Celsius boiled the water at sea level. At higher altitudes the boiling point of water is lower than 212°F or 100°C, as explained in Sec. 6.5.

ature stays below 32°F for awhile. Let's take a closer look at the way heat affects the states of matter by doing the following experiment.

If we place a thermometer in a dish, pack some cracked ice around it, and put the dish in a freezer at −25°C, we note that the temperature of the ice falls until it reaches −25°C. We then bring the dish and ice into a warm room and note the temperature periodically. On a graph we plot temperature against time and get the results shown by the line *ab* in Fig. 5.2. During the time between *a* and *b*, the temperature of the ice is slowly rising, but only solid ice is present.

When the temperature reaches 0°C, it stops rising, and we note that some ice is melting to form liquid water. Now we put the dish and the ice water on an electric hotplate. We find that no matter how high we turn the heater, as long as some ice remains the temperature stays at 0°C, although more and more ice melts. Between the times *b* and *c* the temperature remains constant, as shown by the line *bc*.

When all the ice has melted, the temperature rises again between times *c* and *d* until it reaches 100°C (if we are working at sea level), as shown by the line *cd*.

At time *d*, the water begins to boil and continues to boil at 100°C until all of it has evaporated at time *e*. Then the temperature of the gaseous water rises to 125°C by the time *f*.

Since we have been adding heat throughout this experiment, we conclude that ice at 0°C contains more heat than ice at −25°; liquid water at 0° contains more heat than solid ice at 0°; liquid water at 100° contains more heat

FIGURE 5.2 Change of temperature with time when heat is added to a sample of ice at a constant rate

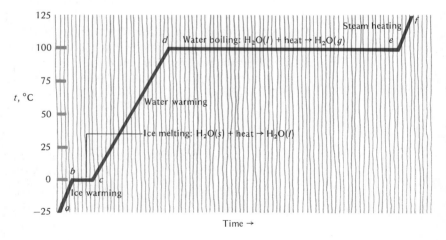

Ice melting: $H_2O(s)$ + heat → $H_2O(l)$

Water boiling: $H_2O(l)$ + heat → $H_2O(g)$

Water warming

Steam heating

Ice warming

t, °C

Time →

vapor

than liquid water at 0°; and gaseous water (water *vapor* or steam) at 100° contains more heat than liquid water at 100°.

Most substances can exist as gases, liquids, or solids. If we heat a solid substance, it melts (liquefies); if we cool a liquid, it freezes (solidifies). If we heat a liquid, it evaporates (changes into a gas); if we cool a gas, it condenses (liquefies). Some gases have to be both cooled and compressed to liquefy them. Some solids, however, are so unstable when heated that they decompose before they melt, and some liquids decompose before they boil.

sublimation

When some solids are heated, they pass into the vapor (gaseous) state without melting. This process is termed *sublimation*. The evaporation of Dry Ice (solid carbon dioxide) and of mothballs are common examples. Snow also sublimes at temperatures below 0°C.

5.4 Measuring Quantity of Heat: Calorimetry

In the experiment described in the preceding section we found that liquid water at 0°C contains more heat than solid ice at 0°C, and liquid water at 100°C more heat than liquid water at 0°C. The next thing we want to know is "*How much* more?" To find this out, we need some definitions, some units, and a device for carrying out experiments.

specific heat

For definitions, we say that the heat absorbed by one gram of a substance with an accompanying increase of one degree in temperature is its *specific heat*. The heat absorbed by one gram of a solid when it melts at a fixed temperature is its *heat of melting* (also called heat of fusion) at that temperature. The heat absorbed by one gram of a liquid when it boils at a given temperature is called its *heat of vaporization* at that temperature.

heat of melting

heat of vaporization
calorie

For our basic unit of heat measurement we use the *calorie*, the amount of heat required to raise the temperature of one gram of water one degree Celsius. When large amounts of heat are involved in a process, we find it convenient to use the *kilocalorie* as the unit of heat; 1 kcal = 1000 cal.

kilocalorie

calorimeter

To measure the amounts of heat involved in different processes we use a *calorimeter*, a simple form of which is shown in Fig. 5.3. In order to keep to a minimum the heat lost to the surroundings in an experiment, the cups are made of foam plastic, a poor conductor of heat, and covered with a piece of cardboard.

EXAMPLE 5.1 What is the specific heat of copper?

ANSWER If we heat a 100-g piece of copper in boiling water at 100°C and quickly transfer it to 50 g water at 25.0°C in the calorimeter, we note that the temperature rises to 36.8°C. Assuming that no heat was absorbed by the plastic cup and that none leaked into or out of the water in it, the heat absorbed

FIGURE 5.3 A simple calorimeter

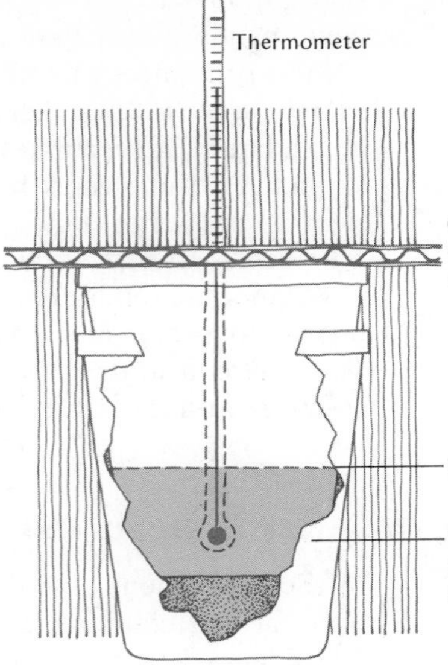

Thermometer

Corrugated cardboard cover

Liquid level

Two nested foam-plastic cups

by the water in warming from 25.0 to 36.8° was liberated by the copper in cooling from 100.0 to 36.8°. Let x be the specific heat of copper.

$$\text{Heat absorbed by water} = \text{Heat liberated by copper}$$

$$\text{Heat absorbed by water} = (50 \text{ g})\left(1.00 \frac{\text{cal}}{\text{g}^\circ\text{C}}\right)(36.8 - 25.0)^\circ\text{C}$$

$$= 590 \text{ cal}$$

$$\text{Heat liberated by copper} = (100 \text{ g})\left(x \frac{\text{cal}}{\text{g}^\circ\text{C}}\right)(100.0 - 36.8)^\circ\text{C}$$

$$= 6320x \text{ cal}$$

$$\text{Heat absorbed} = \text{Heat liberated} \quad \text{so} \quad 590 = 6320x$$

$$x = \frac{590}{6320} \quad \text{or} \quad 0.093$$

The specific heat of copper is 0.093 cal/g°C

The preceding calculations were based on the assumption that no significant amount of heat flowed from the water into the plastic cup or thermometer. That is, we assumed that the specific heats of the cup and thermometer were negligible. If we had taken these into account, our answer would have been changed very slightly. The specific heats of some common substances are given in Table 5.1.

The heat of melting of ice is determined by placing a known weight of ice in a known weight of water and noting the drop in temperature.

EXAMPLE 5.2 What is the heat of melting of ice?

ANSWER We put 100 g water into a calorimeter and note that its temperature is 30.2°C. When 24.6 g ice is added to the water, the temperature drops to 8.5°C.

> Heat absorbed by melting of ice
> > plus
>
> Heat absorbed by warming the cold water so formed
> > equals
>
> Heat liberated by cooling the warm water

Let L be the heat of melting of ice in cal/g.

> Heat absorbed by melting of ice $= (24.6 \text{ g ice})(L \text{ cal/g})$

> Heat absorbed by warming the cold water so formed
> $$= (24.6 \text{ g water})\left(1.00 \frac{\text{cal}}{\text{g}°\text{C}}\right)(8.5 - 0)°\text{C} \quad \text{or} \quad 209 \text{ cal}$$

> Heat liberated by cooling the warm water
> $$= (100 \text{ g water})\left(1.00 \frac{\text{cal}}{\text{g}°\text{C}}\right)(30.2 - 8.5)°\text{C} \quad \text{or} \quad 2170 \text{ cal}$$

> $24.6 L \text{ cal} + 209 \text{ cal} = 2170 \text{ cal}$

> $$L = \frac{2170 - 209}{24.6} \quad \text{or} \quad \frac{1961}{24.6} \quad \text{or} \quad 79.7$$
> The heat of melting of ice is 79.7 cal/g

TABLE 5.1 *Specific Heats of Some Common Substances*

Substance	cal/g°C	Substance	cal/g°C
Water, $H_2O(l)$	1.00	Sulfur, S	0.18
Ice, $H_2O(s)$	0.50	Salt, NaCl	0.20
Steam, $H_2O(g)$	0.48	Sugar, $C_{12}H_{22}O_{11}$	0.30
Aluminum, Al	0.21	Ethyl alcohol, C_2H_5OH	0.58

The heat of condensation of water may be determined by passing dry steam at 100°C into a known weight of cold water in a calorimeter until the temperature rises a convenient amount. Reweighing the calorimeter cup and its contents indicates the amount of steam condensed to water. The observed value for the heat of condensation is 539 cal/g at 100°C.

The absorption of heat by a substance during a change of state is completely reversible. When 1 g ice melts at 0°C, 79.7 cal is absorbed from the surroundings; when 1 g water freezes at 0°C, 79.7 cal is liberated. For this reason, a mixture of pure ice and water will maintain a constant temperature of 0°C as long as both states are present. If heat is added to the system, some ice melts, but if heat is removed, some water freezes. These processes maintain the constant temperature of 0°C so well that the temperature of pure water in contact with pure ice is an excellent fixed point on our thermometric scale.

When 1 g water condenses at 100°C, 539 cal is liberated; when 1 g water changes to steam at 100°C, 539 cal is absorbed. This reversibility makes the boiling point of pure water a good fixed point on the thermometric scale.

Because so much heat is liberated when steam condenses to water, steam at 100°C is much more dangerous than boiling water. If a gram of steam comes into contact with your skin, it liberates 539 cal by condensing to water and 63 cal more as the condensed water cools from 100 to 37°C, your body temperature. This total of 602 cal is almost 10 times the 63 cal of heat liberated on your skin when a gram of boiling water falls on it. A steam burn is likely to do 10 times as much damage to your flesh as scalding with an equal amount of water.

The effects of heat on water noted in the preceding discussions are not peculiar to water. Each pure substance has its own specific values for melting point, boiling point, specific heat, heat of melting, and heat of vaporization.

5.5 How Changes in Temperature Affect the Volumes of Liquids, Solids, and Gases

In Sec. 5.2 we noted that mercury in the bulb of a thermometer when heated expands and pushes up into the capillary tube. Any liquid in such a bulb expands when heated, but the amount of expansion with a given increase in temperature varies according to the particular liquid. It is 0.04 %/°C for water and 0.02 %/°C for mercury.

If we carefully measure the volume of a solid at different temperatures, we find that it, too, expands slightly as the temperature rises. Copper expands 0.002 %/°C and Pyrex glass 0.0004 %/°C.

We can measure changes in the volume of a gas by enclosing it in a cylinder under a piston (Fig. 5.4) that floats on the enclosed gas and is free to slide up or down as the volume of the gas changes. First, we set the cylinder in a mix-

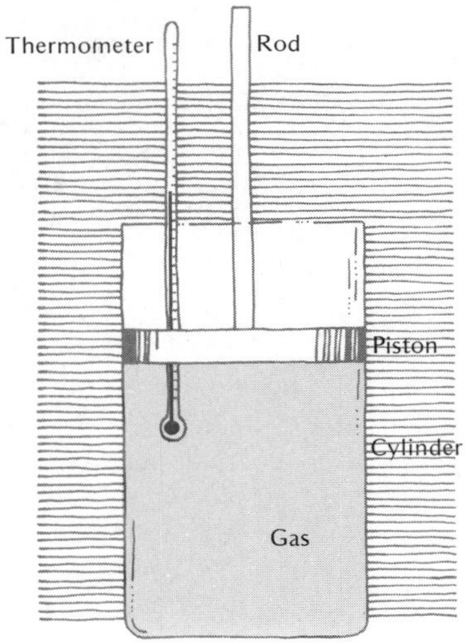

FIGURE 5.4 A sample of gas enclosed in a cylinder with a movable piston

ture of ice and water so that the temperature of the gas is 0°C, and we measure the distance from the bottom of the cylinder to the bottom of the piston. Next we set the cylinder in water at 1°C, measure the distance from the bottom of the cylinder to the bottom of the piston, and find that it has increased by $\frac{1}{273}$. That is, if the bottom of the piston was 273 cm above the bottom of the cylinder at 0°C, this distance becomes 274 cm at 1°C. If we raise the temperature to 10°C, the piston rises $\frac{10}{273}$, and is 283 cm above the bottom of the cylinder. If we place the cylinder in boiling water so that the gas heats to 100°C, the piston rises $\frac{100}{273}$, and is 373 cm above the bottom of the cylinder. Since the volume of the gas is directly proportional to the height of the piston (doubling the height doubles the volume), we conclude that the volume of a sample of gas increases $\frac{1}{273}$ with every degree Celsius rise in temperature.

If we cool the gas to −1°C, the volume decreases $\frac{1}{273}$ (that is, the piston drops to 272 cm above the bottom of the cylinder). If the gas continued to decrease in volume with decreased temperature, at −273°C it would have zero volume! But, as we noted in Sec. 5.3, gases liquefy when cooled enough, and liquids do not decrease in volume at the same rate as gases. But it is convenient to think of −273°C as the lowest possible temperature, since we cannot conceive of a gas having a volume less than zero. Hence −273°C is called *absolute zero*.

absolute zero

extrapolation

Measuring volumes over a given range of temperatures and predicting what would happen at lower temperatures is called *extrapolation* of the data. In Fig. 5.5 the dotted line has been extrapolated (extended beyond the region covered by experimental data) from the solid line. The volume of the gas is

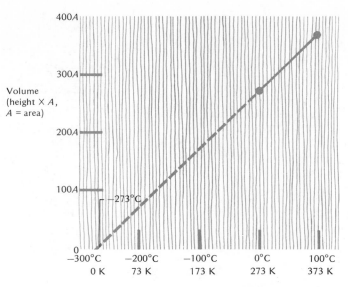

FIGURE 5.5 The effect of changing temperature on the volume of a sample of gas

the *height* to the bottom of the piston multiplied by the *area A* of cross section of the cylinder in Fig. 5.4. The volume then becomes height times *A*. This value is plotted on the vertical axis and the temperature is plotted in °C on the horizontal.

For dealing with the effects of temperature changes on the volume of a sample of gas, it is handy to set the zero on the temperature scale to correspond with the hypothetical volume of zero. We then have the absolute or *Kelvin scale*.

Kelvin or absolute scale of temperature

The Kelvin temperature (K) is obtained by adding 273 to the Celsius temperature:

$$K = °C + 273 \tag{5.1}$$

Figure 5.5 shows the temperatures in K under the corresponding temperatures in °C. The degree symbol is not used with the Kelvin temperature.

Many studies of temperature-volume relationships for different gases led to the conclusion that all gases expand by $\frac{1}{273}$ or 0.366%/°C over long ranges of temperature. This is about 10 times the percentage expansion of liquids and 100 times that of solids. Furthermore, all gases expand by the same percentage per degree Celsius, but all liquids and solids expand by different percentages. Apparently gases are much more alike in physical properties than are liquids or solids.

When we measure temperature and volume with the apparatus shown in Fig. 5.4, we assume no change in atmospheric pressure during the experiment. (Atmospheric pressure is discussed in Sec. 5.8.) We vary the temperature and observe changes in volume. We say that temperature and volume are the variables in the experiment, and that atmospheric pressure is a constant. From Fig. 5.5 we see that when temperature *increases*, volume *increases*. If we double the temperature of the gas in degrees Kelvin, we double the volume it occupies. We call this *direct proportionality*. The French scientists Jacques Charles and Joseph Gay-Lussac made extended studies of temperature-volume relationships in gases and summarized the results in this natural law: *The volume of a sample of gas at constant pressure is directly proportional to the absolute temperature.* This is often called *Charles' law.*

direct
proportionality

Charles' law

5.6 The Densities of Solids, Liquids, and Gases

density

One of the characteristic properties of a substance is its *density*, defined as mass per unit volume:

$$d = \frac{m}{v} \tag{5.2}$$

Densities of solids and liquids are usually reported in grams per cubic centimeter (g/cm^3) because this unit gives convenient values for most densities, which range from about 1 to 20 g/cm^3. Because gases are so much less dense, we usually report their densities in $g/1000\ cm^3$ (or g/liter), giving values that range from about 0.1 to 7 g/liter. All substances change somewhat in volume when the temperature changes, so 1 g of any substance will have a different volume at a different temperature and thus a different density. A numerical value for density should always be accompanied by a specified temperature (Table 5.2).

5.7 Pressures Exerted by Solids and Liquids

If we place a piece of a solid on a table, we don't have to push in on the sides to keep them from bulging; we say that a solid exerts pressure only on the surface that supports it. If we put liquid in a vessel with holes in its sides and bottom, the liquid leaks out. If the holes are the same size, it leaks out fastest through the bottom and more and more slowly up the sides. There is no leakage through a hole just at the surface of the liquid. We conclude that a liquid exerts pressure on the bottom and on the sides of its container, the pressure on the sides decreasing from the bottom upward and becoming zero at the top.

TABLE 5.2 *Densities of Some Common Substances*

Solids	Density, g/cm³	Liquids	Density, g/cm³	Gases	Density, g/1000 cm³
Cork	0.21	Gasoline	0.7	Hydrogen	0.0899
Lithium	0.53	Ethyl alcohol	0.79	Ammonia	0.77
Ice	0.92	Water (4°C)	1.000	Nitrogen	1.25
Sugar	1.59	Glycerine	1.26	Air (dry)	1.29
Salt	2.16	Carbon tetrachloride	1.60	Oxygen	1.43
Granite	2.6	Ethyl iodide	1.93	Hydrogen chloride	1.639
Aluminum	2.70	Carbon tetrabromide	3.42	Carbon dioxide	1.98
Zinc	6.92	Mercury	13.55	Chlorine	3.21
Iron	7.90				
Copper	8.89				
Lead	11.3				
Gold	19.3				
Platinum	21.4				
Osmium	22.5				

All densities of liquids and solids are at 20°C unless otherwise indicated. Densities of gases are at 0°C and 1 atm.

5.8 Pressure Exerted by the Atmosphere

atmospheric pressure

We live at the bottom of an ocean of air. Our atmosphere is a gaseous envelope around our planet, held by gravitational attraction between the earth and the air, i.e., by the weight of the air. At sea level the weight of the column of air above 1 in.² of the earth's surface averages 14.7 lb. We say the *atmospheric pressure* is 14.7 lb/in.².

It is easy to demonstrate atmospheric pressure. Place a glass tube open at both ends in a vertical position with the lower end beneath the surface of some mercury in an open dish. The level *l* of the mercury inside the tube is the same as in the dish, as in Fig. 5.6a. Now connect a pump that can remove air from the tube. When the pump is started, the mercury rises in the tube as in part *b*. But when the top of the mercury column inside the tube is about 76 cm above the surface in the open dish as in part *c*, the level ceases to rise even though the pump is still running.

From these observations we conclude that when the top of the tube is open to the air (or connected to the pump before it is started), the air inside the tube pushes downward on the mercury at *l* just as hard as the air pressure on the mercury in the open dish pushes mercury up into the tube at *l*. The two tendencies are just balanced, and so the level remains fixed at *l*. When we start the pump, air is removed from inside the tube, and the pressure it

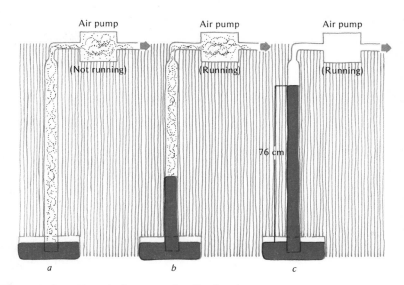

FIGURE 5.6 Pumping air from a tube dipping in mercury

exerts downward at *l* becomes less than the pressure upward caused by the air pushing on the mercury surface in the open dish, so that the mercury is then pushed up into the tube as in Fig. 5.6*b*. As pumping continues, air pressure in the tube is further decreased, and the mercury is pushed up further by the external air pressure. Finally, when the weight of the column of mercury inside the tube presses downward just as hard as the external air pressure pushes mercury upward at *l*, the column will rise no further. The pressure of air above the mercury in the tube is now zero, and the height of the column of mercury is a measure of the atmospheric pressure. When the height is 76 cm, we say the pressure is 1 *atmosphere* (1 atm).

atmosphere (unit)

5.9 Measuring Atmospheric Pressure

A device to measure atmospheric pressure is called a barometer. A mercury barometer is easy to make (Fig. 5.7). Seal one end of a glass tube about 1 m long and fill the tube with mercury to displace the air as in part *a*. Holding a finger tightly against the open end, invert the tube, putting the open end under the surface of mercury in an open dish. When the finger is removed, some mercury runs out of the tube as in part *b*.

The top of the mercury remaining in the tube will be about 760 mm above the level of that in the open dish if we are working at sea level. Since there is no gas above the mercury in the tube, the pressure there is zero. The pressure of the atmosphere on the mercury in the open dish may then be stated as

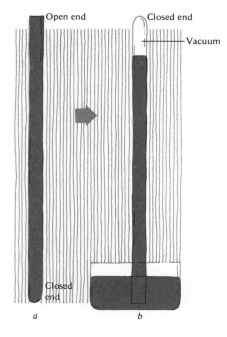

FIGURE 5.7 A mercury barometer. Tube filled with mercury (*a*), inverted with open end in dish (*b*), will come to rest with mercury column 76 cm high, at 1 atm pressure.

760 mm mercury. Since the density of mercury decreases as temperature increases, the column of the liquid that balances the atmospheric pressure will be longer as the mercury gets warmer. We define one *standard atmosphere* of pressure as 760.0 mm mercury at 0°C. The unit of pressure, one millimeter of mercury, is often called one *torr* in honor of Torricelli, who was the first to construct an apparatus like that shown in Fig. 5.7. Weather reports often give barometric pressures in inches of mercury.

standard
atmosphere
torr

5.10 How Changes in Pressure Affect the Volumes of Liquids, Solids, and Gases

Let us carry out a series of experiments in a laboratory at constant temperature. If we put liquid in a cylinder with a pressure gauge (Fig. 5.8*a*) and insert the piston, we can increase the pressure on the liquid by pushing down on the rod. If we let the piston float on the liquid, the pressure shown on the gauge is about 1 atm. If we push down hard on the piston, the pressure shown by the gauge increases, but the piston does not move down perceptibly (Fig. 5.8*b*). Since the volume of the liquid is virtually unchanged by pressure, we say that a liquid is incompressible. If we put a piece of a solid into the cylinder (Fig. 5.8*c*), cover it with a liquid, insert the piston, and again push down hard on the rod, we see that the pressure increases, but that again the volume

Torricelli's experiment with a barometer

FIGURE 5.8 Apparatus for testing the compressibility of liquids and solids. Note that the volume of a liquid or a solid does not change with pressure.

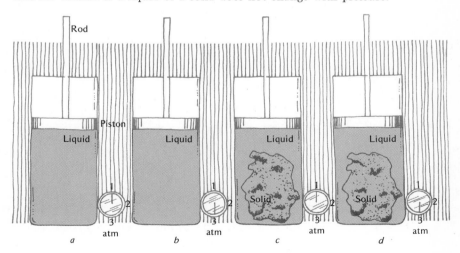

of the liquid and of the solid is virtually unchanged by pressure (Fig. 5.8*d*). We conclude that solids, like liquids, are incompressible.

If we fill the volume under the piston with a gas, we find a very different situation. As we push down harder and harder on the rod, the volume decreases as the pressure increases. If we start with 12 liters in the cylinder at a pressure of 1 atm (Fig. 5.9*a*), and push down the rod until the pressure is 2 atm, we note that the volume is 6 liters, as indicated in Fig. 5.9*b*. If we push down hard enough to increase the pressure to 3 atm, the volume decreases to 4 liters (Fig. 5.9*c*). We varied the value of one observable quantity to see the effect on the value of another observable quantity. The quantities that change are variables. The variables in this experiment are pressure and volume. The temperature does not vary but is a constant. We note that when the pressure on a gas *increases*, the volume *decreases*. We call this *inverse proportionality*. If doubling one variable results in halving the other, we say one is inversely proportional to the other.

If we call one variable x and the other y, we may express this inverse variation as a proportionality,

$$x \propto \frac{1}{y}$$

FIGURE 5.9 The change in volume of a sample of a gas when the pressure on the gas is increased. Note the amount of change.

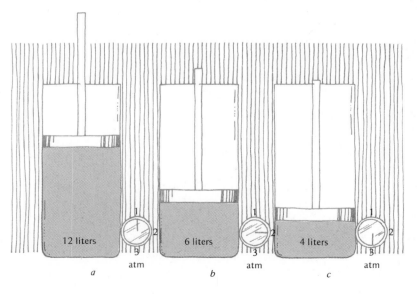

which we read:

> x varies inversely as y
>
> or   x is inversely proportional to y
>
> or   x varies as the reciprocal of y
>
> or   x is proportional to the reciprocal of y

proportionality
constant

To convert a proportionality into a mathematical equation, we insert what we call a *proportionality constant*. Thus

$$x = k\left(\frac{1}{y}\right)$$

where k is the proportionality constant. We can solve this equation for k by multiplying each side of the equation by y. (If equals are multiplied by equals, the results are equal.)

$$xy = k \tag{5.3}$$

In our experiment the two variables are pressure P and volume V. Equation (5.3) tells us that $PV = k$ (that the product of pressure times volume is constant). Table 5.3 verifies this relationship. We see that PV is indeed constant as predicted by Eq. (5.3).

Boyle's law

Robert Boyle discovered the relationship shown in Eq. (5.3), and *Boyle's law* is a description of this observed relationship. A simple statement of the law is: When the temperature of a given sample of gas is held constant, the volume varies inversely with the pressure.

5.11   Standard Temperature and Pressure

From Charles' law and Boyle's law we see that the volume occupied by a given weight of gas depends on its temperature and pressure. Consequently, it is customary to give both the temperature and pressure when we make a statement about the volume of a gas with which we are dealing. For convenience in comparing volumes of gases we often state that such and such a volume is measured at STP, *standard temperature and pressure*. Standard temperature is 0°C and standard pressure is 1 atm (760 torr).

standard
temperature
and pressure (STP)

TABLE 5.3   *Variation in Volume of Gas as Pressure Is Varied*

P (atm)	1	2	3
V (liters)	12	6	4
PV (liter-atm)	12	12	12

Robert Boyle, 1627–1691

Summary

Temperature is a measure of the degree of hotness of a substance. Scientists use the Celsius temperature scale, on which 0°C is the freezing point of water and 100°C is the boiling point of water at sea level.

A mixture of ice and water maintains a temperature of 0°C. Heating the mixture does not raise the temperature, it just melts some ice. Cooling the mixture does not lower the temperature, it just freezes some water. The temperature will change only when either the ice or the water has disappeared.

The unit for measuring amount of heat is the calorie. One calorie is the amount of heat required to raise the temperature of one gram of water one degree Celsius. This amount of heat is equal to the amount of heat liberated when one gram of water *cools* one degree Celsius.

When 1 g ice melts, it absorbs about 80 cal. When 1 g water freezes, it liberates about 80 cal. When 1 g water boils off to steam at 100°C, it absorbs 539 cal. When 1 g steam at 100°C condenses to water, it liberates 539 cal.

The volume of a sample of gas at constant pressure is directly proportional to the absolute temperature. The volume of a sample of gas at constant temperature is inversely proportional to the pressure. The standard temperature and pressure for measuring gas volumes is 0°C and 1 atm.

Atmospheric pressure is measured by a mercury barometer. The pressure is proportional to the length of the column of mercury in the barometer. The *average* atmospheric pressure at sea level is 760 mm Hg (760 torr). This pressure is called one atmosphere.

New Terms and Concepts

ABSOLUTE ZERO: The temperature at which the volume of a gas would become zero if its rate of decrease of volume with decreasing temperature at ordinary temperatures were constant as the temperature became lower and lower.

ATMOSPHERE (UNIT): A pressure that balances a column of mercury 760 mm high at 0°C; 1 atm = 760 torr.

ATMOSPHERIC PRESSURE: The pressure exerted by the weight of the atmosphere.

BOYLE'S LAW: At constant temperature, the volume of a given mass of gas varies inversely as its pressure; $V \propto (1/P)$.

CALORIE: One calorie is the amount of heat required to raise the temperature of one gram of water one degree Celsius.

CALORIMETER: A device for measuring the amount of heat produced or consumed during a given process.

CELSIUS SCALE OF TEMPERATURE: The scale with 0° for the freezing point of water and 100° for its boiling point at a pressure of 1 atm.

CHARLES' LAW: At constant pressure, the volume of a given mass of gas varies directly as its absolute (Kelvin) temperature; $V \propto T$.

DENSITY: The ratio of the mass of a sample of a substance to the volume it occupies; $d = m/v$.

DIRECT PROPORTIONALITY: If the doubling of one variable in a system causes the doubling of a second variable, the second is said to be directly proportional to the first.

EXTRAPOLATION: Inferring or estimating unknown information by extending or projecting known information.

FAHRENHEIT SCALE OF TEMPERATURE: The scale with 32° for the freezing point of water and 212° for its boiling point at 1 atm pressure.

HEAT OF MELTING: The number of calories of heat absorbed when 1 g of a solid substance changes to the liquid state. Also called heat of fusion.

HEAT OF VAPORIZATION: The number of calories of heat absorbed when 1 g of a given material changes from the liquid to the gaseous state.

INVERSE PROPORTIONALITY: If the doubling of one variable in a system causes the halving of a second variable, the second is said to be inversely proportional to the first.

KELVIN OR ABSOLUTE SCALE OF TEMPERATURE: The scale on which 0° is the temperature at which the pressure of a gas becomes zero and 273° is the freezing point of water, $T K = t°C + 273$.

KILOCALORIE: The amount of heat required to raise the temperature of one kilogram of water one degree Celsius.

PRESSURE: The application of continuous force by one body upon another that it is touching.

PROPORTIONALITY CONSTANT: The number that converts a statement of proportionality into an equation.

SPECIFIC HEAT: The number of calories required to raise the temperature of one gram of a substance one degree Celsius.

STANDARD ATMOSPHERE: 760 mm mercury at O°C; 760 torr.

STANDARD TEMPERATURE AND PRESSURE (STP): Temperature is 0°C, and pressure is 1 atm (for measurements involving gases).

SUBLIMATION: The vaporization of a solid to a gas without melting.

TEMPERATURE: The degree of hotness or coldness of a body.

TORR: A unit of pressure equal to the pressure exerted by a column of mercury 1 mm in height; 1 atm = 760 torr.

VAPOR: A substance in the gaseous state but at a temperature only slightly above that at which it would condense to a liquid.

Testing Yourself

5.1 How can you explain the fact that when you mix ice and water, the temperature falls to 0°C but no further? What happens in terms of heat flow if you heat the mixture? If you cool it?

5.2 How can you explain the fact that when you boil water at sea level, the temperature doesn't go above 100°C?

5.3 Are the phenomena discussed in your answers to questions 5.1 and 5.2 peculiar to water, or do they apply to other substances?

5.4 If you are boiling potatoes, will they cook faster if you have them boiling madly over a high flame than if you let them boil gently over a small flame?

5.5 Dry Ice is solid carbon dioxide. It is called "dry" for what reason? How can we explain the dryness? The iciness? What name do we give to the process taking place when Dry Ice disappears?

5.6 Farmers who have trouble with their fruits and vegetables freezing in underground storage rooms on a cold night sometimes put several large tubs of cold water in the storeroom in the evening. How does this help prevent the produce's freezing?

5.7 Suppose that 100-g weights (masses) of water, copper, and aluminum are placed in a hot oven together. Which gets hot most quickly? Most slowly?

5.8 How much heat is required to convert 50 g ice at $-10°C$ to steam at $110°C$?

5.9 What makes a liquid go up a straw when you suck it?

5.10 When a car is supported on a jack so that no weight is borne by the casing of the tire on a wheel, the volume of the uninflated casing is nearly the same as when it is fully inflated. As more and more air is forced into the tire, what happens to the pressure? What does this tell you about the relation between the weight (mass) of air contained in a given volume and the pressure the air exerts?

The photograph shows water in three states: solid, liquid, and gas. Note the molecular structure of water in its liquid state, as shown in the drawing.

6. The Kinetic-Molecular Theory: A Mental Model

When a scientist has gathered an array of facts and figures he can often understand them better if he constructs a theory to account for his observations. Such a theory often predicts relationships that have not yet been observed and thus stimulates the investigator to make further experiments. If the predicted relationships are verified by the results of these experiments, the theory is upheld and may lead to further predictions. If the data from these experiments do not agree with the predictions, then the theory has to be modified to take the new facts into account.

In this chapter we shall develop a theory to explain the observations noted in Chap. 5 and also to explain some further data. The theory with which we shall be concerned has proved of great value in helping chemists and others understand many phenomena in our environment, phenomena that seemed quite inexplicable without the theory.

Because our theory involves visualizing actions that we shall never be able to see because molecules are so small, we call it a *mental model* for the behavior of matter.

6.1 Summary of the Behavior of Gases

1 A gas does not settle out in a container; it completely fills a closed container and exerts the same pressure on the top, bottom, and sides.
2 All gases mix completely with all other gases.
3 They mix spontaneously as one gas diffuses through others. (A gas with an odor released in one part of a room can soon be smelled everywhere in the room.)
4 If the pressure on a sample of gas held at a constant temperature is doubled, the volume is halved.
5 If a sample of gas is heated at constant pressure, the volume increases in direct proportion to the absolute temperature.
6 If a sample of gas is kept in a container insulated against heat flow, its temperature and pressure remain constant.

6.2 A Theory to Explain the Behavior of Gases

We can explain the above observations by constructing the following mental model for gases.

1 Gases are composed of tiny particles that we call molecules.
2 The molecules are far apart so that the volume occupied by a gas is mostly empty space.
3 The molecules are in constant and rapid motion in all directions.
4 When the molecules collide with one another or with the walls of a containing vessel at the same temperature, they do not lose their energy of motion.
5 When a gas is heated, its molecules move faster; the energy of their motion is directly proportional to the absolute temperature.

kinetic-molecular
theory

We call our model the *kinetic-molecular theory* (the Greek word for "moving" is "kinetikos").

6.3 How Does Our Mental Model Account for the Behavior of Gases?

OBSERVATION	INTERPRETATION
A gas does not settle out; it completely fills a closed container.	Rapidly moving molecules rebound from collisions with one another or with container walls with no net loss of motion. They move in all directions so that they fill any container.

OBSERVATION	INTERPRETATION
A gas exerts the same pressure on the top, bottom, and sides of a closed container.	The pressure of a gas is produced by the bombardment of the containing walls by the rapidly moving molecules. They move equally in all directions so that they exert pressure equally on all containing walls.
All gases mix spontaneously with each other.	Since the molecules in a gas are far apart, there is plenty of room between the molecules of one gas for the molecules of another gas. The rapid motion in all directions of all the molecules causes them to mix completely.
At constant temperature the pressure exerted by a given weight of gas varies inversely as the volume it occupies.	The pressure exerted by a gas upon its container walls is due to the bombardment of the walls by the rapidly moving molecules. If the volume is halved, the number of molecules per unit volume is doubled, so that they bombard the walls twice as frequently, thus producing twice the pressure.
At constant pressure, the volume occupied by a sample of gas varies directly as its absolute temperature.	When a gas is heated, its molecules move faster and bombard the container walls more frequently and more vigorously. The energy of molecular motion is directly proportional to the absolute temperature. To keep the pressure (bombardment of the container walls) unchanged, the volume must be increased; thus the molecular bombardment on the walls is kept constant.
If a sample of gas is in a closed container insulated to prevent heat from flowing into or out of the gas, the pressure and temperature of the gas remain constant.	When moving molecules collide with each other or the walls of the container, there is no net loss of energy of motion. (If they lost energy upon collision, they would slow down and the temperature and pressure would decrease.)

6.4 Individual Gas Pressures in Mixtures of Gases

Air contains about 20 percent oxygen, 79 percent nitrogen, and 1 percent other gases. Do the oxygen and nitrogen in air make equal contributions to the atmospheric pressure? Does water vapor in the air have any effect on atmospheric pressure? What happens to the pressures of gases when they are mixed? To answer these questions, we can use two cylinders like that shown in Fig. 5.9 but with a gas cock in the bottom of the cylinder wall so that gas can be put into or taken out of the cylinder without removing the piston.

Put 1 liter of oxygen in the first cylinder at a pressure of 1 atm, and 1 liter of nitrogen in the second cylinder at a pressure of 1 atm. Connect the gas cocks to each other, open both, and push down on the piston in the first cylinder to force the oxygen into the second. Keeping the temperature constant, we find a volume of 2 liters of the oxygen-nitrogen mixture in the second cylinder with the pressure remaining at 1 atm.

According to Boyle's law, if the oxygen alone occupied the volume of 2 liters, it would exert a pressure of half an atm; and the same would be true for the nitrogen. The total pressure of the mixture of gases is 1 atm. Apparently, in the mixture, each gas exerts the same pressure it would exert if it alone filled the entire volume, and the total pressure is the sum of these two pressures.

Many different samples of gases with known volumes at known temperatures and pressures have been mixed and measured. In all cases: In a mixture of gases, each gas exerts the same pressure that it would if it alone occupied the container, and the total pressure of the mixture is the sum of the partial pressures of the components. John Dalton was the first to suggest this relationship. It is often called Dalton's *law of partial pressures.*

law of partial pressures

Our kinetic-molecular model for gases helps us understand partial pressures. When gases are mixed, the molecules of each gas still bounce around with the same motion, and so each kind of gas exerts its pressure by bombarding the container walls without being affected by the other kinds of gas.

6.5 The Evaporation of Liquids

If we place water in an open dish, we note that it gradually evaporates, and it does so more rapidly when it is hot than when it is cold. If we blow air across its surface, the water is cooled by the evaporation. Our bodies are kept cool by the drying of perspiration from our skins. A spray of liquid ethylene chloride directed at a spot on the body evaporates from the skin so rapidly that it freezes the flesh and so serves as a local anesthetic.

We can measure the tendency of a liquid to evaporate by using the apparatus in Fig. 5.9 modified with a gas cock, as described in Sec. 6.4. First, fill the cylinder with air and lock the piston in place to maintain a constant volume in the cylinder. Then, force a few cubic centimeters of alcohol, C_2H_5OH, into the cylinder through the gas cock. Some of the alcohol evaporates, and the pressure in the cylinder increases slightly. If we then force in a little more alcohol, there is no further increase in pressure. Since we are dealing with a mixture of gases (air and alcohol vapor), we assume that the law of partial pressures is operative. Therefore the gaseous alcohol in the cylinder is exerting a pressure, and this pressure is measured by the increase noted when the liquid evaporated in the cylinder. We call this pressure exerted by the vapor

vapor pressure

from the alcohol the *vapor pressure* of alcohol. If we immerse the apparatus in a mixture of water and ice, the observed vapor pressure decreases. If we immerse it in boiling water, the vapor pressure increases.

If we now empty the apparatus and repeat the experiment by putting in acetic acid, CH_3COOH, through the gas cock, we find that the vapor pressure of this substance at any given temperature is lower than that of alcohol.

If we fill the apparatus with dry air and admit liquid water through the gas cock, we find that the vapor pressure of water at any given temperature is intermediate between the values for alcohol and acetic acid (Table 6.1 and Fig. 6.1).

When we heat any liquid, the vapor pressure increases. When it equals atmospheric pressure, the liquid boils. If the liquid and its vapor are enclosed at a high pressure, the boiling point is higher; if at a low pressure, the boiling

normal boiling
point

point is lower. Consequently, the *normal boiling point* of a liquid is the temperature at which it boils at a pressure of 1 atm (760 torr).

TABLE 6.1 *Vapor Pressures of Liquids at Various Temperatures*

°C	Vapor pressure in torr		
	Ethyl alcohol	Water	Acetic acid
0	12.2	4.6	3.5
10	23.6	9.2	6.4
20	43.9	17.5	11.8
30	78.8	31.8	20.1
40	135.3	55.3	34.2
50	222.2	92.5	56.3
60	352.7	149.4	88.3
70	542.5	233.7	137.9
80	812.6	355.1	202.3
90	1187	525.8	292.7
100	1693	760	417.0
110	2360	1074	580.8
120	3055	1489	794.0

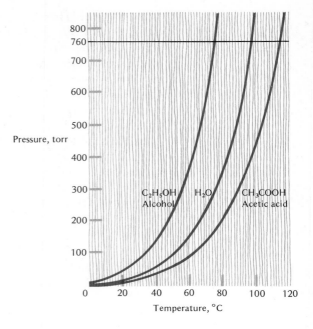

FIGURE 6.1 Variation of vapor pressures of several liquids with temperature

Pressure, torr

Temperature, °C

Water in an open dish boils at 100°C at sea level, where the atmospheric pressure is 760 torr. At an altitude of 5000 ft, where the atmospheric pressure is less than 760 torr, the water boils at about 96°C. On the top of Pike's Peak, Colorado (altitude about 14,000 ft), the boiling temperature is so low that outdoorsmen say, "You can't cook beans on Pike's Peak!" But if you put your beans and water in a stout pot with a tight-fitting lid, the pressure builds up so that the boiling temperature is much higher than 100°C; at 2 atm the boiling point is 121°C. In this "pressure cooker" the beans will cook much faster than in the open air.

6.6 Extending the Kinetic-Molecular Model to Include Liquids

The concepts of the kinetic-molecular model for matter in the gaseous state are helpful in understanding the nature of liquids. We have assumed that gaseous matter consists of tiny molecules very far apart and moving at very high speeds. When a sample of gas is heated in a container with a constant volume, the pressure increases. We interpret this increase as due to an increase in the speed of the molecules at a higher temperature, causing them to bombard the walls of the container more frequently and with greater impact. We assume, therefore, that the temperature of a gas is related to the *kinetic energy* (energy of motion) of its molecules.

kinetic energy

When water vapor at a temperature above 100°C under a pressure of 1 atm is cooled to 100°, it begins to condense to form liquid water at 100°. Since the gaseous and the liquid water have the same temperature, we assume that the kinetic energy of the water molecules in the liquid phase is the same as that in the gaseous phase, i.e., that the speed of the molecular motion is the same in the two phases. How, then, can we account for the liberation of 539 cal of heat for each gram of water condensed from the gaseous to the liquid phase at 100°C?

When 1 g liquid water is vaporized at 100°C, it absorbs 539 cal of heat, and the volume it occupies increases from about 1 cm³ as a liquid to about 1700 cm³ as a gas. It seems sensible, then, to assume that the 539 cal of heat absorbed must have been required to push back the atmosphere to make room for this large increase in volume at the constant temperature of 100°C and to overcome forces of attraction between the molecules. The corresponding contraction in volume during condensation liberates 539 cal. We conclude, therefore, that liquid water consists of molecules that are much closer together and much more strongly attracted to one another than molecules in gaseous water. Though these molecules at 100°C are moving at the same high speeds as in the gas at 100°C, they are so close together in the liquid that they just jostle around in the very limited space available to them.

When liquid water gives up heat (i.e., cools to a lower temperature), the volume does not decrease much. It seems likely that the molecules continue to occupy about the same space but jostle around in it with less kinetic energy, i.e., at lower speeds.

Earlier we noted that a liquid in an open dish gradually evaporates, and it does so more rapidly when it is hot than when it is cold. Why don't all the water molecules evaporate at once? Bearing in mind our kinetic-molecular concept of matter, we conclude that not all molecules in liquids move at the same speed. Some move faster than others, and the hot (high-speed) molecules tend to escape more frequently than the others. The average speed of the molecules is thus reduced by evaporation, and the liquid cools. When we are overheated, our skin excretes sweat that reduces our temperature by evaporative cooling. At higher temperatures, more molecules in a liquid have speeds enabling them to escape, so that the vapor pressure increases and evaporation is more rapid.

6.7 Extending the Kinetic-Molecular Theory to Include Solids

When water cools to 0°C, solid ice begins to form. There is little change in volume, and the temperature remains constant during the freezing process. It seems logical to conclude that the small change in volume and the solidity of the ice arise from the fact that the water molecules are still separated by the same distance as in liquid water but are less free to move about in the solid state—that they are held in a definite geometric pattern. We can see this pattern in snowflakes, crystals of solid water, all of which have a hexagonal pattern. The movement associated with the kinetic energy of the molecules in a solid probably takes place as a vibration about a fixed point rather than in random jostling as in a liquid. Thus a solid cannot flow, whereas a liquid can because its molecules are free to slip around one another.

6.8 Relations between the Volumes of Gases Involved in Chemical Reactions

In Sec. 3.12 we found that there are definite relationships between the *weights* of reactants and products in a chemical reaction. Note the relationships in the following reaction.

$$
\begin{array}{ccccc}
\text{hydrogen} & + & \text{chlorine} & \longrightarrow & \text{hydrogen chloride} \\
H_2(g) & + & Cl_2(g) & \longrightarrow & 2HCl(g) \\
1 \text{ molecule} & + & 1 \text{ molecule} & \longrightarrow & 2 \text{ molecules} \\
1 \text{ mole} & + & 1 \text{ mole} & \longrightarrow & 2 \text{ moles} \\
2 \times 1.0 \text{ g} & + & 2 \times 35.5 \text{ g} & \longrightarrow & 2 \times (1.0 + 35.5) \text{ g} \\
2.0 \text{ g} & + & 71.0 \text{ g} & \longrightarrow & 73.0 \text{ g}
\end{array}
$$

The same kind of relationships can be observed in the following.

$$
\begin{array}{ccccc}
\text{hydrogen} & + & \text{oxygen} & \longrightarrow & \text{water} \\
2H_2(g) & + & O_2(g) & \longrightarrow & 2H_2O(g) \\
2 \text{ molecules} & + & 1 \text{ molecule} & \longrightarrow & 2 \text{ molecules} \\
2 \text{ moles} & + & 1 \text{ mole} & \longrightarrow & 2 \text{ moles} \\
4.0 \text{ g} & + & 32.0 \text{ g} & \longrightarrow & 2 \times (2.0 + 16.0) \text{ g} \\
4.0 \text{ g} & + & 32.0 \text{ g} & \longrightarrow & 36.0 \text{ g}
\end{array}
$$

A third example involves the formation of ammonia.

$$
\begin{array}{ccccc}
\text{hydrogen} & + & \text{nitrogen} & \longrightarrow & \text{ammonia} \\
3H_2(g) & + & N_2(g) & \longrightarrow & 2NH_3(g) \\
3 \text{ molecules} & + & 1 \text{ molecule} & \longrightarrow & 2 \text{ molecules} \\
3 \text{ moles} & + & 1 \text{ mole} & \longrightarrow & 2 \text{ moles} \\
6.0 \text{ g} & + & 28.0 \text{ g} & \longrightarrow & 2 \times (14.0 + 3.0) \text{ g} \\
6.0 \text{ g} & + & 28.0 \text{ g} & \longrightarrow & 34.0 \text{ g}
\end{array}
$$

We can summarize the preceding figures thus:

2.0 g hydrogen + 71.0 g chlorine ⟶ 73.0 g hydrogen chloride
4.0 g hydrogen + 32.0 g oxygen ⟶ 36.0 g water
6.0 g hydrogen + 28.0 g nitrogen ⟶ 34.0 g ammonia

Clearly there are definite relationships between the *weights* of reactants, but the relationships do not involve simple numbers. What about the relationships between the *volumes* of reactants?

Knowing that the volume of a sample of gas depends on its temperature and pressure, we realize that to be compared volumes must be measured at the same temperature and pressure. When we collect data, astonishingly simple relationships are observed:

1 liter hydrogen + 1 liter chlorine $\longrightarrow$ 2 liters hydrogen chloride
2 liters hydrogen + 1 liter oxygen $\longrightarrow$ 2 liters gaseous water
3 liters hydrogen + 1 liter nitrogen $\longrightarrow$ 2 liters ammonia

From many experiments like these we conclude that the volumes of gases (measured at the same temperature and pressure) produced or consumed in chemical reactions are in the ratio of simple whole numbers. This natural law is called Gay-Lussac's *law of combining volumes*.

law of combining volumes
Avogadro's hypothesis

To explain these ratios, the Italian chemist *Avogadro* made the *hypothesis* that at the same temperature and pressure, equal volumes of all gases contain the same number of molecules.

To understand this hypothesis, we can visualize reactions between tiny volumes of hydrogen and chlorine, each containing 9 molecules (Fig. 6.2).

Amedeo Avogadro, 1776–1856

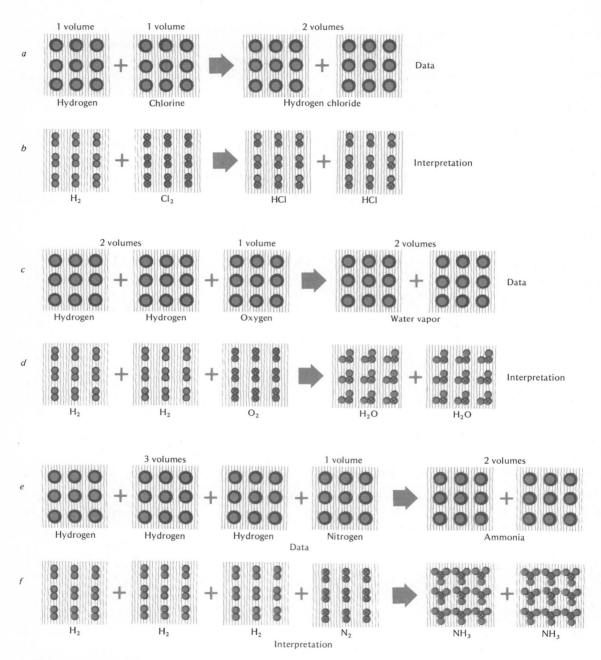

FIGURE 6.2 Combining volumes of gases: data and Avogadro's interpretation

In part *a* we see that 18 molecules of HCl are produced, and so there must be 18 atoms of hydrogen involved in the reaction. Since we start with only 9 molecules of hydrogen, these must contain 18 atoms of hydrogen, or 2 per molecule. The same must be true of chlorine, since 9 Cl_2 (9 molecules) produce 18 HCl molecules, each containing 1 Cl atom (Fig. 6.2*b*). Figure 6.2*c* and *d* shows that oxygen also exists in *diatomic* molecules, and Fig. 6.2*e* and *f* shows the same for nitrogen. Similar experiments and interpretations indicate that gaseous fluorine, bromine, and iodine are also diatomic; their formulas are F_2, Br_2, and I_2 in the gaseous state.

diatomic

6.9 The Relation between Weights and Volumes of Gases Involved in Chemical Reactions

At the beginning of Sec. 6.8 we noted this reaction:

$$H_2(g) \; + \; Cl_2(g) \longrightarrow 2HCl(g)$$
$$1 \text{ mole} + 1 \text{ mole} \longrightarrow 2 \text{ moles}$$
$$2.0 \text{ g} \; + \; 71.0 \text{ g} \longrightarrow \; 73.0 \text{ g}$$

What volumes do each of these gases occupy? In Table 5.2 we see that the density of hydrogen is 0.0899 g/liter at 0°C and 1 atm pressure. The volume occupied by 1 mole or 2.02 g hydrogen is thus

$$\frac{2.02 \text{ g}}{0.0899 \text{ g/liter}} \quad \text{or} \quad 22.4 \text{ liters}$$

The volume of 1 mole of H_2 at 0°C and 1 atm pressure is 22.4 liters.

The density of chlorine is 3.21 g/liter at 0°C and 1 atm. The volume occupied by 1 mole or 71.0 g Cl_2 is

$$\frac{71.0 \text{ g}}{3.21 \text{ g/liter}} \quad \text{or} \quad 22.1 \text{ liters}$$

The volume of 1 mole of Cl_2 at 0°C and 1 atm is 22.1 liters.

The density of hydrogen chloride is 1.639 g/liter at 0°C and 1 atm. The volume occupied by 2 moles or 73.0 g HCl is

$$\frac{73.0 \text{ g}}{1.639 \text{ g/liter}} \quad \text{or} \quad 44.6 \text{ liters}$$

The volume of 2 moles of HCl is 44.6 liters at 0°C and 1 atm. The volume of 1 mole of HCl at 0°C and 1 atm is 22.3 liters.

Here we see a simple relationship between weights and volumes of gases. The volume of 1 mole of these gases at 0°C and 1 atm is about 22.4 liters. If we measure the volume occupied by 1 mole of many different gases at 0°C and 1 atm pressure, we come out with an average value of 22.4 liters. We may summarize these relationships: The volume of 1 mole (the *molar volume*) of any gas is approximately 22.4 liters at STP.

It is much easier to measure the volumes of gases than to measure their weights. The value of 22.4 liters/mole of gas at STP is very useful in calculating volumes of gases consumed or produced in a chemical reaction.

EXAMPLE 6.1   What volume of CO_2 at STP will be produced by burning 1 ton (900 kg) of coal, which is essentially pure carbon?

ANSWER

$$C(s) \quad + \; O_2(g) \longrightarrow \quad CO_2(g)$$
$$1 \text{ mole} \qquad\qquad \longrightarrow \quad 1 \text{ mole}$$
$$12.0 \text{ g} \qquad\qquad \longrightarrow \; 22.4 \text{ liters}$$

We know that 900 kg is 900,000 g. When 900,000 g/(12.0 g/mole) or 75,000 moles of carbon are burned, 75,000 moles of CO_2 are produced. Then

$$75,000 \text{ moles } CO_2 \times 22.4 \, \frac{\text{liters}}{\text{mole}} = 1,680,000 \text{ liters } CO_2$$

EXAMPLE 6.2   What is the volume occupied by 1 mole of a gas at 100°C and 1 atm pressure?

ANSWER   In Sec. 5.5 we concluded that the volume of a sample of any gas held at constant pressure is directly proportional to the absolute (Kelvin) temperature. One mole of a gas at 0°C (273 K) occupies a volume of 22.4 liters. If we heat it to 100°C (373 K), its volume increases by the fraction expressing the ratio of temperatures:

$$\text{Volume at 373 K} = 22.4 \text{ liters} \times \frac{373 \text{ K}}{273 \text{ K}} \quad \text{or} \quad 30.6 \text{ liters}$$

If you like setting up and solving proportions, you might express the relationship in this way:

$$V{:}22.4 = 373{:}273$$
$$273V = 22.4 \times 373$$
$$V = 22.4 \times {}^{373}\!/_{273} \quad \text{or} \quad 30.6 \text{ liters}$$

At 100°C 1 mole of a gas occupies a volume of 30.6 liters at 1 atm pressure.

EXAMPLE 6.3   What is the increase in volume when 1 g liquid water at 100°C boils off to produce 1 g steam (water vapor) at 100°C and 1 atm pressure?

ANSWER   From Table 5.2 we see that 1 g water at 4°C has a volume of 1 cm^3. In Example 6.2 we found that 1 mole of water (which is 18 g) in the gaseous state has a volume of 30.6 liters or 30,600 cm³. One gram of water vapor then has a volume of $^{30,600}/_{18}$ or 1700 cm³. Thus 1 cm³ liquid water at 100°C boils off to produce 1700 cm³ steam at 100°C and 1 atm.

Summary

The kinetic-molecular model of gases helps us understand the behavior of these substances in our environment. Gases are composed of tiny molecules flying around with a lot of empty space between them. When we compress a gas, we reduce its volume by reducing the amount of empty space. Because of their rapid motion, molecules exert pressure on the walls of a containing vessel by bombarding the walls and rebounding from them. In a smaller volume they exert a greater pressure if the temperature remains constant. When we heat a gas, we increase the average speed with which the molecules are moving and thus increase the pressure (the molecular bombardment on the container walls). When one gas is added to another, the molecules added just slip in between the molecules already there, and each gas exerts the same pressure as it would if it alone occupied the container.

The kinetic-molecular model is also helpful in explaining the behavior of liquids. When a gas condenses to a liquid, the volume of the liquid is much less than that of the gas; i.e., the molecules get much closer together. The molecules are still in rapid motion, but they are just jostling about and slip around one another so that the liquid can flow and take the shape of its container. Liquids tend to evaporate, and so we know that some of the molecules must be knocking about faster than others and fly off into the space above the liquid; i.e., they evaporate, producing a vapor pressure. At higher temperatures the molecules in the liquid are moving faster, and evaporation and vapor pressure increase. When we heat a liquid hot enough, it boils; i.e., some of the molecules in the liquid move fast enough to push back the liquid and form bubbles of vapor within the liquid. At the boiling point, the vapor pressure of the liquid equals the pressure of the atmosphere above it.

The kinetic-molecular model is also helpful in explaining the behavior of solids. When a liquid freezes, there is little change in volume, but fluidity (ability to flow) is lost. The molecules in the solid must be about as close together as in the liquid, but they must be in fixed positions that prevent flow. The molecules are still in rapid motion, but each is vibrating about a fixed point. As we cool the solid, the intensity of vibration decreases; as we heat the solid, the intensity of vibration increases. When we heat a solid to its melting point, the intensity of vibration is so great that the molecules break out of their fixed positions in the structure of the solid and move around one another; i.e., they flow, and we have a liquid.

From the fact that the volumes of gases involved as reactants and products in chemical processes are always in ratios of simple whole numbers, we conclude that equal volumes of gases at the same temperature and pressure contain the same number of molecules. The volume containing 1 mole (6.02×10^{23} molecules) of a gas at STP is 22.4 liters. This is called the molar volume. This molar volume applies only to gases—*not* to liquids and solids.

New Terms and Concepts

AVOGADRO'S HYPOTHESIS: At the same temperature and pressure, equal volumes of all gases contain the same number of molecules.

DIATOMIC: Composed of two atoms.

KINETIC ENERGY: Energy associated with motion.

KINETIC-MOLECULAR THEORY: All gases are composed of molecules that are in rapid motion, that do not lose energy of motion when they collide, and whose volume is very small compared to the total volume of the space occupied by the gas.

LAW OF COMBINING VOLUMES: The volumes of gases (measured at the same temperature and pressure) produced or consumed in chemical reactions are in the ratio of small whole numbers.

LAW OF PARTIAL PRESSURES: In a mixture of gases, each gas exerts the same pressure that it would if it alone occupied the container, and the total pressure of the mixture is the sum of these partial pressures of the components.

MOLAR VOLUME: The volume occupied by 1 mole of gas at STP; 22.4 liters.

NORMAL BOILING POINT: The temperature at which the vapor pressure of a liquid becomes 760 torr.

VAPOR PRESSURE: The pressure exerted when a solid or liquid is in equilibrium with its own vapor.

Testing Yourself

6.1 In terms of the kinetic-molecular model for matter, account for the following facts.
   a. Solids have a definite shape, liquids take the shape of their container, and gases must be kept in a closed vessel.
   b. Liquids and gases are fluids (flowables), but solids are not.
   c. All gases are completely miscible (mixable), but this is not true for all liquids.
   d. Water in an open dish evaporates gradually, not all at once.
   e. When water reaches its boiling point, it gradually boils away; it does not all turn to steam at once.
   f. In cold weather a pond freezes slowly, not all at once.

g. Water boils at 95°C at an altitude of 5000 ft but at 120°C in a pressure cooker at 2 atm.

h. One liter of hydrogen will combine with 1 liter of chlorine to form 2 liters of hydrogen chloride, but 1 liter of hydrogen will combine with only ½ liter of oxygen to form 2 liters of water vapor, all these volumes being measured at the same temperature and pressure.

i. Boyle's law.

j. Charles' law.

k. Dalton's law of partial pressures.

6.2 Air is 20 percent oxygen, 79 percent nitrogen, and 1 percent other gases. What is the partial pressure of oxygen in air at sea level? At an elevation of 10,000 ft the air pressure is about 500 torr. What is the partial pressure of oxygen in air at this altitude? Why does one have difficulty breathing at this altitude?

6.3 Deep-sea divers breathe a mixture of oxygen and helium when they are under great pressure in deep dives. If a diver is working at a depth where the pressure is 10 atm, what percentage of oxygen should be present in the helium-oxygen mixture he breathes if the partial pressure of oxygen is to be that in air at sea level?

6.4 Can the boiling point of water be raised above 100°C? How? Can the boiling point of water be lowered below 100°C? How?

6.5 What would happen if a sample of water were placed in a closed container and heated to a temperature above 100°C?

6.6 How many molecules of gas are there in 1 liter of air at STP? How many molecules of oxygen?

6.7 How many molecules of water are present in 1 cm^3 of liquid water? In 1 cm^3 of gaseous water at 100°C and 1 atm?

6.8 What is the weight of a mole of gas if at STP it has a density of 2.00 g/liter?

6.9 The gas sulfur dioxide combines with oxygen to form the gas sulfur trioxide, thus: $2SO_2(g) + O_2(g) \longrightarrow 2SO_3(g)$. What ratio would you expect for the following?

a. $\dfrac{\text{Number of } SO_3 \text{ molecules produced}}{\text{Number of } O_2 \text{ molecules consumed}}$

b. $\dfrac{\text{Volume of } SO_3 \text{ gas produced}}{\text{Volume of } O_2 \text{ gas consumed}}$

6.10 If the air pressure in the tires on a car is measured on a cool morning and the car is then driven all through the day, which is hot and sunny, what happens to the air pressure in the tires? Explain in terms of the kinetic-molecular model of a gas.

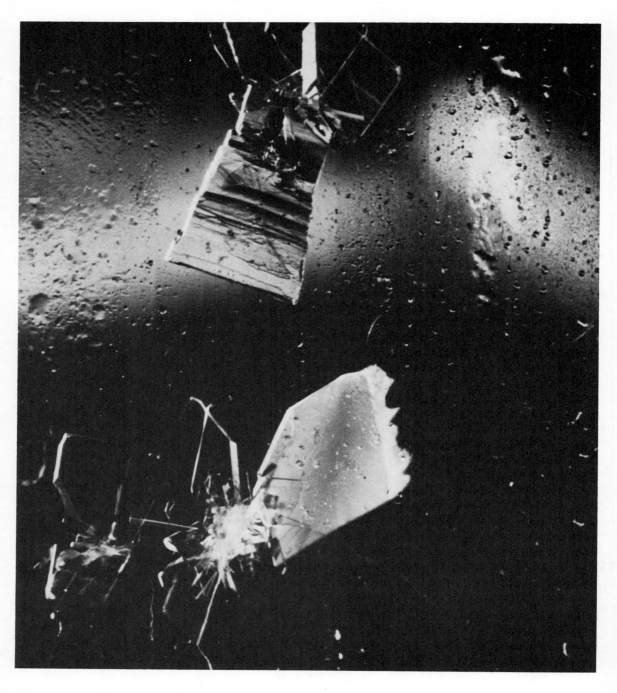

Nitrogen crystals found in frozen automobile exhaust are seen in the photograph. Nitrogen is also a pollutant in the form of nitrogen dioxide whose molecular structure is shown.

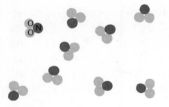

7. Air Pollution

The kinetic-molecular model gives us many insights into the behavior of matter. In this chapter we shall use these insights to help us understand the nature of the atmosphere around our earth—what substances are present in it, how they interact with one another, how they interact with light from the sun, and how the atmosphere purifies itself naturally. We shall also consider the kinds of pollutants that are discharged into our air, where they come from, the effects they produce in the air, where they go, what harm they do to us, and how we can reduce their menace.

7.1 What Is the Problem?

Have you ever tasted rain as sour as vinegar? Have you felt your eyes smart as you drove along a freeway? Have you flown over a city hidden under a layer of dirty gray smog? What color is the smoke from the factory chimneys in your home community? Does the United States have a monopoly on air pollution?

In all the industrialized areas of the world and in all great cities, man is suffering from polluted air. In industrialized regions some of the most damaging pollutants form *acids* in the air which are then washed from the atmosphere in sour rains. As we shall see in Chap. 10, any sour solution is an acid. All but the very weakest acids are detrimental to plants and animals as well as to manmade structures like bridges, railways, and buildings.

acid

113

Rain in Sweden is sometimes as sour as weak lemonade. Swedish scientists have found a two-hundred-fold increase in the acidity of the rain in some parts of Scandinavia since 1965. The rainfall in the Netherlands and Belgium is the most acidic in the world, apparently because this area is centrally located in the European complex of metallurgical and chemical industries polluting the air. This area includes the eastern regions of England, the valley of the Meuse in Belgium and France, and the valleys of the Ruhr and the Rhine in Germany and the Netherlands. Winds carry the acid air as far west as Ireland, to the North Cape of Norway, to the southern boundaries of Switzerland and Austria, and eastward into the Balkans.

Acid rain is injurious to plant and animal life. Some streams in southern Norway have such high acidity that salmon eggs can no longer develop in the water, and salmon runs up these streams have ceased. It is estimated that by the end of this century the acid rain in Sweden will have reduced the crop of trees for lumber by 10 to 15 percent. Acid rain also rapidly corrodes metals and stone; it is costing our country $1.5 billion each year in damage to buildings and bridges. The faces on some stone statues in Oxford, England, have become almost unrecognizable in the last 50 years. Michelangelo's statue of David that stands in the Piazza Signoria, a public square in Florence, Italy,

Air pollution ravaged this statue at the Château de Versailles.

is a copy. The original is safely housed in a museum where it is protected from the polluted rain and air of modern Florence.

Worldwide problems arising from air pollution are but extensions of those we encounter in the United States. We shall consider the general problem as it is reflected in our country, because data for the United States are more readily accessible than for the world at large.

7.2 Air Pollution in the United States

Rain in New England is increasingly acidic. Acid content in precipitation there is now 10 to 100 times higher than in the pure air found over wilderness areas. Equally acid rain and snow have been found in the central Finger Lakes region of New York State. Studies are now under way to determine whether the acidity of our lakes and streams is being increased by acid rain and snow.

Acidity in the atmosphere comes from pollution by oxides of sulfur contained in the stack gases from domestic and industrial furnaces, and by oxides of nitrogen in the exhaust fumes from buses, trucks, and cars. In addition to oxides of nitrogen and sulfur, air contains other pollutants, summarized in Table 7.1.

Because the weight of CO (carbon monoxide) dumped into the atmosphere each year is more than half the weight of all pollutants together, we might

TABLE 7.1 *Air Pollutants Produced in the United States in 1969*

Kind of pollutant	Symbol	Tons/yr, in millions
Nitrogen oxides (nitric oxide, NO, and nitrogen dioxide, NO_2, are generally lumped together and symbolized NO_x)	NO_x	24
Sulfur oxides (sulfur dioxide and trioxide, SO_2 and SO_3, are generally lumped together and symbolized SO_x)	SO_x	33
Particulates (tiny particles too small to be seen with the naked eye but visible with a microscope)	particulates	35
Hydrocarbons (dozens of compounds, composed of H and C only, are generally lumped together and symbolized HC)	HC	37
Carbon monoxide	CO	151
Total		280

TABLE 7.2 *Estimated Magnitude of Concern for Various Air Pollutants*

Pollutant	Particulates	SO_x	NO_x	HC	CO	Total
Percent of total concern	50	27	12	7	4	100

tolerance level

residence time

particulates

conclude that CO is the most dangerous of all. But three other factors besides quantity create the potential harmfulness of pollutants. (1) The *tolerance level* is the maximum concentration of a pollutant that can be present in the air we breathe without unfavorable effects on our health. (2) The *residence time* is the average time it takes for a pollutant to be removed from the atmosphere by natural purification processes. (3) Interactions with other pollutants determine whether one pollutant tends to increase, decrease, or have no influence on the effects of other pollutants. The tolerance level for CO is 40,000 $\mu g/m^3$ (micrograms per cubic meter), whereas that for *particulates* is 375. (A microgram is one-millionth of a gram.) On this score, particulates are much more dangerous than CO. The residence time for CO is only a few hours (Sec. 7.8), while that for tiny particulates is months. Again particulates are more dangerous. CO has no known effect on other pollutants, whereas

FIGURE 7.1 Decreasing density of air with increasing altitude

Atmosphere: the gaseous shell
at least 700 mi thick which
completely envelopes the earth

particulates greatly increase the damage done to lung tissue by inhaled oxides of sulfur (Sec. 7.13). So again particulates are more dangerous.

From similar consideration of other pollutants, some atmospheric scientists have arrived at the ratings of concern shown in Table 7.2. Other scientists have suggested different orders of concern, but all agree that particulates are especially dangerous and that CO is a minor problem.

How can we keep our air fit to breathe? First, we need to know something about the components of unpolluted air and about the structure and behavior of the atmosphere.

7.3   The Atmosphere

Because of the opposing effects of the upward diffusion of the gases in the atmosphere and the earth's gravitational pull downward, there is no sharp line above the earth where the atmosphere ends. Its density gradually decreases at higher and higher altitudes (Fig. 7.1) until, somewhere between 250 and 900 mi above the earth, the air thins out to empty space. Pressure decreases with increasing altitude (Fig. 7.2).

Temperature drops, then begins to rise again (Fig. 7.3) at an altitude of about 7 mi. The layer below this break in the temperature drop is called the *troposphere*, and that immediately above it is called the *stratosphere*. Al-though 95 percent of the total air mass is concentrated in a 12-mi layer above

troposphere
stratosphere

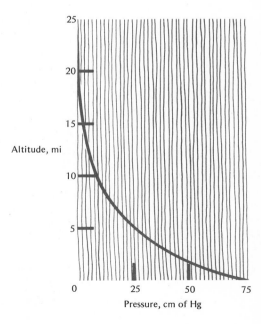

FIGURE 7.2   Change of atmospheric pressure with altitude

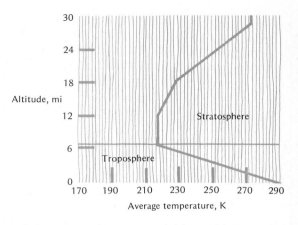

FIGURE 7.3 Change in temperature with altitude

the earth, only a small portion of the troposphere is available to dilute or disperse pollutants. During the summer this may amount to several thousand feet. In the winter, when less heat is received from the sun and there is less atmospheric circulation, the mixing layer over the United States extends on the average only 750 to 2500 ft above the ground.

7.4 Absorption of Solar Energy by the Earth and Its Atmosphere

If there were no atmosphere, the energy received on the surface of the earth from the sun when it is directly overhead would be 2 cal/cm²-min. The atmosphere reduces this rate. When the sun is not directly overhead, its slant-

FIGURE 7.4 Equatorial regions receive more direct—hence more intense— sunlight than other areas.

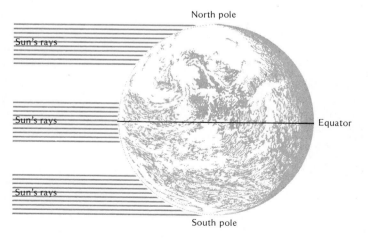

FIGURE 7.5 Circulation of air along the earth's surface toward the equatorial zone, and upward movement of air at the equator

Temperate zone   Equatorial zone   Temperate zone

ing rays bring less heat per cm²-min. The sun's rays (Fig. 7.4) strike the earth more directly near the equator than at surfaces north or south of this region. Much of the sunlight reaching the earth is absorbed by the atmosphere. Consequently the air heats up more in equatorial regions tnan in temperate and polar regions where the sun's rays strike the surface of the earth at an angle of less than 90°.

As we noted in Sec. 6.3, when any gas is heated, it tends to expand because of the faster movement of its molecules. A given mass of gas then occupies a larger volume at a given pressure, and its density decreases with increasing temperature. The warm air of equatorial regions rises through the atmosphere until its density is equal to that of the surrounding air mass. This thinner air produces a pressure at the earth's surface that is less than it would be without such heating. Cooler, denser air from the temperate regions then flows along the earth's surface toward the equator to take the place of the warm air rising. If the earth did not rotate, its atmosphere would circulate as shown in Figs. 7.5 and 7.6.

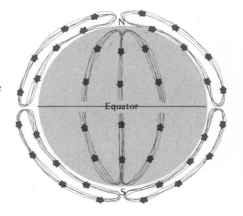

FIGURE 7.6 How the earth's atmosphere would circulate if it were heated uniformly at the equator and the earth were not rotating

But the rotation of the earth distorts this circulation, producing the pattern shown in Fig. 7.7. Localized distortions in the general circulation arise from the uneven heating of air at a given latitude. Air gets hotter over barren ground than over vegetated areas, and it is cooler over water than over land. Clouds or haze absorb more heat than clear air, and mountains and canyons deflect air flow. The combined effects of these factors make the flow of air very complicated.

Of the energy in the sunlight falling on the whole earth, about 30 percent is reflected back by the atmosphere, by clouds, and by the surface of the earth. About 50 percent finally reaches the ground or the ocean, where it is absorbed as heat. Of the remaining 20 percent, a small amount (1 to 3 percent), which is *ultraviolet light*, is absorbed by layers of oxygen (O_2) and ozone (O_3) in the upper atmosphere. Most of the rest of the 20 percent is absorbed in the atmosphere by water vapor, dust, and water droplets in clouds. Although there have been long-term changes in climate, there is no indication that the earth is now undergoing any appreciable net long-term heating or cooling. Therefore the earth must radiate an amount of energy equal to that which it receives from the sun. Though the radiation from the sun is largely in the form of *visible light* and ultraviolet light, the radiation from the earth is *infrared light*, often called heat rays. Much of the sunlight absorbed at the earth's surface is used to evaporate water, which then rises into the atmosphere. The flow of energy through the system comprised of the earth and its atmosphere is schematized in Fig. 7.8.

ultraviolet light

visible light
infrared light

FIGURE 7.7 Idealized pattern of circulation of the atmosphere due to equatorial heating and earth's rotation; zones of high and low pressure.

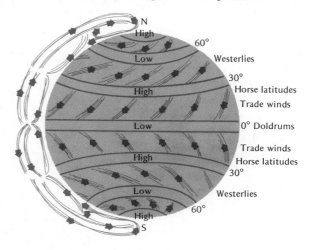

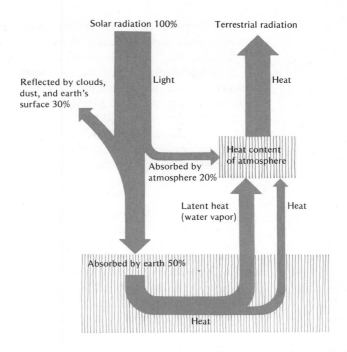

FIGURE 7.8 The flow of the sun's energy through the atmosphere and the earth

7.5 The Natural Pollution and Purification of the Atmosphere

Even if man and his technology did not exist, the air would be polluted by various natural processes. Hydrogen sulfide (H_2S), sulfur dioxide (SO_2), carbon monoxide (CO), dust, and ashes are injected into the atmosphere by volcanic action. Methane (CH_4), ammonia (NH_3), and H_2S are produced extensively from the decay of dead plants and animals under water, which excludes oxygen from the process. Carbon dioxide (CO_2) is produced by the respiration of plants and animals. Oxides of nitrogen (N_2O, NO, and NO_2) are produced by electrical discharges in thunderstorms. Forest and grass fires produce smoke, CO, and CO_2. All these natural pollutants are circulated through the troposphere and are gradually washed out by rain, snow, and hail. Because living plants and animals are continually producing large amounts of CO_2, it is always present in the atmosphere in significant amounts even though it, too, is washed out of the air by rain.

Because the air circulates all around the world, substances introduced into it at one place are transported to all others. When the island of Krakatoa in the straits between Sumatra and Java blew up in 1883, the sound of the major explosion was heard 1500 mi to the north in Burma and 2000 mi to the southeast in Australia. The cloud of dust and ashes produced total darkness for 24 hr, 100 mi west of the volcano. The great dust cloud was observed all around the world near the equator within 2 weeks. Its passage into the skies above the temperate zones produced brilliantly colored sunrises and sunsets for two years. Rains of mud fell 50 mi from the eruption and floating islands of pumice circled about in the Indian Ocean. The pollution of the atmosphere by Krakatoa was worldwide, but within 3 years of this most violent cataclysm in the history of man, natural processes had restored the air to its normal purity.

7.6 The Chemical Composition of the Atmosphere

parts per million

Earth's atmosphere has a mass of 6 thousand trillion (6,000,000,000,000,000) tons and is a mixture of many gases. Analyses of many samples of air collected far from sources of pollution are given in Table 7.3. Concentrations of gases in mixtures are conveniently expressed in *parts per million* (ppm). Thus, if we mix 1 ml oxygen with 999 ml nitrogen, the mixture contains one part of oxygen per thousand of mixture. If we mix 1 ml oxygen with 999,999 ml nitrogen, the mixture contains one part of oxygen per million of mixture, or 1 ppm.

7.7 The Impact on the Environment of CO_2 from Burning Fuels

Each year millions of tons of carbon dioxide are discharged into the atmosphere by the respiration of plants and animals and by the burning of fuels containing carbon. The combustible part of coal is principally carbon, and it

TABLE 7.3 *Composition of Clean, Dry Air near Sea Level*

Component	Formula	Concentration, ppm	Component	Formula	Concentration, ppm
Nitrogen	N_2	780,900	Krypton	Kr	1
Oxygen	O_2	209,400	Nitrous oxide	N_2O	0.5
Argon	Ar	9,300	Hydrogen	H_2	0.5
Carbon dioxide	CO_2	315	Xenon	Xe	0.08
Neon	Ne	18	Nitrogen dioxide	NO_2	0.02
Helium	He	5.2	Ozone	O_3	0.01–0.04
Methane	CH_4	1.0–1.2			

reacts with the oxygen of the air thus:

$$C(s) + O_2(g) \longrightarrow CO_2(g)$$

Octane, C_8H_{18}, one of the chief components in gasoline, reacts with oxygen in the following way:

$$2C_8H_{18}(l) + 25O_2(g) \longrightarrow 16CO_2(g) + 18H_2O(g)$$

hydrocarbons

Other *hydrocarbons* (compounds containing only carbon and hydrogen) likewise burn to produce CO_2 and H_2O. The sugar we eat ($C_{12}H_{22}O_{11}$) combines with oxygen to produce CO_2. Although this process takes place in many consecutive reactions, the overall equation is

$$C_{12}H_{22}O_{11} + 12O_2 \longrightarrow 12CO_2 + 11H_2O$$

We emit the CO_2 from our lungs as we exhale.

Before we began to burn large amounts of coal, gas, and oil in our homes and industries, the concentration of CO_2 in air was fairly constant at about 280 ppm. In the 100 years between 1860 and 1960 the figure rose to 315 ppm. It is estimated that in the year 2000 the CO_2 concentration will be about 400 ppm. A study sponsored by the United Nations indicates that by the year 2000 the combustion of fuels will have produced an amount of CO_2 equal to one-fifth of the total amount in the atmosphere today. This is about 100 times the CO_2 produced by volcanoes and 10 times that produced by the breathing of all living organisms. Although this increase will have no harmful effects on the lives of most organisms, it *may* bring about important changes in climate.

greenhouse effect

Carbon dioxide in the atmosphere creates what is known as the *greenhouse effect*. The glass in a greenhouse allows visible sunlight to pass through to heat the air and the soil around growing plants, but it traps the infrared heat rays emitted from the warm soil and air inside the greenhouse. Likewise, CO_2 permits the passage of ultraviolet and visible light from the sun down to the surface of the earth, but it readily absorbs infrared light, keeping it from passing up through the atmosphere and thus raising the temperature of the atmosphere.

As the average concentration of CO_2 has risen during the past century, so has the average temperature of the earth. If the average temperature were to rise 9°F, part of the polar ice caps could melt and the level of the oceans could rise several feet. This in turn could adversely affect man's activity on land near sea level. Harbors could be flooded and fertile valleys drowned with salt water. It must be emphasized that we *do not know* what the effect of increasing CO_2 will be; we *do know* that many areas which today are dry land were shallow seas or swamps when the average temperature of the earth was higher than it is today. It is imperative that we continue and intensify our studies of the amounts of CO_2 in the earth's atmosphere and the accompanying changes in climate.

7.8 The Carbon Monoxide Problem

When coal (which is largely carbon) is burned with less than the amount of oxygen needed to produce CO_2, it forms carbon monoxide, CO:

$$2C(s) + O_2(g) \longrightarrow 2CO(g)$$

When any substance containing carbon is burned with insufficient oxygen, CO is produced. In an automobile engine the burning of the hydrocarbons in gasoline produces exhaust gas containing 10,000 to 40,000 ppm CO.

Carbon monoxide is a particularly insidious poison, because we cannot see, taste, or smell it, and it does not irritate our eyes, nasal passages, or lungs. It passes unchanged through the walls of our lungs into the blood. There it combines with hemoglobin, the substance in red blood corpuscles that carries oxygen (in the form of oxyhemoglobin, O_2Hb) to all the tissues of the body. The combination of CO with hemoglobin produces carboxyhemoglobin (COHb). Because CO is about 200 times more strongly bound than oxygen to hemoglobin, all CO present in the lungs will be absorbed to form COHb before much O_2Hb can be formed. The resulting decrease in the transport of oxygen from the lungs to the tissues may cause the latter to suffer from oxygen deprivation. The two tissues most sensitive to lack of oxygen are the heart and the brain.

If you rest quietly and breathe air containing 10 ppm CO for 10 hr, then 2 percent of your hemoglobin is converted to COHb. Further exposure produces no additional conversion. If you are doing hard physical work and breathing 10 ppm CO, you reach the maximum 2 percent COHb in 3 hr. It takes several hours of rest (or less time doing heavy work) while breathing pure air to remove the COHb from your blood. If you are similarly exposed to air containing 30 ppm CO, you convert 5 percent of your hemoglobin to COHb. Levels of 2 to 5 percent COHb in your blood impair sharpness of vision, discrimination of brightness, and judgment of time intervals. Similar exposure to air containing 50 or more ppm CO will result in headache, fatigue, drowsiness, even coma and death. Never run a car engine in a closed garage. Never use a gas space heater without a proper vent to remove the products of combustion. Never drive a car with a leaky exhaust pipe or muffler. Never drive in a car with all the windows and air vents closed.

In slow, heavy traffic on a freeway you may breathe air containing 50 or more ppm CO. In a traffic jam, the CO content of air may rise to 140 ppm. On a smoggy day in Los Angeles, there may be 30 ppm CO in the air for 8 hr or more. In downtown New York City, levels of CO are 15 ppm all day long. People who are continuously exposed to city air in heavy traffic (traffic policemen, newsstand vendors, and drivers of buses, trucks, and taxis) are the chief sufferers from CO poisoning. The commuting suburbanite who spends an hour or more driving in heavy traffic every day also suffers appreciably. CO is not present in air over the open countryside because it is rapidly removed by the metabolism of microorganisms (principally fungi) found in almost all soils.

Of the 151 million tons of CO dumped into our air annually, 60 percent comes from gasoline engines in cars and trucks, and the other 40 percent comes from the burning of brush, straw, and weeds, from forest fires, industrial processes, and the burning of solid wastes in dumps or incinerators. Clearly, the number-one contributor to air pollution by CO is the automobile, and strenuous efforts are now being made to modify its operation so that it will generate less CO. Understanding the nature of these modifications requires a knowledge of several basic principles of chemistry not yet considered in this book. For this reason, abatement of CO pollution will be discussed in Chap. 11.

7.9 Oxides of Sulfur

When sulfur burns in the open air, a gas is produced that is highly irritating to the eyes and respiratory passages of the nose, throat, and chest. If this gas is collected and cooled to room temperature, about 4 percent of it condenses to a liquid. Analysis of the uncondensed gas shows that it has the formula SO_2. The formula of the liquid is SO_3. The equations for the formation of these two oxides are

$$S(s) \ + \ O_2(g) \longrightarrow \ SO_2(g)$$
$$2S(s) \ + \ 3O_2(g) \longrightarrow 2SO_3(g)$$

If we pass the gaseous SO_2 through water, it dissolves to give a sour solution—an acid. If SO_3 is added to water, it too dissolves to give an acid solution. Analysis of the products shows that the following reactions take place:

$$SO_2(g) \ + \ H_2O(l) \longrightarrow H_2SO_3(l) \qquad \text{(sulfurous acid)}$$
$$SO_3(g) \ + \ H_2O(l) \longrightarrow H_2SO_4(l) \qquad \text{(sulfuric acid)}$$

SO_2 in the atmosphere is gradually changed to SO_3. Thus

$$2SO_2(g) \ + \ O_2(g) \longrightarrow 2SO_3(g)$$

Most of the acidity in rain is due to sulfuric acid.

When SO_2 is inhaled, it dissolves in the moisture on the mucous membrane lining the respiratory tract, and the sulfurous acid so formed damages the tissue. Sulfuric acid deposited in respiratory membranes by the inhalation of SO_3 or H_2SO_4 droplets is even more damaging. Fortunately, the odor of SO_2 is so noticeable that one is immediately warned of its presence and can get out of danger before damaging concentrations are reached. The odor is detectable at 3 to 5 ppm in air. Prolonged exposure to 20 ppm is unpleasant but bearable; it is the maximum concentration allowable for persons working in confined atmospheres. The principal effect is difficulty in breathing, but there is no evidence of respiratory disease in healthy persons. Respiratory diseases already in existence, however, are aggravated by exposure to SO_2, and deaths from such diseases increase in periods of intense air pollution.

About three-fourths of the SO_x pollutants in our atmosphere come from the combustion of coal and oil containing sulfur compounds. When these are burned, most of the sulfur is converted to SO_2. Much of the remaining one-fourth of the SO_x is emitted by various industrial processes. For this reason, eliminating SO_x pollution depends largely on adequate controls in large plants burning coal and oil for heat or to generate electricity. Such controls can then be modified to reduce SO_x emissions from industrial processes. The chemistry needed to see how we can reduce SO_x will be developed in Chap. 10. Methods of abatement are discussed in Chap. 11.

7.10   Oxides of Nitrogen

The nitrogen and oxygen in air do not combine at ordinary temperatures, but burning fuels with air at temperatures of 2000 to 3000° F produces significant amounts of nitric oxide, NO:

$$N_2(g) \ + \ O_2(g) \longrightarrow 2NO(g)$$

About 10 percent of the NO produced in combustion is converted to nitrogen dioxide, NO_2, by the reaction

$$2NO(g) \ + \ O_2(g) \longrightarrow 2NO_2(g)$$

About 50 percent of the NO and NO_2 pollution comes from electric generating plants, 40 percent from cars, trucks, and buses, and the remainder from forest fires and other open burning.

The average residence time of NO and NO_2 in the atmosphere is 3 or 4 days. Ultimately, these oxides are converted to nitric acid, HNO_3. This reacts with various substances in the air to produce solids that are removed from the atmosphere by rainfall. The chemistry of these processes is developed in Chap. 10.

smog

The concentrations of NO and NO_2 in polluted air are not toxic to man, but they react with other pollutants to produce *smog*, which is becoming the number-one pollution problem in large cities. Abatement of NO_x pollution will be discussed in Chap. 11.

7.11   Photochemical Smog

The term "smog"—a combination of the words "smoke" and "fog"—was coined to name visibly polluted air over a city. Smog is a complex mixture of

Pollution in Los Angeles makes downtown buildings scarcely visible.

many gases and tiny particles of liquids and solids. Its combined effect on the eyes and the respiratory system is far more irritating than the effects of its separate ingredients.

photochemical

When NO_2 is exposed to sunlight, a *photochemical* reaction takes place:

$$NO_2(g) + \text{sunlight} \longrightarrow NO(g) + O(g)$$

The O atoms combine with O_2 molecules to form ozone, O_3:

$$O(g) + O_2(g) \longrightarrow O_3(g)$$

If no other pollutants are present, the ozone reacts with NO to produce NO_2 and O_2:

$$O_3(g) + NO(g) \longrightarrow NO_2(g) + O_2(g)$$

This cycle of nitrogen-oxygen reactions is shown in Fig. 7.9.

Internal combustion engines emit large quantities of hydrocarbons. In the air these react with the oxygen atoms released by the decomposition of NO_2 to form compounds called *hydrocarbon free radicals* (unusually reactive fragments of a normal molecule) which quickly react with NO to produce NO_2 again. This process removes NO from the air and stops the reaction in the cycle that uses up ozone. Ozone, which is highly irritating to the eyes and the respiratory passages, is one of the most active components of smog. The free radicals also react with O_2 and NO_2 to form a series of compounds known as *peroxyacylnitrates*, commonly abbreviated PAN. These are also very irritating and contribute greatly to the effects of smog. Many other irritants are produced by the interaction of hydrocarbon free radicals and oxygen. The disruption of the NO_2–O_3 cycle caused by the hydrocarbons is shown schematically in Fig. 7.10.

In a large city there is less smog in late afternoon than at noon. At dawn, NO, NO_2, and O_3 concentrations are low. During the morning rush hours, NO and the hydrocarbons build up. The latter absorb sunlight and speed the conversion of NO into NO_2 by oxygen molecules, and so the concentration of NO decreases and that of NO_2 rises. It is a brown gas, plainly visible in heavy smog. Then the sunlight interacts with NO_2 to produce ozone. The disruption of the NO_2–O_3 cycle by hydrocarbons builds the concentrations of ozone, PAN, and other irritants to a maximum around noon. The evening rush-hour traffic again builds up NO and the hydrocarbons, but without bright sunlight the latter cannot facilitate the conversion of NO into NO_2 by O_2. The accumulated NO then reacts with ozone to produce NO_2, and the concentrations of all irritants decrease. Smog disappears during the night. The key to

hydrocarbon
free radicals

peroxyacylnitrates
(PAN)

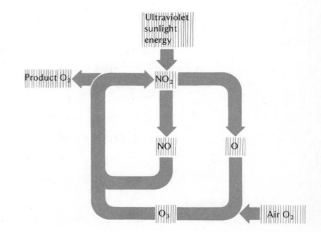

FIGURE 7.9 The nitrogen dioxide–ozone cycle

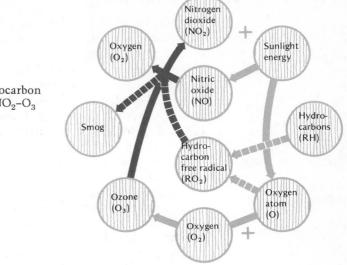

FIGURE 7.10 Hydrocarbon disruptions of the NO_2–O_3 cycle

reducing photochemical smog is to keep motors from giving off NO_x and hydrocarbons. Ways of doing this are discussed in Chap. 11. The evaporation of hydrocarbons into the air when gasoline is being pumped from one tank into another is also a problem. New ways for eliminating such evaporation are now being developed.

7.12   Lead in Polluted Air

Another air pollutant is lead. A record of its concentration in the atmosphere has been kept in the ice of northern Greenland. There lead is removed from the air by falling snow and accumulates in sheets of ice. By sampling the ice at different depths, scientists can show clearly that the lead concentration in the atmosphere has risen sharply since 1940, when "leaded" gasolines were introduced. Today the use of these gasolines accounts for the emission into the atmosphere of 181,000 tons/yr of lead, more than 98 percent of the total lead emitted from all sources. Though the effects of airborne lead compounds on our health are not well known, no one doubts that we would be better off without them. Modern methods of petroleum refining can produce unleaded gasolines that burn as well as those containing lead. Although we have already decreased the lead content in some gasolines, we should eliminate it completely. The barrier to this step forward is economic, not technological. It is estimated that to convert our nation's refineries to the production of unleaded gasolines would cost some $2 billion. Who will pay this price?

7.13   Particulates as Pollutants

As noted in Sec. 7.2, pollution by particulates is probably our worst air problem. About a third of the total mass of polluting particulates comes from electric power generating plants, a third from open burning (forest fires, agricultural burning, refuse coal burning), a quarter from industrial processes, and the remainder from vehicles—cars, trucks, buses, railroad locomotives, ships, and airplanes.

Particulates are droplets of liquids or bits of solids. Droplets are spherical, but solid particles are very irregular in shape. Nevertheless we speak of the "diameter" of a solid particle, meaning an average of its length, breadth, and thickness. The diameters of particulates range from 0.0002 micron (the size of a small molecule) to 500 microns (the size of a grain of sand.) A micron is $\frac{1}{1,000,000}$ m, or $\frac{1}{1000}$ mm. The smallest particulates take weeks to settle out of the air (mostly by rainfall) and are carried by winds through the lower troposphere. During the drought on the western plains in the 1930s, fine dust blown from parched farmlands darkened the skies in New York City. The largest particulates come down almost immediately, landing close to the source of pollution. The ground near a typical portland cement factory is blanketed with layers of coarse dust from the cement kilns and grinders.

For convenience, we classify particulates into two categories: (1) those smaller than 0.1 micron and (2) those larger than 0.1 micron. In a typical sample of polluted air, 95 percent, by count, of the particles are smaller than 0.1 micron; of the total weight of the particles present, 95 percent, by weight, is in particles larger than 0.1 micron.

For air over a forested wilderness where there has been no fire for some months, 2 to 5 μg of particulates per m^3 is typical. Air over farmland runs about 10 to 15 μg/m^3. In a large city, a typical sample of air contains about 100 to 200 μg/m^3. In a heavily polluted area the weight of particulates may be as much as 2000 μg/m^3.

A sample of air taken over the Atlantic Ocean contained 300 particles per cm^3 (cubic centimeter), most of these probably from ocean spray. A sample of air collected over Chicago at 6 a.m. contained 44,000 particles per cm^3; at 8 a.m. the figure had risen to 116,000.

A typical automobile discharges 100 billion particles per sec in its exhaust gases. Typical air in a conference room or classroom of medium size has 70,000 particles per cm^3.

The damage done to us by particles in the air we breathe depends on three factors: (1) their size, (2) their chemical and physical characteristics, and (3) the extent to which they have absorbed gaseous pollutants and hold them in contact with tissues in the lungs. When we inhale, air travels through the nose, the trachea, the bronchi, and the bronchioles into the *alveoli*, the tiny

alveoli

sacs whose walls permit the transfer of oxygen from the air into the blood and of carbon dioxide from the blood into the air (Fig. 7.11).

cilia

The hairs in the nose remove coarse particles; smaller ones are caught on the mucous membranes that line the respiratory tract. Tiny hairs (*cilia*) move the mucus up into the throat, where it is swallowed. Particles larger than 0.5 micron are caught and removed before they get to the alveoli. Particles smaller than 0.5 micron may penetrate the alveoli and remain there for a considerable time. They may even be absorbed into the blood. High concentrations of particulates heighten the effect of SO_2, which is apparently absorbed onto the surface of the particles and held in contact with lung tissue longer than it would be held in the gaseous state, thus causing greater damage.

7.14 The Effects of Particulates on Sunlight

prism

In addition to causing direct damage to bodily tissues, particulates in the air have another effect—they scatter sunlight. If we pass a beam of sunlight through a *prism* (Fig. 7.12), the white light is split up into the colors of the rainbow, which we can see on the screen. If we reflect this rainbow back

FIGURE 7.11 (*a*) The respiratory system. (*b*) The alveoli.

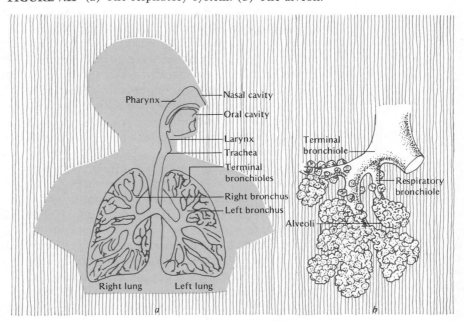

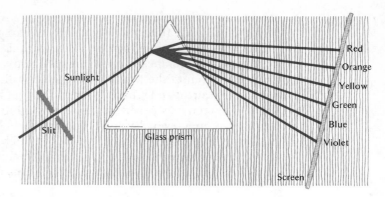

FIGURE 7.12 The splitting of white light into colors

through another prism to mix the colors, we see white light again. If we block out the blue and violet from white light, we see reddish light; if we block out the red and orange, we see bluish light.

When a ray from the sun passes through air, the particulates present scatter some of the light out of the ray's path. We know that both red and blue light are scattered appreciably, since we can look at the sun at sunset but not at noon. Blue is scattered about 16 times as much as red. When the sun is overhead, the whole sky looks blue because the strongly scattered blue light comes to our eyes from all directions, but most of the less strongly scattered red light continues in the direction of the rays from the sun. When we look directly at the sun at sunrise or sunset, it is red. The light is traveling through much greater thicknesses of air than when the sun is directly overhead, and most of the blue light has been scattered out. The red light in the sun's rays comes straight through to our eyes with little loss by scattering. The presence of particulates in the air increases scattering. This is most noticeable at sunset, because the blue light from the sun is so completely scattered that the red seems much more intense.

Particles in the atmosphere not only scatter light but also absorb it. When we look at an object, we see it by virtue of the light it reflects to our eyes. As the concentration of particulates in air increases, there is a decrease in both the intensity of light reflected from an object and the intensity of the background. Visibility is decreased; our eyes have difficulty distinguishing an object from its background. An object visible at a distance of 25 mi through clean air over the countryside may be barely detectable at a distance of 1 mi in heavily polluted air. This condition creates serious hazards to aircraft approaching and leaving airports.

In heavily polluted areas, scattering and absorption of light by particulates may reduce the amount of sunlight falling on the ground by one-third or more. A study in Leningrad showed that in winter the city received 70 percent less sunlight per unit area than the surrounding countryside. Under these circumstances, the need for artificial lighting increases, more electric energy is consumed, power plants generate more air pollution, and a dangerous cycle is created. The reduction of solar energy reaching the surface of the earth may lead to a worldwide decline in average air temperature. As we saw, the increase in CO_2 in the air may have the opposite effect (Sec. 7.7). We do not yet have enough data to know which effect will predominate, or whether they will cancel one another.

High concentrations of particulates encourage the formation of clouds, rain, and snow. Rainfall induced over cities may be lost to surrounding farming areas. A study in the state of Washington showed an increase of 30 percent in average precipitation over long periods of time as a result of air pollution in an area affected by pulp and paper mills.

7.15 Temperature Inversions

When winds and currents of rising air carry air pollutants away, conditions in major cities and industrial areas are made tolerable if not enjoyable. But winds are decreased by high buildings and by mountains. In a large city or an industrial area adjacent to mountains, pollutants are dispersed principally by rising air currents. If these fail, pollution builds up and its toxic effects become acute.

As shown in Fig. 7.3, the temperature of the air immediately above the earth normally decreases with altitude. As the air at the surface is warmed and expands, it becomes less dense than the air above it. It rises through the cooler air, which then flows down to replace it. This cooler air is then warmed by the earth and rises in its turn, carrying pollutants with it.

inversion layer

When a large mass of cool air flows into an area at a low altitude (down the sides of nearby mountains or inland from the sea), it pushes the warm surface air up 1500 to 3000 ft. Air temperature then decreases in normal fashion from the ground up to some altitude, say 2000 ft. But where the lower cool air touches the warm air above it, the air temperature increases with altitude. This region of rising temperatures is called an *inversion layer*. The cooler, denser, polluted air below this warm layer cannot rise; the inversion layer serves as a kind of "lid," and trapped pollutants build up to intolerable levels (Fig. 7.13).

In Donora, Pennsylvania, in 1948 an inversion lasting 4 days caused a buildup of pollutants which killed 20 persons and made half the population of 14,000 sick. During a 4-day inversion over London in December of 1952, 4000 more deaths occurred than normal. An inversion over New York City in 1963 caused an intense smog during which there were 300 more deaths

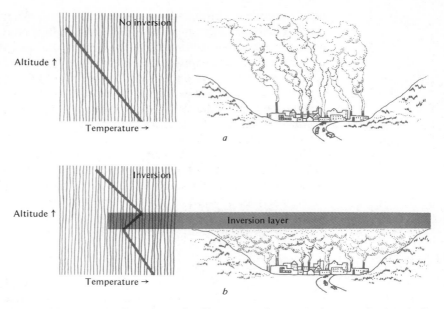

FIGURE 7.13 (*a*) Dispersion of pollutants under normal atmospheric conditions. (*b*) Trapping of pollutants under an inversion layer. (From P. R. Ehrlich and A. H. Ehrlich, *Population, Resources, Environment: Issues in Human Ecology*, W. H. Freeman and Company, San Francisco, 1972.)

temperature
inversion

than normal. *Temperature inversions* occur frequently in Los Angeles and greatly aggravate the formation of smog.

Summary

Clean air is composed of a mixture of many gases, chiefly nitrogen (79 percent), oxygen (20 percent), argon (a little less than 1 percent), and water vapor, carbon dioxide, and several other gases in very small quantities. Because we live at the bottom of this sea of air, we are subjected to atmospheric pressure arising from the weight of the air above us. This pressure decreases with altitude and finally becomes zero at some indefinite distance above the earth where the atmosphere peters out. The average pressure of the atmosphere at sea level is 760 torr, often called normal or standard atmospheric pressure.

The atmosphere influences the amount and quality of the sunlight we receive on earth. It filters out the ultraviolet light in sunlight and reduces the amount of infrared light radiated from the earth into space. The warming of the atmosphere by sunlight causes winds to blow, and these serve to keep the air mixed up. The cooling of warm, moist air produces rain, snow, or hail. These wash and purify the air as they fall to earth.

The chief pollutants in our air are the oxides of carbon, the oxides of sulfur, the oxides of nitrogen, hydrocarbons, gaseous compounds of lead, and tiny solid or liquid particles. Carbon dioxide is not damaging to health, but increasing amounts in the atmosphere may bring about a slow rise in average yearly temperature which could result in melting the polar ice caps and raising the level of the oceans. Carbon monoxide is very dangerous in heavy city traffic, but once it is dispersed away from the city into the country, it is quickly removed by organisms living in the soil. Sulfur dioxide and sulfur trioxide form droplets of sulfuric acid in the atmosphere. Rain, snow, and hail bring these to earth, where the acid is very damaging to plants, animals, and manmade structures. The oxides of nitrogen and the hydrocarbons are harmful chiefly because they produce photochemical smog in contaminated city air. Lead compounds are poisonous, but we do not yet know their effects on the air we breathe. Particulates are the most dangerous of all air pollutants because they increase the damage done by other pollutants in our respiratory passages and are very difficult to remove completely from the atmosphere once they are in it.

New Terms and Concepts

ACID: A sour substance.

ALVEOLI: Air cells in the lungs.

CILIA: Tiny hairs in nasal passages and bronchi.

GREENHOUSE EFFECT: The heating of the earth's atmosphere by sunlight due to the fact that the atmosphere transmits visible and some ultraviolet light from the sun but absorbs the infrared light which is generated on the earth's surface by sunlight and then radiates upward.

HYDROCARBON FREE RADICALS: Very reactive fragments of molecules containing only carbon and hydrogen.

HYDROCARBONS: Compounds containing only carbon and hydrogen.

INFRARED LIGHT: Radiation with wavelengths longer than those of visible light and shorter than radio waves. Heat waves.

INVERSION LAYER: A region in the atmosphere where the temperature rises with increasing altitude.

PARTICULATES: The general term for any small bodies of matter circulating in the atmosphere, usually small solid particles. But the term is also sometimes applied to droplets of liquid.

PARTS PER MILLION: If 1 g of a given solid, liquid, or gas is mixed with 999,999 g of other substances, the concentration of the first substance is said to be 1 part per million (ppm).

PEROXYACYLNITRATES (PAN): Complex compounds of oxygen and nitrogen.

PHOTOCHEMICAL: Relating to the effect of light in causing chemical changes.

PRISM: A three-sided glass or crystal object that breaks up light into a rainbow.

RESIDENCE TIME: The average length of time a substance remains in the atmosphere; the time between its emission into the atmosphere and its removal by some process.

SMOG: An irritating mixture of atmospheric pollutants arising from the presence of hydrocarbons, ozone, and oxides of nitrogen in the air; tiny particles of solids and liquids are also generally present.

STRATOSPHERE: The portion of the atmosphere beginning about 7 mi above the surface of the earth and extending to about 17 mi.

TEMPERATURE INVERSION: An increase in temperature with increasing altitude in the atmosphere.

TOLERANCE LEVEL: The maximum concentration of a pollutant that can be present in the air we breathe without unfavorable effects on our health.

TROPOSPHERE: The portion of the atmosphere extending from the earth's surface upward about 7 mi.

ULTRAVIOLET LIGHT: A kind of radiation in sunlight that cannot be seen; it causes sunburn.

VISIBLE LIGHT: Radiation that can be seen by the human eye.

Testing Yourself

7.1 Why does atmospheric pressure decrease with altitude?

7.2 When a weather balloon is being inflated to ascend into the stratosphere (more than 7 mi above the surface of the earth), it is only partially filled when it is inflated on the ground. Why should it not be filled full?

7.3 Why are airplanes that fly at high altitudes built with "pressurized" cabins?

7.4 Why does the barometric pressure decrease in an air mass that increases in temperature?

7.5 Pike's Peak is about 14,000 ft above sea level. What is the average barometric pressure on the peak? At what temperature would water boil on Pike's Peak?

7.6 What is the greenhouse effect of CO_2 in the atmosphere? How does the presence of particulate pollutants in air counteract the greenhouse effect? What might happen to the level of the oceans if the particulate effect dominated? If the greenhouse effect dominated?

7.7 Why is carbon monoxide such a dangerous pollutant? How does nature remove it from air?

7.8 Why are the oxides of sulfur so obnoxious as air pollutants? How does nature remove them?

7.9 The concentrations of NO and NO_2 found in polluted air are not toxic to man. Why, then, are these oxides so undesirable in air?

7.10 Why are particulates such dangerous pollutants, even though they may be chemically inert solids?

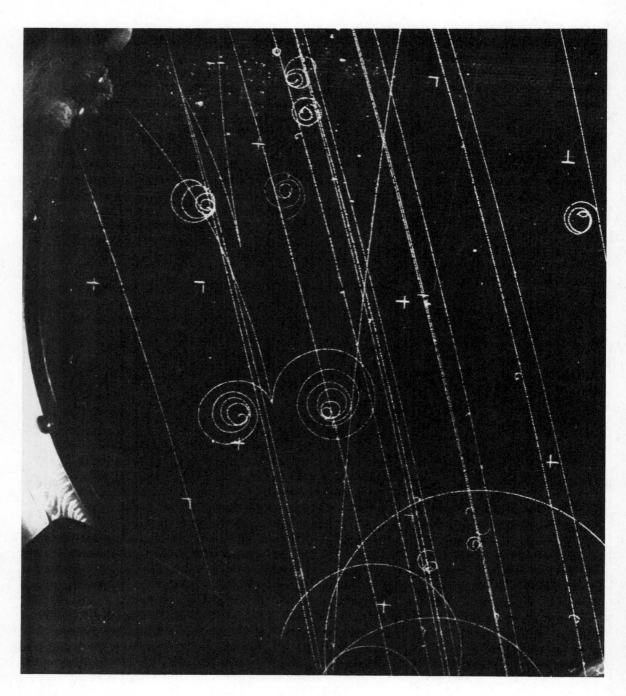

The drawing is a schematic representation of a photon splitting into an electron-positron pair;
the photograph illustrates phenomena such as the splitting of a photon (shown by the double-spiral track).

γ ray

e^+ e^-

8. What Are Atoms Made of?

We have greatly deepened our understanding of our environment and the processes within it through two mental models for matter: the atomic theory (Chap. 3) and the kinetic-molecular theory (Chap. 6). Once Dalton's atomic theory was generally accepted, scientists began to wonder what atoms are made of, whether they can be broken down into smaller, *subatomic* particles, and whether there are kinds of matter more fundamental than the chemical elements.

subatomic

William Prout, an English physician, pointed out in 1815 that most atomic weights, compared to weight 1 for hydrogen, were either whole numbers or very nearly so. He proposed that all atoms might be composed of aggregates of hydrogen atoms—that hydrogen is the really fundamental stuff of matter. But more and more precise determinations of atomic weights did not always yield simple multiples of the atomic weight of hydrogen, so Prout's hypothesis was put aside.

We now know that Prout was 150 years ahead of his time. By the middle of the twentieth century we came to realize that hydrogen is indeed the stuff of which many elements are made, as we shall see in Chap. 22. This realization evolved from the discovery of the fundamental importance of electric charges in atoms. With the discovery of the nature of electricity came the discovery of subatomic particles within atoms. In this chapter we shall consider these discoveries and how they led to another great mental model—the nuclear atom.

8.1 Matter with an Electric Charge

If you shuffle across a heavy carpet on a dry winter's day and touch someone, a spark jumps between you, and your nerves twitch where the spark jumps. If you comb your dry hair in dry weather, it is attracted to the comb and stands on end. If you rub your comb on a piece of woolen cloth, the comb will pick up threads and bits of paper and dust.

As early as 600 B.C. the Greeks knew that a piece of wool rubbed on amber caused it to attract small bits of thread, chaff, and dust. The Greeks did not explain this phenomenon, and no systematic studies were made of it for 2000 years, until William Gilbert, physician to Queen Elizabeth I of England, discovered that many substances exhibit this "amber effect." Gilbert called these substances "electrics" from the Greek word "elektron" for amber. By the beginning of the eighteenth century the word "electrification" was accepted as a name for the property shown by some substances when they are rubbed. In the eighteenth century it was discovered that this property could be conducted along metallic wires but not along silk threads. It was also observed that electrified objects attract some objects and repel others. Today we say that an object electrified by rubbing has been "charged." When its electric properties have been removed, we say it has been "discharged."

Several important aspects of the behavior of electricity can be demonstrated quite simply by a series of observations of a simple apparatus (Fig. 8.1).

1 Figure 8.1*a* shows a device consisting of two small balls made of very lightweight pith from a plant stalk, covered with tinfoil to give them surfaces that readily conduct an electric charge; the pithballs are suspended by silk threads.
2 In part *b* a rod of plastic has been rubbed with a wool cloth and brought close to the hanging pithballs; they are attracted toward the rod.
3 Part *c* shows that after the balls are allowed to touch the rod, they are immediately repelled from it and from each other.
4 When the rod is removed and the pithballs are handled, they resume the position shown in part *a*; they have been discharged. Touching them with a wire connected to the earth also discharges them; we then say they have been grounded.
5 When a glass rod that has been rubbed with a silk cloth is brought close to the hanging pithballs, they are attracted to the rod (part *d*).
6 If the balls touch the rod (part *e*), they are immediately repelled from it and from each other. Again they can be discharged by handling or grounding.
7 When a plastic rod is rubbed with wool and brought near the balls charged as in observation 6, both balls are attracted to the plastic rod (part *f*).

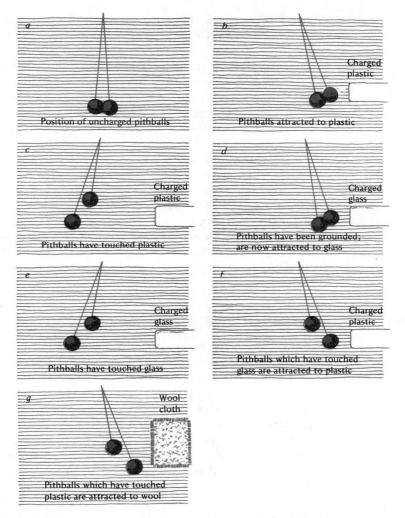

FIGURE 8.1 Interactions between electrically charged objects

8 In part g the wool cloth has been rubbed on a plastic rod, and the pithballs have been charged by touching the plastic rod; when the cloth is brought near, the balls (which repel one another) are attracted to the cloth.

The plastic rod (observation 2) is charged (electrified) by rubbing with wool. Some charge (observation 3) is transferred from the rod to the pithballs. The repulsion between the two balls, and between the rod and the balls, indicates that like charges repel one another. Charges (observation 4) can be removed from objects by handling them or connecting them to the earth. Observations 5 and 6 lead us to the same conclusions as 2, 3, and 4. Observation 7 shows us that the charge on a glass rod that has been rubbed with silk is different from the charge on a plastic rod that has been rubbed with wool. These two different or unlike charges attract one another. Observation 8 shows that when a plastic rod is charged by rubbing it with a wool cloth, the cloth acquires a charge unlike that on the rod.

It is awkward to describe a charge as "that which is produced on a glass rod by rubbing it with a silk cloth" or "that which is produced on a plastic rod by rubbing it with a wool cloth." So we call the charge on the glass rod positive and that on the plastic rod negative. Positive charges repel each other; negative charges repel each other; positive charges attract negative charges. Like charges repel; unlike charges attract. These repulsions and attractions are strong when the charged objects are close together, but they decrease rapidly as the charged objects are separated. When two objects are electrified by rubbing them together, one becomes positively charged and the other negatively charged. As we shall see, this simple fact is of profound importance to the structure of the atom.

8.2 Subatomic Particles with a Negative Charge: Electrons

The next step in the investigation of electricity was the design of machines to produce much larger charges than could be obtained by rubbing rods with bits of cloth. In the latter part of the nineteenth century, electricity was a popular novelty. A favorite pastime was to connect a source of electricity to various pieces of apparatus to see what would happen. When a source delivering several thousand volts is connected to a glass tube containing two wires sealed through its ends, and the tube is partly filled with a gas, beautifully colored discharges appear between the wires. Displaying these phenomena in tubes of many sizes and shapes filled with different gases became a popular form of parlor entertainment.

electrode

William Crookes, a British physicist, shifted the observation of electric discharges from the parlor to the laboratory. He built a straight tube with a metal disk called an *electrode* at each end connected to a wire sealed through the glass. He also added a side tube connected to a vacuum pump (Fig. 8.2). When a source of about 50,000 V (volts) was connected to the electrodes, there was no discharge when the tube was filled with a gas at atmospheric pressure. But when the pressure was reduced to a few torr, an electric discharge took place through the gas, producing a uniform glow inside the tube.

The color of this glow depends on the gas in the tube. With neon, the discharge is bright orange-red; with sodium it is yellow; with argon, blue. Almost any color can be produced by a suitable mixture of gases. As the pressure is

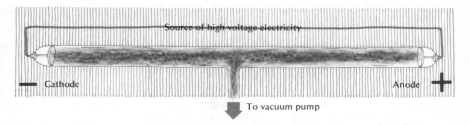

Source of high-voltage electricity

Cathode

Anode

To vacuum pump

FIGURE 8.2   Flow of electricity through a gas at low pressure

reduced, the discharge in the tube changes. At 0.001 torr, the original color disappears (although electricity is still flowing through the gas), and a greenish glow appears on the surface of the glass near the positive electrode. This glow is called *fluorescence*. The color of the discharge through the tube at higher pressures depends on the gas, but the same greenish glow appears at very low pressures regardless of what gas is in the tube. The color of the fluorescence depends on the kind of glass.

fluorescence

When a gaseous discharge tube of the form shown in Fig. 8.3 is used and the pressure is a few torr, the path followed by the glowing discharge depends on which electrodes are connected. Since there are three *anodes* (positive electrodes), the discharge can follow the path *ab*, *ac*, or *ad*.

anode

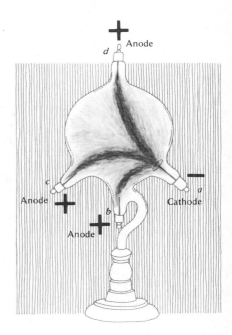

FIGURE 8.3   Three possible paths of electric discharge through a gas at moderately low pressure (moderate vacuum)

cathode

cathode ray

Figure 8.4 shows that when the pressure is 0.001 torr or less, there is no visible path for the discharge between the electrodes, but the glass surface of the evacuated tube directly opposite the *cathode* (negative electrode) fluoresces, no matter whether the positive electrode connected is *b*, *c*, or *d*. Apparently something is streaming from the cathode directly across the evacuated space and colliding with the glass, causing it to fluoresce. This something we call a *cathode ray*.

The apparatus in Fig. 8.5 shows that cathode rays carry considerable energy. The cup-shaped cathode focuses these rays—which seem to be emitted at right angles to the surface of the cathode—upon a piece of platinum, which becomes red-hot as soon as a high voltage is impressed across the electrodes.

Still more information about cathode rays comes from the discharge in the tube shown in Fig. 8.6. When an object made of metal foil is placed in the path of the rays, it casts a sharp shadow on the fluorescent screen, which consists of a coating of zinc sulfide, ZnS, on the inside of the glass. Zinc sulfide fluoresces readily, producing a bright-green color.

If a stream of negatively charged particles is passed between two charged plates, the path of the particles is bent toward the positively charged plate. If a stream of such particles is passed between the poles of a magnet, the magnet neither attracts nor repels the particles, but causes them to move in a curved path perpendicular to the line drawn between the poles of the magnet.

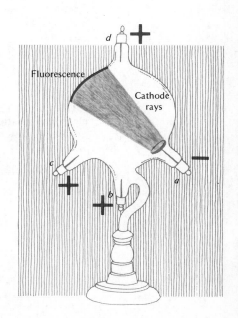

FIGURE 8.4 Path of electric discharge through a gas at very low pressure (high vacuum). Cathode rays travel in straight lines at a right angle to a flat cathode.

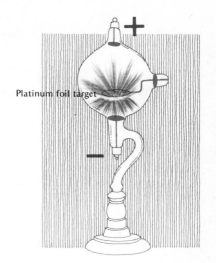

FIGURE 8.5 Cathode rays carry energy.

Platinum foil target

In Fig. 8.7 cathode rays pass through a hole in the metal anode lined up with a similar hole in another metal disk, so that a pencil of rays strikes the zinc sulfide coating at the far end of the evacuated tube. When there is no charge on the metal plates P and P' and the magnet is removed, the cathode rays travel a straight line and produce a bright spot at point O. But when the plates are charged (with the positive one above the pencil of rays), the bright spot is deflected from O to O'. If the plates are uncharged and the magnet is in place, the spot is deflected from O to O''. *The cathode rays behave like the stream of negatively charged particles* described in the preceding paragraph. We conclude, therefore, that *a cathode ray is a stream of negatively*

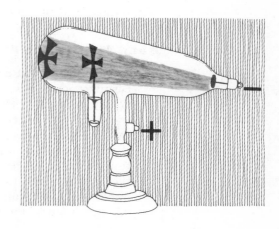

FIGURE 8.6 The straight-line path of cathode rays

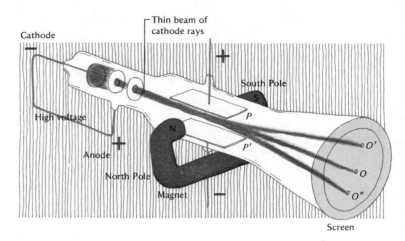

FIGURE 8.7 Deflection of cathode rays by effects of electric and magnetic fields

electron

charged particles. These particles were called *electrons* by G. J. Stoney as early as 1891.

A television tube operates on the principle shown in Fig. 8.7. The picture is produced by deflecting an electron beam over the fluorescent face of the tube horizontally by one field and vertically by the other, while the strength of the beam is varied to make each little region of the face glow bright or stay dark in response to signals from the TV camera. You can easily demonstrate the deflection of electron beams by electric and magnetic influences. When the atmosphere is dry, if you rub the face of the picture tube (or even the plastic protective window in front of it) with wool, silk, or nylon, the picture is distorted by the electric effect from the charges you produce. If you bring a strong magnet near the face of the tube, it also distorts the picture.

8.3 The Charge and Mass of the Electron

The next important development in the nineteenth-century investigation of electricity came from the famous Cavendish Laboratory at Cambridge University in England. Here J. J. Thomson, using an apparatus much like that shown in Fig. 8.7, determined the ratio of the charge to the mass of what he called the *cathode corpuscles.* The beam of electrons was first allowed to pass through the tube in the absence of both the electric and the magnetic fields, thus producing a spot at the point *O*. The magnetic field was then applied and the beam deflected to the point *O″*. Then, with the magnet still in place, an electric field was built up between the plates *P* and *P′* until it was just strong enough to return the beam of electrons to its original position, striking the

cathode corpuscle

screen at O. In these circumstances, the force exerted downward upon a moving electron by the magnetic field is then exactly equal to the force exerted upward by the electric field. From his knowledge of the effects of magnetic and electric fields on the motion of a particle with known mass and electric charge, Thomson was able to determine e/m, the ratio of the charge e to the mass m of a cathode corpuscle, soon called an electron by most scientists.

charge of
the electron

Ten years later, R. A. Millikan, a physicist at the University of Chicago, measured the *charge of the electron* (Fig. 8.8). In his apparatus, oil is sprayed into the air above the upper metal plate A, and a few tiny droplets of it find their way down through the small hole in this plate. Illuminated from the side and viewed with a short focus telescope, these droplets appear as tiny bright spots against a dark background. When such a droplet falls freely under the influence of gravity, it first accelerates, then acquires a constant speed because the friction produced by its passage through the air increases with the speed of its fall. When the upward force of friction equals the downward force of gravity, it falls at a constant speed. This can be measured by timing the fall through a known distance measured by a scale in the eyepiece of the telescope.

When oil is sprayed into the upper chamber of the apparatus, the rubbing of the oil on the sprayer gives the droplets an electric charge, sometimes positive and sometimes negative. The upper plate can be made negative to attract positive droplets upward against the force of gravity; the plate can also be made positive to attract negative droplets. By balancing the electric force pulling the droplet upward against the gravitational force pulling it downward, Millikan was able to determine the charge on the droplet.

By measuring the charge on droplets of various sizes formed by various liquids, Millikan found that the charge always had a certain minimum value or an integral multiple of it. He concluded that electricity comes in unit packets and that the ultimate unit of negative charge is the minimum value he had determined. This is now called the charge of the electron. A droplet

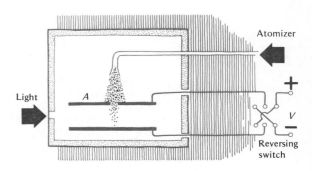

FIGURE 8.8 Millikan's apparatus for measuring the charge of the electron

may acquire 1, 2, 3, or more times the unit (minimum) charge, but not some fraction of the charge. If one or more electrons were rubbed *off* a droplet during atomization, its charge would become positive, since matter in bulk is electrically neutral (uncharged or with equal numbers of positive and negative charges). If one or more electrons were rubbed *onto* a droplet, its charge would become negative. In either case, the magnitude of the charge would be e or some integral multiple of e.

From Thomson's determination of e/m for cathode corpuscles and Millikan's determination of the ultimate charge, it is easy to calculate the mass of the electron. It turns out to be about $\frac{1}{1836}$ of the mass of a hydrogen atom.

8.4 Positive Rays

Thomson was able to release cathode corpuscles in an evacuated tube by impressing a high voltage across the electrodes, by heating the cathode, or by illuminating it with ultraviolet light. These results indicated strongly that electrons are part of a fundamental stuff (*prima materia*) of which atoms are composed. Since matter in bulk is electrically neutral, it must have some positive component in amounts sufficient to neutralize the negative charges on the electrons in a given atom. Thomson set about to find this positive stuff by modifying his apparatus.

In 1886 Eugen Goldstein had constructed a Crookes tube with a perforated cathode mounted at some distance from one end of the tube (Fig. 8.9). When the tube was partially evacuated and a source of high voltage was connected to the electrodes, streaks of light appeared on the side of the cathode away from the anode. These streaks originated at the holes in the cathode and were self-luminous, unlike cathode rays, which are invisible and can be traced only by producing luminous fluorescence in substances they strike. Also, unlike cathode rays, which are independent of the nature of the gas in the tube, the color of the streaks depends on the gas remaining in the tube: yellowish in air, rose-colored in hydrogen, yellowish-rose in oxygen, and green-

FIGURE 8.9 Goldstein's gaseous discharge tube showing production of luminescent streaks from the side of the cathode away from the anode

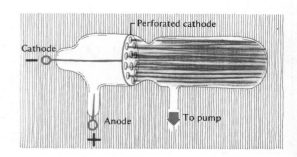

ish-gray in carbon dioxide. If the cathode is constructed of a cylinder a few centimeters long with holes bored through its length, the streaks become clearly defined rays which Goldstein termed *canal rays*. By passing these rays between highly charged plates or the poles of a magnet (as in Fig. 8.7) it became evident by the direction in which the rays were deflected that they were streams of positively charged particles—positive rays.

Thomson modified Goldstein's discharge tube (Fig. 8.10) to produce a beam of these positive rays that could be subjected to electric and magnetic fields of known magnitude, so that the ratio of charge to mass for the particles could be determined.

He found that not all the particles in a positive ray produced from a given gas with a given applied voltage had the same speed, and so the fluorescent pattern where the beam struck the end of the tube was a line, not a point. In spite of this difficulty he was able to determine the ratio of charge to mass for the positive particles produced from several different gases. He assumed that these positive particles were produced by knocking one electron off an atom or a molecule. The particle with the largest ratio of charge to mass was produced from hydrogen. This value was equal to the charge on the electron divided by the mass of a single hydrogen atom (obtained by dividing the atomic weight of hydrogen by Avogadro's number, 6.02×10^{23}). Thomson therefore concluded that this positively charged particle is a hydrogen atom bearing one positive charge. It was called a hydrogen ion or a *proton* and was symbolized by H^+. An atom or a group of atoms bearing an electric charge is called an *ion*. Apparently, then, a hydrogen atom consists of one proton and one electron bound tightly together. Only $\frac{1}{1836}$ of the mass of this atom is due to its electron.

proton

ion

8.5 The Nuclear Atom

In the last decade of the nineteenth century and the first of the twentieth, when researches with cathode ray tubes and other electric apparatus were revealing the nature of cathode radiation, a different kind of radiation was found to be emanating from the ores of uranium and thorium. By subjecting this radiation to electric and magnetic fields, it was found that one component

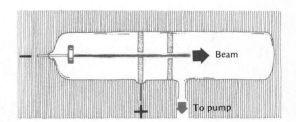

FIGURE 8.10 Thomson's gaseous discharge tube for producing a beam of positively charged particles

alpha (α) particle
beta (β) particle
gamma (γ) ray

was a beam of positively charged α (*alpha*) *particles*, another was a beam of negatively charged β (*beta*) *particles*, and a third was a beam not deflected by either electric or magnetic fields. These were called γ (*gamma*) *rays*. When particles were allowed to accumulate in a closed vessel, helium gas could be shown to be present. Studies of the charge-to-mass ratios for α particles indicated that they were doubly charged ions of the element helium, He^{2+}. Apparently α particles can pick up stray electrons to become helium atoms: $He^{2+} + 2e^- \longrightarrow He$. Likewise β particles are electrons traveling at a wide range of speeds. And the behavior of γ rays showed that they are invisible light rays.

radioactivity

The emission of α, β, and γ rays was called *radioactivity*.

Ernest Rutherford, a pupil of J. J. Thomson while the latter was working on the charge-to-mass ratios of electrons and positive ions, became interested in the radiation from uranium and thorium ores at about the same time that Pierre and Marie Curie were doing their work in Paris that led to the discovery of radium and other radioactive elements. After studying the rates at which radioactive elements disintegrate and the way α and β particles are absorbed or scattered by various substances, Rutherford decided to examine the scattering of α particles in more detail. He describes his work in a vivid passage:

> I would like to use this example to show how you often stumble upon facts by accident. In the early days I had observed the scattering of alpha par-

Ernest Rutherford, 1871–1937

ticles, and Dr. Geiger in my laboratory had examined it in detail. He found, in thin pieces of heavy metal, that the scattering was usually small, of the order of one degree. One day Geiger came to me and said, "Don't you think that young Marsden, whom I am training in radioactive methods, ought to begin a small research?" Now I had thought that, too, so I said, "Why not let him see if any alpha particles can be scattered through a large angle?" I may tell you in confidence that I did not believe that they would be, since we knew that the alpha particle was a very fast, massive particle, with a great deal of energy, and you could show that if the scattering was due to the accumulated effect of a number of small scatterings the chance of an alpha particle's being scattered backwards was very small. Then I remember two or three days later Geiger coming to me in great excitement and saying, "We have been able to get some of the alpha particles coming backwards. . . ." It was quite the most incredible event that has ever happened to me in my life. It was almost as incredible as if you fired a 15-inch shell at a piece of tissue paper and it came back and hit you. On consideration, I realized that this scattering backwards must be the result of a single collision, and when I made calculations I saw that it was impossible to get anything of that order of magnitude unless you took a system in which the greater part of the mass of the atom was concentrated in a minute nucleus. It was then that I had the idea of an atom with a minute massive center carrying a charge. I worked out mathematically what laws the scattering should obey, and I found that the number of particles scattered through a given angle should be proportional to the thickness of the scattering foil, the square of the nuclear charge, and inversely proportional to the fourth power of the velocity. These deductions were later verified by Geiger and Marsden in a series of beautiful experiments.*

The Geiger-Marsden experiments and Rutherford's interpretation of what happens when α particles bombard a thin foil of a heavy metal are indicated in Fig. 8.11a and b. Only those particles that pass very close to the tiny nucleus of a metal atom (*atomic nucleus*) are deflected—most of the particles pass through as though the metal foil were empty space. Rutherford concluded that the nucleus carries a positive charge and is very small—with a radius of the order of $\frac{1}{100,000}$ of the atom as a whole! An atom has a diameter of about 10^{-8} cm; the diameter of the nucleus is approximately 10^{-13} cm.

atomic nucleus

8.6 Subatomic Particles with a Positive Charge: Protons

Rutherford and his associates discovered that the atom has a nucleus in 1911 and continued studying the interactions between α particles and various forms of matter for several years. Their technique was to use a beam of α particles from a radioactive material and to set up a screen coated with a flu-

*"The Development of the Theory of Atomic Structure," in Needham and Pagel, Eds., *Background to Modern Science*, Macmillan, Toronto, 1940.

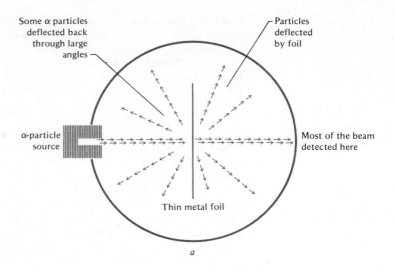

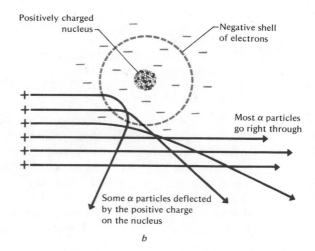

FIGURE 8.11 (*a*) Rutherford's experiment with alpha particles: the particles emerge from a source at the left and strike a thin metal foil; most of the particles go right through the foil; a few are deflected through various angles; some are even deflected backward. (*b*) Rutherford's interpretation: the regions which deflect the positive alpha particles must bear all the positive charge of the atom and most of its mass; only those alpha particles which come very close to a nucleus of positive charge are deflected; most of the volume of the atom is occupied by the negative electrons.

orescent material such as zinc sulfide to detect the presence of α particles after they had struck a metal foil or other material. The screen was viewed through a microscope so that an individual flash of light could be seen for every α particle that struck the screen. When Rutherford passed α particles through gases, he found that the number of flashes decreased with the distance the particles passed through the gas, indicating that the particles were slowed down by a series of collisions with gas molecules until they lost so much kinetic energy that they would not produce a flash when they struck the zinc sulfide. This absorbent effect was clearly observable in oxygen and carbon dioxide, but in air there were *more* flashes than in a vacuum! It soon became apparent that it was the nitrogen in the air that produced this remarkable phenomenon. When α particles passed through nitrogen in a magnetic field, the position of the flashes showed that new particles were produced with a positive charge and a mass 1 or 2 times that of the hydrogen atom. Since the mass of an α particle is 4 times that of a hydrogen atom, Rutherford felt sure that either the α particle was breaking a fragment off the nitrogen nucleus or vice versa. Since α particles are ejected from many radioactive substances, they are probably quite stable. And since the increase in flashes was found only with nitrogen, it rather than the α particles was the probable source of the new fragments. Rutherford suspected that these were *protons*, nuclei of H atoms.

proton

Rutherford's conclusion was soon confirmed by the work of a physicist, C. T. R. Wilson, who devised an ingenious apparatus called a *cloud chamber* for detecting the paths of moving charged particles. In Fig. 6.1 we saw that as the temperature of a liquid is lowered, its vapor pressure decreases. If a gas is saturated with gaseous water at a given temperature, a sudden lowering of temperature will cause the gaseous molecules of water to condense as droplets of liquid water. Such droplets form most readily around a nucleus. Wilson reasoned that a charged particle moving at high speed through the gas would tear electrons off the gaseous molecules and produce positively charged ions that would serve as condensation nuclei along the path of the particle. Figure 8.12 shows the essentials of Wilson's cloud chamber. A cylinder, painted black inside, is closed at the top by a glass plate and at the bottom by a tightly fitting black piston. A small amount of water is placed on top of the piston, which is then pushed upward, slightly warming the compressed gas. A short time is allowed for the water vapor to saturate the gas at the higher temperature, and then the piston is suddenly jerked downward. The gas suddenly expands in volume and suddenly decreases in temperature, and droplets of liquid water immediately condense on any available nuclei. With light coming from the side, the path of a high-speed particle is a white streak of fog droplets against the black background.

cloud chamber

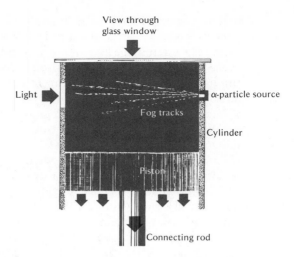

FIGURE 8.12 Wilson's cloud chamber. (Figure adapted from G. S. Christiansen and Paul H. Garrett, *Structure and Change: An Introduction to the Science of Matter*, W. H. Freeman and Company, San Francisco, 1960.)

Figure 8.13 shows some typical cloud tracks. In part *a*, α particles are passing through oxygen saturated with water vapor. Most of the particles go straight through the gas; only occasionally does one come close enough to the nucleus of an oxygen atom to be deflected. The branched track reveals a collision. Because the oxygen atom has 4 times the mass of an α particle, it rebounds a

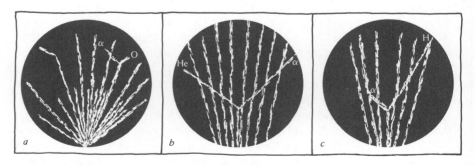

FIGURE 8.13 Fog tracks of alpha particles colliding with (*a*) an oxygen atom, (*b*) a helium atom, and (*c*) a hydrogen atom. (Figure adapted from G. S. Christiansen and Paul H. Garrett, *Structure and Change: An Introduction to the Science of Matter*, W. H. Freeman and Company, San Francisco, 1960.)

shorter distance after collision than does the α particle. In part b, the α particles are passing through helium gas. Since a helium atom and an α particle (which is a helium atom that has lost two electrons) have essentially the same mass, the two particles produce similar tracks after collision. In part c, the α particles are passing through hydrogen gas. Because the hydrogen atom has only one-fourth the mass of an α particle, it rebounds over a much longer path after collision than does the α particle. Also, because of its small mass, the hydrogen atom does not produce so many gaseous ions along its path, and so its fog track is noticeably narrower than that of the α particle.

P. M. S. Blackett, another physicist using the cloud chamber, reported that out of more than 20,000 cloud chamber photographs with 400,000 tracks of α particles through nitrogen, only eight showed the characteristics of the track shown in Fig. 8.14. Since the lightly *ionized* track goes backward, it is clear that more than a simple collision has taken place. The lightly fogged, long branch of the backward fork looks like the trail of the hydrogen atom in Fig. 8.13c; the heavily fogged, short branch of the fork extending forward must have been made by a heavy ion, since it looks like the track of the oxygen ion in Fig. 8.13a. The α particle seems to have vanished—an astonishing result!

These tracks indicate that some new process is taking place. There must be some interaction between the colliding nuclei that produces two new par-

ionize

FIGURE 8.14 Fog track record of the collision of an alpha particle with a nitrogen atom. (Figure adapted from G. S. Christiansen and Paul H. Garrett, *Structure and Change: An Introduction to the Science of Matter*, W. H. Freeman and Company, San Francisco, 1960.)

ticles out of the two there before the collision. By photographing the tracks produced by such a collision when the cloud chamber is mounted between the poles of a powerful magnet, it is possible to identify the mass and charge of the particles involved. It turns out that the head-on collision of a particle with a nitrogen nucleus produces a hydrogen ion (proton) and an oxygen ion. This *nuclear reaction* may be symbolized thus:

nuclear reaction

$$\text{}^{4}_{2}\text{He} + \text{}^{14}_{7}\text{N} \longrightarrow \text{}^{1}_{1}\text{H} + \text{}^{17}_{8}\text{O} \tag{8.1}$$

mass number

atomic number

The prefixed superscript, called the *mass number*, is the integer nearest to the atomic weight of the nucleus (determined by mass spectrometry, Sec. 3.10); the subscript shows the *atomic number*, which is the charge on the nucleus (determined by Rutherford's interpretation of the scattering of α particles by solid elements).

Equation (8.1) is translated as follows: When a helium nucleus with a mass number of 4 and a positive charge of 2 units collides with a nitrogen nucleus with a mass number of 14 and a positive charge of 7 units, they react to form a hydrogen nucleus with a mass number of 1 and a positive charge of 1 unit, and an oxygen nucleus with a mass number of 17 and a positive charge of 8 units. The atom $^{17}_{8}\text{O}$ is an isotope of oxygen. The total of the superscripts to the right of the arrow equals the total of those to the left, and the same is true for the subscripts.

Rutherford and James Chadwick were able to eject protons from the nuclei of a number of other elements by bombarding them with α particles. By studying the length of the tracks produced by these protons, they found that some protons have more energy than the original α particles. This observation indicates that the proton is not just knocked off the nucleus, but that the nucleus absorbs the α particle and then ejects the proton, whose energy is largely determined by the instability of the intermediate nucleus. This production of protons by the interaction of α particles with the nuclei of various atoms indicates that these nuclei contain protons, and that the charge of a nucleus is due to their presence.

In the preceding section we quoted Rutherford as stating that the number of α particles scattered by a metal foil through a given angle should be inversely proportional to the square of the nuclear charge. By studying the scattering of α particles by gold, silver, copper, and platinum, Chadwick found the charge on the nucleus for each of these kinds of atoms. If it was assumed that each nucleus contained a number of protons equal to its positive charge, the total mass of that number of protons was only about half the measured mass of the atom. It was assumed, therefore, that the nucleus must contain additional particles with mass enough to make up the difference. Rutherford

156   *What Are Atoms Made of?*

and Chadwick postulated the existence of a subatomic particle with no electric charge, possibly a proton-electron pair.

8.7 Subatomic Particles with No Electric Charge: Neutrons

neutron

By 1932 the masses of the known isotopes of the known elements had been measured accurately and found to be very nearly integral multiples of the mass of the hydrogen atom. It seemed reasonable to assume, then, that the nuclei were made up of enough protons to give the known positive charge on the nucleus plus enough proton-electron pairs to bring the mass up to that found for the isotope. Chadwick called the electrically neutral proton-electron pair a *neutron* and postulated that its mass would be about that of a hydrogen atom. As so often happens, confirmation of this postulate came from a wholly unexpected source.

In 1930 Walther Bothe and Josef Becker in Germany bombarded beryllium with α particles and produced an extraordinarily penetrating radiation. In 1932 Irene Joliot-Curie and Frederic Joliot—the daughter and son-in-law of Pierre and Marie Curie—were studying the absorption of this radiation. When sheets of carbon, aluminum, copper, silver, and lead were placed in the beam of radiation, there was practically no absorption. But when the beam passed through a sheet of paraffin wax (which contains only compounds of C and H), its activity in producing ions was *twice as great* as it was before. When the radiation passed through any substance containing hydrogen atoms, its ionizing power was enhanced. By measuring the tracks produced after the beam had passed through the wax, the Joliots showed that the radiation from the α-particle–beryllium source was ejecting protons from the wax or other hydrogen-containing target.

When Chadwick heard of these phenomena, he undertook similar experiments and was able to show that all the effects observed when the "Bothe-Becker rays" impinged on hydrogen atoms could be explained by assuming that these rays were a stream of uncharged particles having the mass of a proton. Chadwick's term—the "neutron"—was soon accepted as the name of these particles. Thus, when an α particle strikes the nucleus of a beryllium atom, it is absorbed into the nucleus, and the unstable system so produced then ejects a neutron. Later it was found that neutrons can be ejected from the nuclei of many elements by a variety of processes, and we now accept the neutron as one of the fundamental particles of which all atoms (except ordinary hydrogen) are composed.

An atom of ordinary hydrogen contains one proton in the nucleus and one electron outside it. The existence of an isotope of hydrogen composed of atoms having a mass twice that of the usual hydrogen atom is explained by assuming that the nucleus of this second kind of hydrogen, called *deuterium*, contains one proton and one neutron, but that this deuterium atom contains only one electron. A third isotope of hydrogen, *tritium*, contains one proton and two neutrons in the nucleus and one external electron. The nucleus for any isotope consists, then, of the number of protons needed to produce the observed positive nuclear charge and enough neutrons to bring the mass number up to the observed value. External to each nucleus there are as many electrons as there are protons in the nucleus, so that the atom as a whole has no net charge.

deuterium

tritium

Summary

Rubbing different kinds of matter produces two different kinds of electric charge, called positive and negative. A positive charge is that produced on a glass rod by rubbing it with silk. A negative charge is that produced on a plastic rod by rubbing it with wool. Like charges repel each other, and unlike charges attract. When one unit of positive charge meets one unit of negative charge, they cancel to produce matter with no net charge.

Atoms are composed of three kinds of subatomic particles:

1 electrons with a mass $\frac{1}{1836}$ times that of a hydrogen atom and a negative charge of 1 unit

2 protons with a mass $\frac{1835}{1836}$ times that of a hydrogen atom and a positive charge of 1 unit

3 neutrons with a mass about equal to that of a hydrogen atom but with no electric charge

And so we construct another great mental model for matter—the nuclear atom, which can be summarized as follows.

1 An atom contains a very tiny nucleus with a diameter only about $\frac{1}{100.000}$ of the diameter of the atom as a whole. This nucleus contains all the atom's positive charge and most of its mass. It is made up of the protons and neutrons in the atom. The electrons are dispersed in the large space around the nucleus.

2 The atomic number of an element is the number of protons in the nucleus of an atom of the element. This is the same as the number of electrons

around the nucleus, so that the atom as a whole is electrically neutral (has no net electric charge).

3 The mass number of an element is the number of protons and neutrons in the nucleus of an atom of an element.

4 Isotopes of a given element have marked differences in atomic weights but very nearly the same chemical properties. All atoms of a given element contain the same number of protons in the nucleus and electrons around it. Different isotopes of the element consist of atoms with different numbers of neutrons in the nucleus and hence different atomic weights.

New Terms and Concepts

ALPHA (α) PARTICLE: A helium ion bearing two units of positive charge.

ANODE: An electrode bearing a positive charge.

ATOMIC NUCLEUS: The tiny central portion of an atom that bears all the positive charge in the atom and most of the mass.

ATOMIC NUMBER: The number of unit positive charges on the nucleus of an atom; the number of electrons surrounding the nucleus.

BETA (β) PARTICLE: An electron moving at a very high speed.

CANAL RAYS: A special name for the positive rays passing through the holes in a perforated cylindrical cathode in a gaseous discharge tube.

CATHODE: An electrode bearing a negative charge.

CATHODE CORPUSCLE: An early name for an electron.

CATHODE RAY: The beam of energy (consisting of a stream of electrons) emitted from the cathode in a highly evacuated tube containing a highly charged anode and cathode.

CHARGE OF THE ELECTRON: The smallest amount of negative electric charge; the unit of negative electric charge.

CLOUD CHAMBER: A device for detecting the paths of fundamental particles through air saturated with water vapor.

DEUTERIUM: The isotope of hydrogen containing one proton and one neutron in the atomic nucleus.

ELECTRODE: A piece of metal bearing an electric charge.

ELECTRON: The smallest bit of matter that bears the smallest negative charge.

FLUORESCENCE: The emission of visible light by a substance when it is excited by the impact of some kind of radiant energy.

GAMMA (γ) RAY: An invisible light ray of great penetrating power.

ION: An atom or a group of atoms with a charge.

IONIZE: To convert or be converted into ions.

MASS NUMBER: The integer that approximates the mass of 1 mole of atoms of an isotope of an element. The mass number of hydrogen is 1, that of deuterium is 2, and so on.

NEUTRON: A particle with the mass of a proton but bearing no charge.

NUCLEAR REACTION: Any reaction that involves change in an atomic nucleus.

PROTON: The nucleus of a hydrogen atom; a hydrogen ion. It bears a positive charge equal to the negative charge on the electron.

RADIOACTIVITY: The emission of α, β, and γ rays.

SUBATOMIC: Pertaining to particles smaller than an atom.

TRITIUM: The isotope of hydrogen containing one proton and two neutrons in the atomic nucleus.

Testing Yourself

8.1 Summarize our present concepts of the structure of an atom with respect to the location of electric charges within it, the relative sizes of its various parts, the relative masses of its various parts, and the location of the various particles present in it.

8.2 Interpret the atomic weight of an atom and its atomic number in terms of the fundamental particles of matter.

8.3 How do we explain the nature of isotopes of elements in terms of the fundamental particles of matter?

8.4 If an α particle collides head-on with a beryllium nucleus and produces a neutron, what other kind of nucleus must be produced by the nuclear reaction:

$$\,^{4}_{2}\text{He} + \,^{9}_{4}\text{Be} \longrightarrow \,^{1}_{0}n + ?$$

The symbol $\,^{1}_{0}n$ stands for a neutron.

8.5 The atomic weight of chlorine as it is found in nature is 35.453 ± 0.001 g/mole. There are two isotopes of chlorine found in nature: ^{35}Cl and ^{37}Cl. What is the composition of the nucleus of each isotope and how can we explain the value of the atomic weight of naturally occurring Cl?

8.6 What experimental evidence supports each of the following statements? (a) An atomic nucleus is made up of several components. (b) An atom is porous. (c) The atomic nucleus is positively charged. (d) The atomic nucleus is small compared to the size of the atom. (e) Electrons in atoms are external to the nucleus. (f) Many elements exist as isotopes in nature.

8.7 What is the composition of the nucleus of each of the following atoms found in nature?

^1_1H ^2_1H ^3_1H $^{16}_8O$ $^{17}_8O$ $^{18}_8O$

Complete the following table on a separate sheet of paper.

Particle	Electron	Proton	Neutron	α particle	Fluorine nucleus	Radium nucleus
Mass number	_____	_____	_____	_____	_____	_____
Charge	_____	_____	_____	_____	_____	_____

8.8 Balance the following nuclear equations by filling in the formula for the missing isotope.

$^2_1H + {}^9_4Be \longrightarrow {}^1_0n + ?$ $^1_0n + {}^{10}_5B \longrightarrow {}^4_2He + ?$

$^{52}_{24}Cr + {}^2_1H \longrightarrow 2\,{}^1_0n + ?$

8.9 The mass of 1 mole (6.02×10^{23} atoms) of gold is 196.967 g. The density of gold is 19.3 g/cm^3. To simplify the arithmetic for the following calculation, round these figures off to 6×10^{23} atoms per mole, 200 g/mole, and 20 g/cm^3. What volume does each atom of gold occupy? If you think of this volume as a cube, what is the diameter of a gold atom, assuming it is spherical and just fits into the cube?

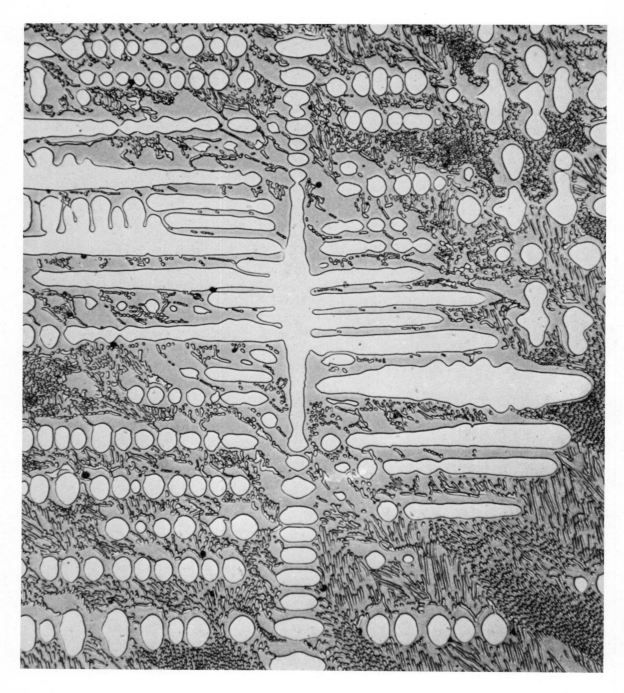

The photograph reveals the microstructure of a cobalt dendrite, one of many radioactive materials used in the treatment of cancer; a typical uranium decay reaction is illustrated schematically.

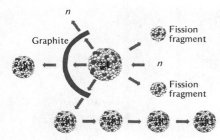

9. Nuclear Reactions and Nuclear Energy

Rutherford's nuclear model of the atom is the key to understanding the radioactive elements which have unstable nuclei that spontaneously disintegrate into other elements. When this occurs, highly penetrating radiations are released. These radiations are very dangerous to living creatures. The nuclei of some elements undergo fission; i.e., they split into nuclei of elements of smaller atomic number. Nuclei of some other elements may fuse together to form nuclei of elements of greater atomic number. Both fission and fusion release tremendous amounts of energy, but fission also yields many different elements that are dangerously radioactive. Nuclear fusion is very difficult to achieve, and progress toward the commercial production of energy from this process is very slow. But, as we shall see in Chap. 11, energy is the basic need of modern civilizations. If our kind of society is to endure, we must harness the energy that can be released by fission and fusion. And if we produce this energy, we must ensure that any harmful products released simultaneously do not damage us or our environment. In this chapter we shall first explore the nature of nuclear reactions and then consider how they may be harnessed to produce the energy we so acutely need.

9.1 Radioactivity of Naturally Occurring Substances

Radioactivity (Sec. 8.5) was discovered by accident in 1896 by Henri Becquerel, a French physicist, who stored some photographic plates wrapped in

Pierre and Marie Curie
in their laboratory

black paper in a drawer with some uranium ore for a few days. When he developed these plates, he was astonished to find heavy blackening of the photographic emulsion where the uranium ore had been lying on top of the wrapped plates. Becquerel later found that uranium and all its compounds are radioactive.

Pierre and Marie Curie, colleagues of Becquerel in Paris, were excited by this new phenomenon and set about to find other radioactive elements. They soon discovered that thorium is radioactive. They also found that certain ores of uranium are more radioactive than the pure element. From this they inferred that the radioactivity of uranium must be producing some new element even more radioactive than its parent. In 1898 the Curies isolated a substance 400 times more radioactive than uranium. They concluded that this substance contained a new element, and they named it polonium in honor of Marie Curie's native land. Six months later the Curies reported the separation of a substance 900 times more radioactive than uranium, with chemical properties quite different from the compounds of uranium, thorium, and polonium. They postulated that this substance contained another new element and named it radium. After 4 years of labor spent in processing several tons of pitchblende, an ore rich in uranium, the Curies were able to produce 0.1 g of radium chloride which showed no contamination by other elements when subjected to the most rigorous methods of analysis then known. The radioactivity of pure radium was found to be more than 1 million times that of pure uranium. The isolation of radium by the Curies was a monumental achievement. By even the most modern methods of extraction, a ton of pitchblende yields only 0.26 g of radium chloride!

By 1903 there was enough experimental data on radioactive transformations to let Rutherford and his colleague, Frederick Soddy, formulate a theory for the process. They concluded that the atomic nuclei of radioactive elements are unstable and tend to disintegrate, ejecting either a high-speed α particle (symbolized as an He ion, He^{2+}, or $_2^4He$) or a high-speed β particle (denoted as an electron, $_{-1}^0e$), thus changing the charge of the nucleus or its charge and mass, and forming a new element. Though these conversions of one element into another can be observed, they cannot be modified. The rates of radioactive processes are unaffected by changes in the temperature, pressure, chemical combination, or any other difference in the conditions under which they take place. In a particular interval of time, a fixed percentage of the atoms present in a pure sample of a radioactive element (free or chemically combined) will disintegrate. This percentage is constant with time but is different for every radioactive isotope. The most convenient way to compare the rates at which the nuclei of various radioactive elements disintegrate to form new elements is to measure the time required for one-half of a given sample of an element to disappear through radioactive decay. This time is called the *half-life* of the element.

half-life

The half-lives of radioactive elements differ enormously. It takes 4.5 billion yr for a piece of uranium $_{92}^{238}U$ to decay enough to lose half its radioactivity. But a piece of polonium $_{84}^{214}Po$ loses half its radioactivity in about $\frac{1}{10,000}$ sec! Thus if we have a million atoms of $_{92}^{238}U$ now, we'll have 500,000 left in 4.5 billion yr. But if we have 1 million atoms of $_{84}^{214}Po$ now, 500,000 will have decayed in $\frac{1}{10,000}$ sec, 250,000 in the next $\frac{1}{10,000}$, 125,000 in the next, etc. Within a second we shall have no $_{84}^{214}Po$ left! The half-life of bismuth $_{83}^{210}Bi$ is about 5 days. If we start with 1 million atoms now, we'll have 500,000 in 5 days, 250,000 in 10 days, 125,000 in 15 days, and so on (Fig. 9.1). A graph showing the rate of decay of *any* radioactive element is given in Fig. 9.2.

The decay of the atomic nucleus of one radioactive element often produces another radioactive element which, in turn, decays to produce a third, and so on. From extensive studies of the chemical properties of elements present in uranium ores, it has been shown that uranium is the parent of a whole series of elements that are formed and, in turn, decay into new elements. The elements in the uranium decay series, the particles the nuclei eject, and the half-lives of the elements are shown in Fig. 9.3. The symbolic statement

$$_{92}^{238}U \xrightarrow[4.5 \times 10^9 \text{ yr}]{\alpha} {_{90}^{234}}Th$$

means: The nucleus of an atom of uranium, with a mass number of 238 and 92 unit positive charges, emits an α particle ($_2^4He$) to produce the nucleus

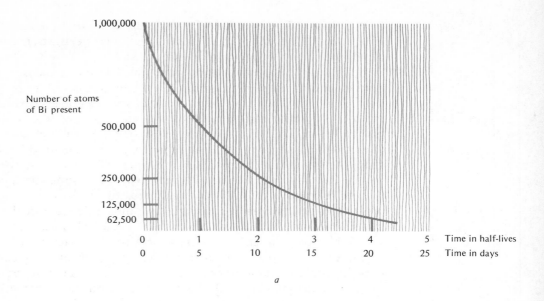

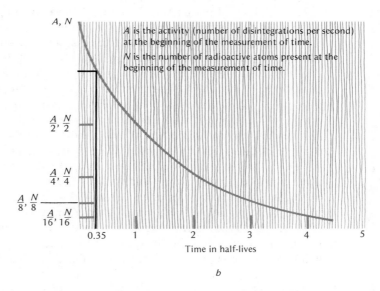

FIGURE 9.1 (*a*) Radioactive decay curve for $^{210}_{83}$Bi. (*b*) Radioactive decay curve for $^{14}_{6}$C (see Problem 9.12).

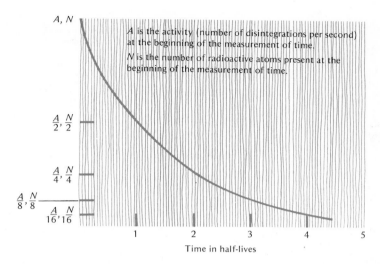

FIGURE 9.2 Characteristic radioactivity curve for one isotope by itself

of an atom of the element thorium, with a mass number of 234 and 90 unit positive charges; the half-life of this process is 4.5 billion yr.

9.2 Radioactive Carbon in Nature

cosmic rays

Neutrons from *cosmic rays* that penetrate the earth's atmosphere react with nitrogen to form radioactive carbon and a proton; the symbol for a neutron is ^1_0n.

$$^1_0n + {}^{14}_7N \longrightarrow {}^{14}_6C + {}^1_1H \tag{9.1}$$

This carbon is eventually incorporated into the CO_2 of the atmosphere and then into plants by photosynthesis. Animals eat the plants and build the radioactive carbon into their tissues. This isotope disintegrates with the emission of a β particle (an electron):

$$^{14}_6C \longrightarrow {}^{\ 0}_{-1}e + {}^{14}_7N \tag{9.2}$$

The half-life of radioactive carbon is about 5760 yr. The rate of formation of $^{14}_6C$ by cosmic rays and the rate of its disintegration by β decay are such that 1 g carbon in the tissues of a living plant or animal contains enough of the radioactive isotope to produce 15.3 disintegrations per minute. When a plant

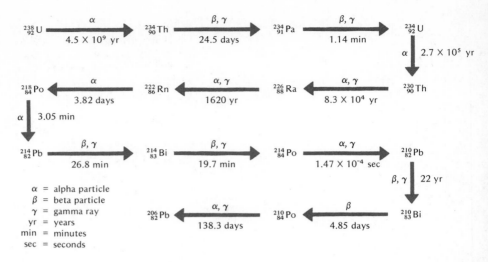

FIGURE 9.3 The $^{238}_{92}$U radioactive decay series

or animal dies, it no longer continues to take up radioactive carbon in its food, and so the amount of this isotope decreases with time. We know that 5760 yr after its death, each gram of carbon in its remaining tissues will produce 7.6 disintegrations per minute; 11,250 yr after its death, 1 g carbon in its tissues will produce 3.8 disintegrations per minute, and so on.

When wood is burned, when papyrus is used to make paper, or when flax is made into linen cloth, the plant is killed and the dead matter begins to lose radioactive carbon by β decay. The observed rate of decay can be used to tell when the charcoal in a cave was left from the fire of a primitive man, or when a piece of papyrus or a strip of linen cloth was made. This radioactive dating technique has been used widely in anthropological and archaeological studies. It is based on two assumptions:

1 that the number of cosmic rays reaching the earth each year has been constant over thousands of years, hence
2 the ratio of radioactive carbon to nonradioactive has remained constant over this period of time

Figure 9.4 shows the relationship between the rate of β particle production from a carbonaceous (carbon-containing) material and its age. It is useful for determining ages of materials between 1000 and 12,000 years old.

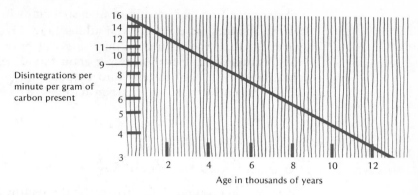

FIGURE 9.4 Rate of disintegration of $^{14}_{6}C$ related to the age of the source

EXAMPLE 9.1 The hair from an Egyptian mummy is found to produce 8 β particles per minute per gram of carbon in it. When did the mummified person die?

ANSWER A line drawn horizontally from 8 on the vertical scale in Fig. 9.4 intersects the slanting line as shown in Fig. 9.5. A vertical line dropped from this intersection meets the horizontal axis at about 5.1 on the horizontal scale. This indicates that the mummified person died about 5100 years ago.

FIGURE 9.5 Rate of disintegration of $^{14}_{6}C$ related to the age of the source

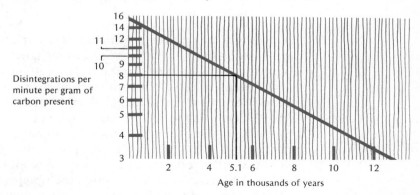

Explosions from the testing of nuclear bombs in 1965 about doubled the normal production of $^{14}_{6}C$ in the atmosphere. Living organisms that incorporated this carbon in succeeding years will have greater activity than the 15.3 disintegrations per minute per gram found before that year. Future archaeologists will have to take into account this marked change in the $^{14}_{6}C$ activity when they use radiocarbon dating.

9.3 Nuclear Reactions Induced by Machines

Blackett's observation that an α particle colliding head-on with a nitrogen nucleus transmutes it into an oxygen nucleus (Sec. 8.6) was the first known example of an artificially induced nuclear reaction. Until 1932, the only sources of high-speed particles of atomic dimensions were radioactive elements. In that year J. D. Cockcroft and E. T. S. Walton, working in Rutherford's laboratory, built a hydrogen discharge tube with electrodes connected to a source of electric charge at 600,000 V (volts). Hydrogen ions with very high speeds are produced by the discharge of electricity through this tube. When these were directed against a lithium target, particles were produced that showed tracks in a cloud chamber like those of α particles. Cockcroft and Walton concluded that the following nuclear reaction was taking place:

$$^{7}_{3}\text{Li} + ^{1}_{1}\text{H} \longrightarrow 2\,^{4}_{2}\text{He} \tag{9.3}$$

cyclotron

deuteron

Shortly thereafter, E. O. Lawrence and M. S. Livingston developed the *cyclotron* at the University of California and bombarded rock salt with high-speed *deuterons* (nuclei of the hydrogen isotope $^{2}_{1}\text{H}$). The reactions observed were

$$^{23}_{11}\text{Na} + ^{2}_{1}\text{H} \longrightarrow ^{24}_{11}\text{Na} + ^{1}_{1}\text{H} \tag{9.4}$$

$$^{24}_{11}\text{Na} \longrightarrow ^{24}_{12}\text{Mg} + _{-1}^{0}e \tag{9.5}$$

Today records are available for thousands of nuclear reactions that have been carried out by the bombardment of various target elements with high-speed electrons, protons, neutrons, and ions from atoms ranging in mass from $^{4}_{2}\text{He}$ to $^{238}_{92}\text{U}$. Many of these bombardments give rise to radioactive decay series producing several different elements by consecutive nuclear reactions.

In 1939, Otto Hahn and Fritz Strassman published a paper in Germany indicating that when uranium was bombarded with neutrons, the nucleus broke apart, yielding fragments much heavier than the α particle which up to that time was the heaviest particle known to be emitted by radioactive atoms. Ten days later Lise Meitner and Otto Frisch prepared a statement in Sweden that "two nuclei of roughly equal size" were produced by the disintegration

nuclear fission

of a uranium nucleus when struck by a neutron; they called this process *nuclear fission*, heralding the birth of the "atomic age."

nuclear fusion

Soon after they split one atomic nucleus to form two of about equal size (nuclear fission), scientists undertook to see whether they could put two nuclei together to form one of larger size (*nuclear fusion*). They found that when atoms are heated to very high temperatures, collisions between fast-moving atoms are so energetic that the electrons surrounding the nuclei are stripped off. This mixture of electrons and nuclei is called a *plasma*. When the hydrogen isotope deuterium, ^2_1H, is in the plasma state at a temperature of around 100 million degrees produced by very powerful magnetic fields, the nuclei can fuse to form the heavier element helium, ^4_2He, as shown by Eqs. (9.6) through (9.9).

plasma

$$^2_1H + ^2_1H \longrightarrow ^3_2He + ^1_0n \ + heat \tag{9.6}$$

$$^2_1H + ^2_1H \longrightarrow ^3_1H \ + ^1_1H \ + heat \tag{9.7}$$

$$^2_1H + ^3_1H \longrightarrow ^4_2He + ^1_0n + heat \tag{9.8}$$

$$^2_1H + ^3_2He \longrightarrow ^1_1H \ + ^4_2He + heat \tag{9.9}$$

This high temperature is generated for a short period of time in a nuclear fission bomb. It was this heat that caused nuclear fusion in what was called the hydrogen bomb or fusion bomb when it was first tested in 1952. Tremendous amounts of energy are liberated in H-bomb detonations. The hydrogen isotope deuterium, ^2_1H, is fairly plentiful in nature, and if controlled fusion could be achieved, mankind would have a vast source of energy from nuclear reactions [Eqs. (9.6) through (9.9)]. But the technical difficulties in the way of controlling fusion are enormous, and commercial production of energy by this process is not expected for several decades.

9.4 The Conservation of Mass and Energy in Physical and Chemical Changes

As we have seen (Sec. 3.4), the concept of the conservation of matter (or mass) was developed from experimental work with accurate balances that precisely measure the masses of reactants and products in chemical changes. By the end of the eighteenth century, scientists working in many European countries using many different chemical reactions took for granted that the total mass of their products was equal to the total mass of the reactants they had used.

Because energy is manifested in so many and such elusive forms—heat, light, electricity, sound, radioactivity, chemical reactivity, mechanical motion—it took about 50 years longer to develop the concept of conservation of energy. The term "energy" came into use in the mid-nineteenth century, and by that time its conservation was generally accepted, largely because of the work of James P. Joule, a wealthy English brewer who practiced science as an avocation. From 1840 to 1878 he devoted much of his time to the study of transformations of one kind of energy into another.

When Joule stirred the water in a calorimeter, he found that the energy of stirring was converted into heat. In his well-insulated calorimeter the amount of heat generated was directly proportional to the amount of mechanical energy (energy of motion) put into the system by stirring. There was no loss of energy in the operation; one kind of energy was transformed into an equivalent amount of another kind. When Joule passed an electric current through his calorimeter, he found that the amount of heat generated was directly proportional to the amount of electric energy put into the calorimeter. Because of Joule's careful and well-publicized studies, the concept of the conservation of energy soon became widely accepted.

law of conservation of energy

Law of conservation of energy: Energy can neither be created nor destroyed; it can only be transformed. Another way of stating the law is: Energy can be transformed from one type to another without loss, and in a system in which no energy leaves or enters, the total amount of energy remains constant.

A steam engine is a machine for converting some of the heat in steam into mechanical energy. Much of the heat is wasted to the surroundings, but a careful determination of the amount of heat put in, the mechanical energy produced, and the heat wasted shows that no energy has been created or destroyed when the engine and its surroundings are considered as a whole.

When gasoline is burned in an automobile engine, some of the chemical energy residing in the arrangement of the atoms in molecules of gasoline and oxygen is converted into heat when the mixture of fuel and oxygen explodes in the cylinders. Some of this heat is converted into mechanical energy that moves the car. Much heat is wasted in this double conversion, but for the system as a whole (the car and its surroundings) the total amount of energy in the fuel and oxygen before combustion is equal to the amount of energy in the moving car plus the heat lost to the surroundings plus the chemical energy of the arrangement of the atoms in the compounds in the exhaust gases.

When a water wheel turns a dynamo, the mechanical energy in the falling water is converted into electric energy. When electricity from the battery in your car passes through the starting motor, electric energy is converted into the mechanical energy required to crank the car engine and start it. When the engine is running, it turns a generator to convert mechanical energy into electric energy. This electric energy is passed through the battery and changes the chemical compounds present into other compounds; that is, the electric energy is converted into chemical energy. We call this process "charging" the battery. When the battery is used to start the car, the chemical reactions are reversed. "Discharging" the battery transforms the chemical energy of some of the compounds in it to other compounds and produces electric energy.

9.5 The Conservation of Mass-Energy in Nuclear Reactions

Radium in a closed container generates so much energy by radioactive disintegration that its temperature is considerably higher than that of the surrounding air. The energy produced by the disintegration of a given number of radioactive atoms is about 1 million times that obtained by burning an equal number of hydrogen atoms to form water. All nuclear reactions involve fantastically greater amounts of energy per atom than ordinary chemical reactions. As early as 1906, when knowledge of radioactivity was still limited, it was asserted that if all the atoms in 1 ton of uranium could be made to decay in 1 yr, the energy released would be as much as that consumed by the entire city of London!

When Cockcroft and Walton bombarded lithium with protons [Eq. (9.3)], they found that considerable energy was liberated, but the most astounding discovery was that the total mass of the products was *less* than the total mass of the reactants! Shortly thereafter it was noted that when nitrogen atoms were bombarded with α particles [Eq. (8.1)], energy was absorbed by the process, and the total mass of the products was *more* than the total mass of the reactants! These two observations seemed to be exceptions to the law of conservation of matter. Clearly the law had to be modified to take these new findings into account.

Fortunately, Albert Einstein had foreshadowed these complexities in his concepts of relativity. Einstein had shown that the mass of a very rapidly moving particle is greater than its mass at rest—that the mass of a body is a measure of its energy content. He derived a remarkably simple mathematical equation for expressing the equivalence of mass and energy:

$$E = mc^2 \tag{9.10}$$

where E stands for energy, m for mass, and c for the velocity of light. We can think of matter as a manifestation of energy and of energy as a manifestation of matter. The energy equivalent to a small mass is extremely large. One gram of mass is equivalent to 22 million cal of heat, an amount produced by burning about 4000 tons of coal. When lithium is bombarded with protons, the nuclear reaction releases particles with such enormous energy that there is a detectable decrease in mass. When α particles bombard nitrogen atoms, the nuclear reaction absorbs energy, and the mass of the products is measurably greater than that of the reactants. Consequently, we find it expedient to combine the laws of conservation of matter and of energy into a single statement, called the *law of conservation of mass-energy*: In a system in which no energy or mass leaves or enters the system, the sum of the masses of all matter at rest and of the mass-equivalents of energy is constant.

law of conservation
of mass-energy

Energy is released by any nuclear reaction that results in products with a mass less than that of the reactants. We now know how to produce energy on a large scale from nuclear reactions involving fission of large nuclei into smaller ones (Sec. 9.6). We are working on ways of producing energy by nuclear reactions involving fusion of small nuclei into larger ones. We believe these fusion reactions are producing huge amounts of energy continuously in the sun and other stars (see Chap. 22), but we have not yet been able to produce them continuously in manmade machines.

9.6 Production of Energy by Nuclear Fission

When uranium from naturally occurring ores is bombarded with neutrons, only a very small fraction of the atoms present undergo nuclear fission. Natural uranium has three isotopes, $^{234}_{92}U$, $^{235}_{92}U$, and $^{238}_{92}U$, and 99.3 percent of the atoms are of the heaviest kind. Since increasing the percentage of $^{235}_{92}U$ increases the rate of fission, it must be this isotope that undergoes nuclear reaction. From careful examination of all the products of uranium fission, it was found that more neutrons were produced than were consumed, and that barium and krypton were formed in considerable amounts. One typical fission reaction is

$$^{235}_{92}U + {}^{1}_{0}n \longrightarrow {}^{141}_{56}Ba + {}^{92}_{36}Kr + 3\,{}^{1}_{0}n + \text{heat} \tag{9.11}$$

The total mass of the fission products at rest is somewhat less than that of the reactants. Consequently, the fission process liberates huge amounts of energy.

Both the barium and the krypton produced by fission are radioactive isotopes that decay into a series of other elements. Furthermore, the fission process produces many pairs of radioactive fragments other than the Ba-Kr pair. For these two reasons, uranium undergoing fission soon becomes contaminated with many different elements, many of them very radioactive.

Harnessing uranium fission to produce useful amounts of energy requires the following conditions.

1 There must be enough $^{235}_{92}U$ atoms so that the neutrons which initiate fission can strike a nucleus before they escape from the metal.
2 One of the neutrons produced by fission must hit another $^{235}_{92}U$ nucleus so that the fission continues as a chain reaction.

3 The flow of neutrons must be controlled. If every neutron released by one atomic fission produces another fission, and every neutron released in this round creates another fission, then the number of released neutrons and the number of fissions will soon multiply uncontrollably, and the entire system will, in a very short time, melt down from the enormous amount of heat generated.

9.7 Nuclear Reactors

nuclear reactor

A device that harnesses nuclear reactions to produce usable energy is called a *nuclear reactor*. A simple reactor (Fig. 9.6) based on the fission of $^{235}_{92}U$ consists of a pile of blocks of very pure carbon (graphite) with channels into which rods of uranium can be inserted. Air circulates through these channels to cool the pile. Rods of steel containing 1.5% boron can be dropped into the spaces between the rods of uranium. Boron atoms are excellent absorbers of neutrons, and so they stop the fission chain when dropped into the pile. When these control rods are not in the pile and only a few uranium rods are inserted, the neutrons produced by the fission of a $^{235}_{92}U$ nucleus are lost by

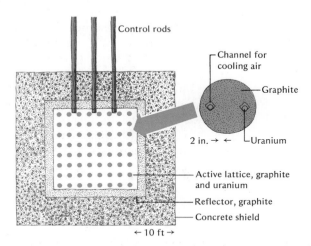

FIGURE 9.6 An early form of nuclear reactor. A large pile of blocks of graphite carries horizontal channels containing rods of uranium with space around them for cooling air. Rods of boron-containing steel can be dropped between the fuel channels to stop the chain reaction by absorbing neutrons. The entire pile is surrounded by thick concrete walls.

escaping from the carbon framework. When more uranium is inserted, the chances increase that a neutron released by the splitting of one $^{235}_{92}U$ atom will collide with another such atom. When just enough uranium is inserted, the reaction will be just self-sustaining. That is, of the three neutrons released by the fission of a $^{235}_{92}U$ nucleus, one will strike another nucleus and carry on the chain reaction. The neutrons produced by the reaction described in Eq. (9.11) have high speeds. If they are slowed down, they will more readily interact with a $^{235}_{92}U$ nucleus. The carbon serves to slow them down and is called a *neutron moderator*. The boron control rods can be dropped quickly into the pile to stop all activity if the reactor needs to be shut down.

neutron moderator

Another kind of reactor utilizes $^{238}_{92}U$ as a fuel. When a neutron strikes a $^{238}_{92}U$ nucleus, the following reactions can take place consecutively:

$$^{238}_{92}U + {}^{1}_{0}n \longrightarrow \qquad ^{239}_{92}U \tag{9.12}$$

$$^{239}_{92}U \longrightarrow {}^{239}_{93}Np + {}^{0}_{-1}e \tag{9.13}$$

$$^{239}_{93}Np \longrightarrow {}^{239}_{94}Pu + {}^{0}_{-1}e \tag{9.14}$$

An atom of $^{239}_{92}U$ is radioactive with a half-life of 23 min, and so it quickly decays into the new element neptunium, Np. This is also radioactive, with a half-life of 2.3 days; it disintegrates to produce another new element, plutonium, Pu. $^{239}_{94}Pu$ is radioactive but its half-life is 24,000 yr, so that it is relatively stable. Because $^{239}_{94}Pu$ is fissionable, a reactor can actually produce more nuclear fuel than it consumes if it is designed so that the extra neutrons not needed for fission of $^{235}_{92}U$ are absorbed by the $^{238}_{92}U$ to produce fissionable $^{239}_{94}Pu$ [see Eqs. (9.12) through (9.14)]. After the reactor has been running for some time, the uranium rods can be removed and the plutonium chemically separated from the unused uranium, since the two metals have quite different chemical characteristics. The main events taking place in a nuclear reactor are shown schematically in Fig. 9.7.

The element thorium provides the fuel for a third type of reactor. When it is bombarded by neutrons, the following reactions can take place:

$$^{232}_{90}Th + {}^{1}_{0}n \longrightarrow \qquad ^{233}_{90}Th \tag{9.15}$$

$$^{233}_{90}Th \longrightarrow {}^{233}_{91}Pa + {}^{0}_{-1}e \tag{9.16}$$

$$^{233}_{91}Pa \longrightarrow {}^{233}_{92}U + {}^{0}_{-1}e \tag{9.17}$$

The half-lives of $^{233}_{90}Th$, protactinium $^{233}_{91}Pa$, and $^{233}_{92}U$ are 22 min, 27 days, and 160,000 yr respectively. This series of reactions thus produces $^{233}_{92}U$ quite rapidly, and since this isotope is readily fissionable, $^{232}_{90}Th$ is an excellent source of nuclear fuel. It can be installed as a blanket around the core of a uranium reactor and produce more fissionable atoms than are consumed by

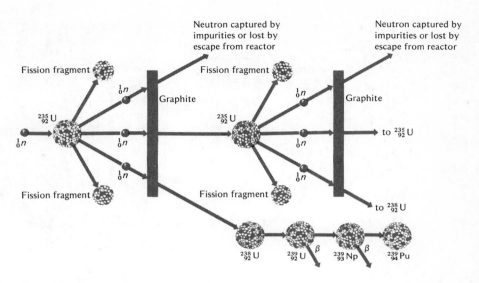

FIGURE 9.7 Main events taking place in a uranium reactor

the disintegration of $^{235}_{92}U$. Such a modification is called a *breeder reactor*.

Many different kinds of nuclear reactors have been built, using different fuels, moderators, control elements, coolants, and breeder design. The fuels may be $^{233}_{92}U$, $^{235}_{92}U$, $^{238}_{92}U$, or $^{239}_{94}Pu$ in the metallic state or combined in various compounds. Scores of different substances have been used as moderators and control rods, and coolants have ranged from molten metals to ordinary water.

A commercial atomic power plant consists of a nuclear reactor designed to produce a maximum of heat that can be efficiently used to generate steam. The steam drives turbines attached to electric generators. A typical design is shown in Fig. 9.8. Elaborate structures are built as envelopes around the reactors to protect the workers in the plant from stray radiation and to minimize air or water pollution should a reactor develop a leak. There is no danger that a reactor may "blow up" like an atomic bomb.

The chief hazard of operating nuclear reactors is the possibility of small but persistent leakages of radioactive gases, liquids, or solids damaging to living organisms. When the products of nuclear reactions in fuel elements have accumulated for some time, efficiency is so reduced that the fuel must be removed and processed chemically to recover unused fuel from waste products. Since the wastes are highly radioactive, fuel elements must be removed, shipped, and processed with extreme care. Waste products separated

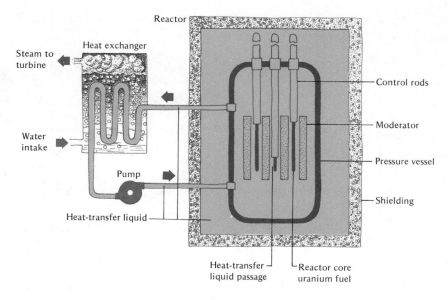

Reactor

Heat exchanger

Steam to turbine

Water intake

Pump

Heat-transfer liquid

Control rods

Moderator

Pressure vessel

Shielding

Heat-transfer liquid passage

Reactor core uranium fuel

FIGURE 9.8 Schematic diagram of a reactor for producing power. (Adapted from an AEC figure, Argonne Laboratory.)

from fuel must be stored. Much of their radioactivity has so short a half-life that it quickly dies down. Storage in containers surrounded by barriers that do not permit the passage of radiation reduces the hazard greatly. Radioactive residues can then be incorporated into solid material that can be stored for hundreds of years inside heavily shielded concrete bunkers. Within a few hundred years the radioactivity of the wastes will be less than we encounter daily from cosmic rays and naturally radioactive compounds around us.

9.8 Damage to Living Organisms from Exposure to Radiation

All radioactive emanations—α, β, and γ rays—injure living organisms because of their ionizing effect. They knock electrons out of the atoms in living tissue, destroying the chemical compounds of which the atoms are a part and impairing the action of the tissue. All living cells contain much water, and when water is irradiated, it splits into H atoms and OH groups. The OH groups combine to form hydrogen peroxide, H_2O_2, which soon decomposes to give H_2O molecules and O atoms. These O atoms are extremely reactive and quickly attack the complex compounds present in cells, thus interfering with the normal operation of cells. If enough healthy cells are thus damaged, a cancer may be started. If testes or ovaries are irradiated, defective sperm or eggs may be produced. These may lead to genetic defects causing the death or malformation of offspring.

Soon after radium was discovered, it was found that when a compound of this element is mixed with zinc sulfide, tiny flashes of greenish-yellow light are continually produced, and the mixture glows in the dark. This light is like the fluorescence produced when cathode rays strike zinc sulfide. A paste made of the mixture was used to paint the numbers and hands of clocks and watches so that they could be seen in the dark. About 40 years ago, the women who painted the clock dials used tiny brushes which they moistened in their mouths to get a fine point. Within a few years many of them developed cancer of the mouth. Only then did medical men become aware of the dangers of overexposure to radioactive compounds. Today we have techniques for handling radioactive substances without danger to the handlers.

We can tolerate a moderate amount of radiation and repair the damage it does. But the damage can readily become irreparable. While the immediate blast from an atomic bomb is frightfully destructive, even worse are the enormous bursts of intense radioactivity it gives off, and the continuing high levels of radiation that linger for many days. Illnesses and deaths from radiation sickness continue for years, and damage to the genes transmitted through ovaries and testes results in birth defects in the next generation.

Radioactive debris from an atomic explosion is driven into the atmosphere and circulates around the world, gradually coming to earth as radioactive fallout. During the 1950s the world became aware of the hazards of atomic bomb tests in the atmosphere, and an international agreement was reached to conduct all future tests underground to confine the radioactive debris to the test site.

At present most of the ionizing radiation we are exposed to comes from medical *x-rays*, which are like γ rays but less energetic and hence less damaging to living tissue. Small amounts come from television screens and luminous dials on watches. About two-thirds of our total exposure comes from manmade radiation and one-third from natural sources. Radiation from the fallout of debris of nuclear explosions has been small. But it is estimated that the reactor in a 100-MW (megawatt) electric generating station will produce each year about the same quantity of long-lived fission products as the detonation of a 1-megaton fission bomb. To prevent radioactive contamination of the environment, radioactive wastes from nuclear reactors are removed, shipped, and stored under the most rigorous safeguards.

Strontium $^{89}_{38}$Sr and $^{90}_{38}$Sr, barium $^{140}_{56}$Ba, cesium $^{137}_{55}$Cs, and iodine $^{131}_{53}$I are particularly dangerous in radioactive debris because they have relatively long

x-rays

half-lives (50 days, 27 yr, 12 days, 30 yr, and 8 days respectively), and because they are readily assimilated by the human body. Strontium and barium are chemically similar to the calcium in bones and easily follow the metabolic paths of calcium into these important structures, where their radioactivity interferes with the production of red blood cells in the bone marrow. Cesium is chemically similar to sodium and potassium, which are widely distributed in blood and muscle tissue; radioactive cesium soon finds its way all over the body and disrupts many vital functions. Ordinary, nonradioactive iodine plays an important role in the thyroid gland; when $^{131}_{53}$I is ingested, it quickly makes its way to the thyroid, where it interferes with the function of that important organ.

The Health Physics Division of the Atomic Energy Commission has ruled that storing a radioactive isotope for 20 half-lives will make it safe for biological exposure. In this length of time, radioactivity will have been reduced to about one-millionth of its original intensity. This rule requires that isotopes like $^{90}_{38}$Sr and $^{137}_{55}$Cs be kept isolated for about 600 yr!

Summary

The nuclei of certain isotopes of some elements are unstable. This instability makes them radioactive. They attain greater stability by ejecting three kinds of radiation: α particles, β particles, and γ rays.

Different radioactive isotopes disintegrate at different rates. The rate of disintegration is conveniently measured as the half-life, the time required for one-half of the isotope to disintegrate. Half-lives vary from a tiny fraction of a second to billions of years.

Radioactive carbon, $^{14}_{6}$C, is produced in nature by the bombardment of our atmosphere by cosmic rays. The half-life of $^{14}_{6}$C is 5760 yr. The amount of $^{14}_{6}$C present in a sample of dead material which was once part of a living organism indicates how long ago the organism died.

In all processes involving the transformation of energy from one form into another, energy is neither created nor destroyed. This is the law of conservation of energy.

The nuclei of certain elements can be split into two fragments of roughly the same size; this disintegration is called nuclear fission. The total mass of the products of fission is less than the total mass of the reactants. This loss of mass releases very large amounts of energy.

Hydrogen nuclei may fuse to form helium nuclei. The mass of the He produced is less than the mass of the H used up. This loss of mass releases very large amounts of energy.

Mass may be considered as a manifestation of energy and vice versa. Amounts of mass and energy are related by the equation $E = mc^2$. In a system in which no energy or mass leaves or enters the system, the sum of the masses of all matter at rest and of the mass-equivalents of energy is constant.

Radiation from radioactive substances damages living tissues. As we build nuclear reactors for producing electric energy, we must use as many safeguards as we can invent to protect the environment from pollution by radioactive isotopes.

New Terms and Concepts

BREEDER REACTOR: A nuclear reactor that produces more fissionable material than it consumes.

COSMIC RAYS: A stream of penetrating radiation entering the earth's atmosphere from outer space.

CYCLOTRON: A device for giving high speed to charged particles by magnetic and electrical means.

DEUTERON: The nucleus of the hydrogen isotope, containing one proton and one neutron.

HALF-LIFE: The time required for the intensity of radiation from a material to decrease to one-half its value at the beginning of the timing period.

LAW OF CONSERVATION OF ENERGY: Energy is recognized in various forms, and when it disappears in one form, it appears in others, in each case according to a fixed rate of exchange. The total quantity of any energy, measured in terms of any one form, is constant whatever forms it may assume.

LAW OF CONSERVATION OF MASS-ENERGY: In a system in which no mass or energy enters or leaves the system, the sum of the masses of all matter at rest and of the mass-equivalents of energy is constant.

NEUTRON MODERATOR: A substance used in a nuclear reactor to slow down fast neutrons.

NUCLEAR FISSION: The disintegration of an atomic nucleus to form neutrons and two nuclei of roughly half the mass of the disintegrating nucleus.

NUCLEAR FUSION: The combination of two atomic nuclei to form the nucleus of a heavier atom.

NUCLEAR REACTOR: A device for producing energy by nuclear reactions and removing the energy as heat.

PLASMA: An electrically neutral, highly ionized gas consisting of ions, electrons, and neutral particles.

X-RAYS: A highly penetrating form of radiation.

Testing Yourself

9.1 On a separate sheet of paper, fill in the following table.

Isotope	Electrons	Protons	Neutrons
Hydrogen	————	————	————
Deuterium	————	————	————
Tritium	————	————	————
$^{12}_{6}C$	————	————	————
$^{14}_{6}C$	————	————	————

9.2 Why are the atomic weights of most elements not whole numbers?

9.3 Assume the existence of a hypothetical isotope with a nuclear charge of 89 and mass of 230. Fill in the table below to show the characteristics of the isotope resulting from the emission of particles from the hypothetical isotope.

Particle(s) lost	α	β	Neutron	α and β
Nuclear charge	————	————	————	————
Nuclear mass	————	————	————	————

9.4 When ordinary nitrogen is bombarded with neutrons, an unstable radioactive isotope is produced that ejects a proton. What is the new element so produced?

9.5 How would you explain to Omits the difference between nuclear fusion and nuclear fission?

9.6 The compounds of uranium in nature contain 1 atom of $^{235}_{92}U$ for every 139 atoms of $^{238}_{92}U$. Why doesn't a nuclear fission chain reaction take place in beds of uranium ore?

9.7 Why is it necessary to build huge and expensive plants to separate $^{235}_{92}U$ from natural uranium for nuclear reactor fuel?

9.8 A neutron production factor of exactly 1 per fission must be maintained in a nuclear reactor. What would happen if the factor exceeded 1? If it fell below 1?

9.9 What fraction of the amount of radioactivity of any isotope is present after: (*a*) 5 half-lives; (*b*) 10 half-lives; (*c*) 20 half-lives?

9.10 A sample of charcoal (practically pure carbon) is found in a cave inhabited by man in prehistoric times. The radioactivity of the carbon is found to be 3 disintegrations per minute per gram. When did man build the fire which produced this charcoal?

9.11 The intensity of radioactivity of a certain sample of $^{128}_{53}I$ indicates that 32 mg (milligrams) of the isotope is present; $2\frac{1}{2}$ hr later, the activity shows the presence of 0.50 mg. What is the half-life of $^{128}_{53}I$?

9.12 A museum possesses a fragment of wood which is claimed to be part of a wood sculpture made 2000 years ago. The production of beta particles from a sliver of the wood is found to be at the rate of 12 disintegrations per gram of carbon per minute. Is the wood old enough to have been carved 2000 years ago? (*Hint*: See Fig. 9.1*b*.)

Helium, shown in the photograph of a model, occurs as the last element of period 1; lithium with its three electrons begins period 2.

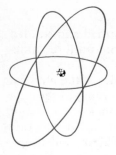

10. Simplification by Classification:
The Periodic Table of the Elements

fly ash

Although we are building more and more nuclear reactors to help meet increasing needs for energy, our fear that such plants may pollute our environment with radioactive poisons is keeping this development from meeting all our unfilled requirements for energy. For at least another 25 years we shall have to keep building more power plants that burn coal to generate steam for turbines that drive dynamos. Burning coal is a dirty business resulting in the pollution of our air by oxides of sulfur and particulates that go up the flues of the steam boiler as *fly ash*. To know how to deal with these pollutants, we need to understand their chemical natures and how they can be removed by chemical reaction with other substances. To understand these particular chemical reactions, we need more knowledge of the general ways in which the chemical elements react. This chapter gives us a general overview of the properties of chemical elements. We use the well-tested method of *simplification by classification* to ease our job of learning these aspects of chemistry and help us apply it to environmental problems. To this end, we examine in this chapter one of the great accomplishments of nineteenth-century chemistry, the periodic table of the elements.

10.1 The Discovery of the Elements

Although the 20 commonest chemical elements were identified and studied before the end of the eighteenth century (Sec. 2.7), it was not until the nineteenth century that elements were discovered in large numbers. This was the era when chemists were heavily engaged in systematic studies of the physical properties and chemical behavior of a host of substances, trying to separate them into their constituent elements. Discovering new elements was one of the most exciting activities in the scientific research of that era. In the years 1800–1825, 22 new elements were isolated: vanadium, niobium, palladium, rhodium, osmium, iridium, potassium, sodium, barium, strontium, calcium, magnesium, boron, iodine, cadmium, selenium, lithium, silicon, zirconium, aluminum, titanium, and bromine. Between 1825 and 1870, beryllium, uranium, ruthenium, cesium, rubidium, thallium, and indium were discovered.

As soon as chemists learned how to isolate a new element, they began to study its physical properties and chemical reactivity and to publish their discoveries. The ensuing flood of papers soon made it evident that some principles for organizing all this factual information would have to be devised, or chemistry would be only a huge conglomeration of facts with no scientific interpretation. Various bases for the classification of the elements were tried. The more successful of these are described in the following pages.

10.2 Classifying the Elements by Their Physical Properties: Metals, Nonmetals, and Metalloids

At room temperature and one atmosphere of pressure (1 atm), 77 of the 90 naturally occurring elements are solid, 2 are liquid (mercury and bromine), and 11 are gaseous (hydrogen, nitrogen, oxygen, fluorine, chlorine, helium, neon, argon, krypton, xenon, and radon). Of the solids 67 are metals—shiny in appearance, malleable (can be hammered into thin sheets), ductile, and good conductors of heat and electricity. Three of the solid elements (phosphorus, sulfur, and iodine) are dull in appearance, brittle, and poor conductors of heat and electricity. Because of these characteristics we call them nonmetals. Seven of the solid elements (boron, carbon, silicon, germanium, arsenic, selenium, and tellurium) exhibit some luster, are moderately malleable, and conduct heat and electricity moderately well. We call these semimetals or *metalloids*. Although mercury is a liquid, its luster and high conductivity of heat and electricity lead us to classify it as a metal. Liquid bromine and all the gaseous elements are clearly nonmetals.

metalloid

10.3 Families of Elements

As nineteenth-century chemists continued their systematic investigation of the physical and chemical properties of the elements then known and the characteristics of the compounds they formed, it became apparent that some elements had many similarities and that they might be grouped into families, thus developing some pattern of organization.

triad

As early as 1817 the German chemist Johann Wolfgang Döbereiner recognized a relationship among the atomic weights of certain well-known *triads* of elements with strikingly similar properties. Thus the atomic weight of strontium is nearly the average of the weights of calcium and barium, and all three react in much the same way. The same is true for bromine with respect to chlorine and iodine. Similar relationships were found in the triads lithium-sodium-potassium and sulfur-selenium-tellurium.

Further attempts to discover such relationships were frustrated by the large amount of erroneous data on atomic weights then being published. The relationship between atomic and molecular weights of elements was not correctly understood until 1860, when the Italian chemist Stanislao Cannizzaro presented a paper entitled "Sketch of a Course of Chemical Philosophy" (based on the interpretation of the law of combining volumes by his teacher Avogadro) to the first International Chemical Congress in Karlsruhe, Germany. Cannizzaro's interpretation cleared up the confusion about the formulas of various compounds and brought order out of the chaos of conflicting determinations of the atomic weights of the elements. Within 10 years, at least four different chemists tried to classify the elements on the basis of atomic weights recalculated in the light of Cannizzaro's interpretation. Of these chemists, the Russian Dmitri Ivanovich Mendeleev was the most successful.

10.4 Mendeleev's Periodic Table

In March 1869 Mendeleev sent to the Russian Chemical Society his first paper on the classification of the elements, entitled "The Relation of the Properties to the Atomic Weights of the Elements." Mendeleev noted that when the known elements were listed in order of increasing atomic weight, elements with certain characteristics in common appeared at regular intervals. The list of the 63 elements (and their atomic weights) known to him appears on page 188. Mendeleev pointed out that members of well-known families of elements that had certain sets of similar properties turned up at regular intervals.

alkali metal

hydroxide

One such family of elements consists of the metals lithium (Li), sodium (Na), potassium (K), rubidium (Rb), and cesium (Cs). In the preceding list, their symbols are enclosed in rectangles made with solid lines. These elements are called the *alkali metals*. All are soft and shiny, tarnish rapidly when exposed to air, and form oxides with the general formula E_2O, where E stands for any member of the family. When dropped into water, these metals react vigorously, producing hydrogen and a dissolved *hydroxide* with the general formula EOH. Lithium hydroxide is LiOH, sodium hydroxide

H 1	Li 7	Be 9.4	B 11	C 12	N 14	O 16	F 19	Na 23	Mg 24	Al 27.3
Si 28	P 31	S 32	Cl 35.5	K 39	Ca 40	Er 56	Yt 60	In 75.6	Ti 50	V 51
Cr 52	Mn 55	Fe 56	Ni 59	Co 59	Cu 63.4	Zn 65.2	As 75	Se 79.4	Br 80	Rb 85.4
Sr 87.6	Zr 90	Ce 92	La 94	Nb 94	Di 95	Mo 96	Rh 104	Ru 104	Pd 107	Ag 108
Cd 112	U 116	Th 118	Sn 118	Sb 122	I 127	Te 128	Cs 133	Ba 137	Ta 182	W 186
Au 197	Pt 197.4	Ir 198	Os 199	Hg 200	Ti 204	Pb 207	Bi 210			

chloride
hydride

NaOH, and so on. The alkali metal hydroxides have very similar properties. The elements form white, crystalline, soluble *chlorides* with the formula ECl, *hydrides* with the formula EH (LiH, NaH, etc.), and have many other properties in common. Mendeleev noted these intervals in his list: Li—6 elements—Na—6 elements—K—16 elements—Rb—18 elements—Cs.

Another family consists of the metals beryllium (Be), magnesium (Mg), calcium (Ca), strontium (Sr), and barium (Ba). In the preceding list their symbols are enclosed in rectangles made with dotted lines. These elements are called the *alkaline earth metals*. They are shiny, somewhat harder than the alkali metals, and not quite so reactive with oxygen and water. Letting E stand for any member of the alkaline earth family, the oxides are EO, the hydroxides are $E(OH)_2$, the chlorides, ECl_2, and the hydrides, EH_2. The compounds of each type have very similar properties. Mendeleev also noted this sequence: Be—6 elements—Mg—6 elements—Ca—16 elements—Sr—18 elements—Ba.

A third family consists of the nonmetals fluorine (F), chlorine (Cl), bromine (Br), and iodine (I). These elements are called the *halogens*; their symbols are enclosed in circles with solid lines in the preceding list. The halogens combine with the alkali metals to form compounds of the type EX, where E stands for an alkali metal and X for a halogen. All these compounds, such as LiF, NaCl, KBr, RbI, CsF, etc., are white, crystalline solids, readily soluble in water. The halogens also form white crystalline solids having the general formula EX_2, where E stands for an alkaline earth metal—compounds like $BeCl_2$, $BaBr_2$, CaI_2, etc. These compounds have similar properties. Halogens combine with hydrogen to form colorless gases with the general formula HX. Mendeleev also noted the sequence: F—6 elements—Cl—16 elements—Br—19 elements—I.

oxygen family

A fourth family consists of the nonmetals oxygen (O), sulfur (S), selenium (Se), and tellurium (Te). The symbols for the elements in the *oxygen family* are enclosed in dotted circles in the preceding list. These elements combine with hydrogen to form H_2O, H_2S, H_2Se, and H_2Te. They also form SO_2, SeO_2, and TeO_2. Here the sequence is: O—6 elements—S—16 elements—Se—19 elements—Te.

Mendeleev was so impressed by these similarities in sequence that he arranged the known elements in a table of horizontal rows and vertical columns to bring out the periodic and family relationships. The form of his periodic table published in 1872 is shown in Table 10.1. The columns headed with Roman numerals contain families or groups of similar elements, and the horizontal rows are called series. Between 1869, when he wrote his first paper,

TABLE 10.1 *Mendeleev's Periodic Table of 1872. E Is a General Symbol for Any Element.*

Series	I E_2O	II EO	III E_2O_3	IV EH_4 EO_2	V EH_3 E_2O_5	VI EH_2 EO_3	VII EH E_2O_7	VIII EO_4	
1	H 1								
2	Li 7	Be 9.4	B 11	C 12	N 14	O 16	F 19		
3	Na 23	Mg 24	Al 27.3	Si 28	P 31	S 32	Cl 35.5		
4	K 39	Ca 40	? 44	Ti 48	V 51	Cr 52	Mn 55	Fe 56	Co 59
								Ni 59	Cu 63
5	(Cu 63)	Zn 65	? 68	? 72	As 75	Se 78	Br 80		
6	Rb 85	Sr 87	Yt 88?	Zr 90	Nb 94	Mo 96	? 100	Ru 104	Rh 104
								Pd 106	Ag 108
7	(Ag 108)	Cd 112	In 113	Sn 118	Sb 122	Te 125	I 127		
8	Cs 113	Ba 137	Di 138?	Ce 140?					
9									
10			Er 178?	La 180?	Ta 182	W 184		Os 195	Ir 197
								Pt 198	Au 199
11	(Au 199)	Hg 200	Tl 204	Pb 207	Bi 208				
12				Th 231		U 240			

and 1872, when he published the form of his table shown in Table 10.1, Mendeleev made changes in the values of the atomic weights for 18 elements. Recalculating earlier data according to the methods worked out by Cannizzaro produced striking changes such as: Er, from 56 to 178; Yt, from 60 to 88; In, 75.6 to 113; Ce, 92 to 140; La, 94 to 180; Di, 95 to 138; U, 116 to 240; Th, 118 to 231. When Mendeleev changed the positions of these 18 elements in his table, he found that each vertical column now consisted of elements in a family with noteworthy similarities. He was so sure of the basic significance of his table that he did not hesitate to transpose I and Te, thus making the halogen and oxygen family sequences internally consistent; Te and I were then in the vertical columns with the other elements of their families. He assumed that the atomic weights of Te and I had been incorrectly determined.

Mendeleev also left vacant spaces where no element was known, since he believed that the missing elements would be discovered. He was so confident of the gradations from element to element that he predicted the properties of the elements that should fill some of the gaps. Those under boron, aluminum, and silicon he named eka-boron, eka-aluminum, and eka-silicon, and he predicted the atomic weights 44, 68, and 72. Within 15 years the new elements scandium, gallium, and germanium were discovered, with atomic weights 45.0, 69.7, and 72.6. With astonishing accuracy Mendeleev also predicted the density of these elements, their atomic volume (the volume occupied by 1 mole of the element in the solid phase), their color, chemical behavior, and the properties of some of their compounds. Their discovery and characterization were a triumph of the periodic concept. The striking agreement between the properties of eka-silicon predicted by Mendeleev and those of the element later named germanium are summarized in Table 10.2.

The periodicity in properties noted by Mendeleev was also recognized by Julius Lothar Meyer, a German chemist and physician. In 1864, when teaching at the University of Breslau, Meyer published a textbook, *Moderne Theorien der Chemie*, in which he had a partially complete periodic table. In 1870 Meyer published an expanded version of his table in an article entitled "Nature of the Elements as Functions of Their Atomic Weights." He based his table on the periodicity he noted in such properties as atomic volume, melting point, volatility, malleability, brittleness, and electrochemical behavior.

In more recent years similar periodicity has been noted from plots of atomic radius, ionic radius, boiling point, compressibility, hardness, electrical conductivity, thermal conductivity, and thermal coefficient of expansion. Compounds of the elements likewise show periodicity in such properties as melting point, boiling point, stability, color, and solubility. Some graphs of the form published by Meyer are presented in Figs. 10.1, 10.2, and 10.3.

TABLE 10.2 *Properties of Eka-silicon Predicted by Mendeleev Compared to Those Found for Germanium*

Property	Eka-silicon	Germanium
Atomic weight	72	72.6
Physical properties	Gray, high melting point, density = 5.5	Gray, melting point = 958°C, density = 5.36
Properties of oxide	XO_2, high melting point, density = 4.7	GeO_2, melting point = 1100°C, density = 4.70
Properties of sulfide	XS_2, insoluble in H_2O, soluble in $(NH_4)_2S$	GeS_2, insoluble in H_2O, very soluble in $(NH_4)_2S$
Properties of chloride	XCl_4, volatile liquid, boiling point < 100°C, density = 1.9	$GeCl_4$, volatile liquid, boiling point = 83°C, density = 1.88

periodic law

The periodic concept was invaluable in codifying enormous amounts of chemical data. In the nineteenth century the concept was stated as the *periodic law*: When the chemical elements are arranged in order of increasing atomic weight, elements with similar sets of properties appear periodically in the list.

One of the great values of the periodic table lay in the impetus it lent to the discovery of new elements. Knowing that elements above and below a vacancy, and to the right and left of it, had similar properties gave hints of what substances might contain the unknown element and how it might be separated from them. Time after time these hints led to success, and the gradations between neighboring elements became clearly established. By 1890 seven elements unknown in 1869 had been discovered and fitted into Mendeleev's table.

10.5 Discovery of the Noble Gases

In the 1880s Lord Rayleigh at Cambridge University undertook an exhaustive study of the densities of gases and the exact volumes that combined to form compounds. Although the samples of oxygen he prepared by three different methods all had the same density, the nitrogen he prepared by removing the dust, oxygen, water vapor, and carbon dioxide from air was 0.5 percent heavier than the nitrogen he prepared by the decomposition of ammonia, NH_3.

In the 1890s Sir William Ramsay became interested in this puzzle and tried combining nitrogen from the atmosphere with magnesium, which readily forms magnesium nitride, Mg_3N_2. After repeated passes over red-hot magnesium, $\frac{1}{80}$ of the original volume of gas remained, and its density was considerably greater than that of nitrogen. Rayleigh and Ramsay concluded that the gas was a new element and named it argon (meaning lazy) because it refused to combine with the hot magnesium. They believed that its atomic

FIGURE 10.1 Melting points of the elements vs atomic weight. The periodic up-and-down fluctuations in the melting points of the elements were noted by Julius Lothar Meyer and led him to the conclusion that elements could be grouped into families with similar physical and chemical properties.

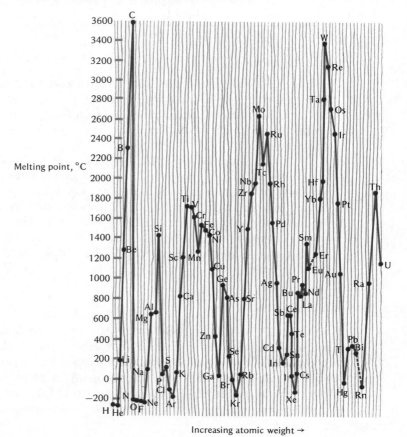

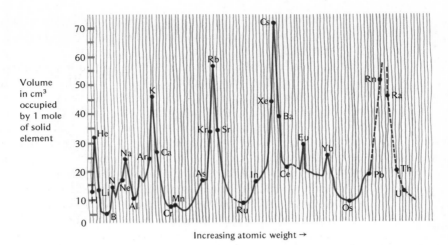

FIGURE 10.2 The volumes occupied by 1 mole of the solid elements vs their atomic weights. Families of elements occupy similar positions on the series of curves. For instance, Na, K, Rb, and Cs are at the peaks in the graph. (Helium, He, was not known in Meyer's time.)

FIGURE 10.3 Energy required to remove an electron from an atom vs atomic weight. Again the elements Na, K, Rb, and Cs occupy similar positions on the series of curves, this time at the lowest points.

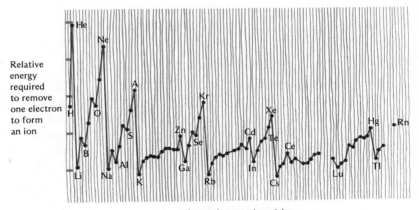

weight would place it between chlorine and potassium in the periodic table and that there might well be a family of elements located between the halogens and the alkali metals.

This exciting suggestion set off a search for such elements. A gas given off from the radioactive mineral uranite was chemically inert and was finally identified spectroscopically as the element helium. (Spectroscopic methods for identifying the elements are discussed in Chap. 16.) Spectroscopic analysis of sunlight in 1868 led to the conclusion that there was an element on the sun that was not present on earth. The name helium was given to this element because "helios" is the Greek word for sun. Its atomic weight placed it between hydrogen and lithium in the periodic table. Ramsay and his coworkers distilled liquefied air in their search for other inert gases in the atmosphere and found one that fitted into the periodic table between bromine and rubidium; they named it krypton (hidden). They later discovered and named neon (new) and xenon (strange); neon fitted between fluorine and sodium, xenon between iodine and cesium. About 1890, a heavy gas produced by the radioactive disintegration of radium was found to be quite inert chemically; it was named radon because of the source from which it was isolated. Radon is the inert gas element having the greatest atomic weight.

Each of these elements fitted into Mendeleev's list of 63 elements after a halogen and before an alkali metal. The new family of elements was fitted into Mendeleev's periodic table (Table 10.1) as a *zero group* to the left of group I, the alkali metals. Because these elements are so unreactive, group zero was often called the inert gases. In the 1960s some chemists succeeded in making a few compounds of some of these elements, so that the name "inert gases" no longer seemed to fit. Early chemists called gold, silver, and platinum, which were rather unreactive with other elements, the noble metals. So it seemed appropriate to call these, by analogy, the *noble gases*. Some chemists prefer the term *rare gases*.

zero group

noble gas
rare gas

10.6   The Long Form of the Periodic Table

Table 10.1 had some inadequacies and inconsistencies. The elements in the even-numbered series, placed to the left in each vertical column (often called the "A subgroup"), had strong family resemblances, but those in the odd-numbered series, placed to the right in each vertical column (called the "B subgroup"), did not so clearly share resemblances. Putting the A and B subgroups in the same vertical column appeared to be forcing the periodic concept too hard. Accordingly, a longer form of the table seemed more appropriate as the relationships among the properties of more and more elements were studied. This long form is shown in Table 10.3, using values accepted today

TABLE 10.3   *The "Long Form" of the Periodic Table*

Group \ Period	IA E_2O EH	IIA EO EH_2	IIIB	IVB	VB	VIB	VIIB		VIII		IB	IIB	IIIA E_2O_3	IVA EO_2 EH_4	VA E_2O_5 EH_3	VIA EO_3 H_2E	VIIA E_2O_7 HE	0
1	1 H 1.008																	2 He 4.003
2	3 Li 6.939	4 Be 9.012											5 B 10.81	6 C 12.01	7 N 14.01	8 O 16.00	9 F 19.00	10 Ne 20.18
3	11 Na 22.99	12 Mg 24.31											13 Al 26.98	14 Si 28.09	15 P 30.97	16 S 32.06	17 Cl 35.45	18 Ar 39.95
4	19 K 39.10	20 Ca 40.08	21 Sc 44.96	22 Ti 47.90	23 V 50.94	24 Cr 52.00	25 Mn 54.94	26 Fe 55.85	27 Co 58.93	28 Ni 58.71	29 Cu 63.54	30 Zn 65.37	31 Ga 69.72	32 Ge 72.59	33 As 74.92	34 Se 78.96	35 Br 79.91	36 Kr 83.80
5	37 Rb 85.47	38 Sr 87.62	39 Y 88.91	40 Zr 91.22	41 Nb 92.91	42 Mo 95.94	43 Tc (99)	44 Ru 101.1	45 Rh 102.9	46 Pd 106.4	47 Ag 107.9	48 Cd 112.4	49 In 114.8	50 Sn 118.7	51 Sb 121.8	52 Te 127.6	53 I 126.9	54 Xe 131.3
6	55 Cs 132.9	56 Ba 137.3	See* below	72 Hf 178.5	73 Ta 181.0	74 W 183.8	75 Re 186.2	76 Os 190.2	77 Ir 192.2	78 Pt 195.1	79 Au 197.0	80 Hg 200.6	81 Tl 204.4	82 Pb 207.2	83 Bi 209.0	84 Po (210)	85 At (210)	86 Rn (222)
7	87 Fr (223)	88 Ra (226)	See** below	104 Ku	105 Ha (260)													

*Lanthanide series	57 La 138.9	58 Ce 140.1	59 Pr 140.9	60 Nd 144.2	61 Pm (147)	62 Sm 150.4	63 Eu 152.0	64 Gd 157.2	65 Tb 158.9	66 Dy 162.5	67 Ho 164.9	68 Er 167.3	69 Tm 168.9	70 Yb 173.0	71 Lu 175.0
**Actinide series	89 Ac (227)	90 Th 232.0	91 Pa (231)	92 U 238.0	93 Np (237)	94 Pu (242)	95 Am (243)	96 Cm (247)	97 Bk (249)	98 Cf (251)	99 Es (254)	100 Fm (253)	101 Md (256)	102 No (253)	103 Lw (257)

The numbers in parentheses are the relative atomic weights of the most stable known isotopes.

The numbers above the symbols are the atomic numbers; those below are the atomic weights. E is a general symbol for any element.

for the atomic weights of all the elements now known. The zero group of elements fits more conveniently after VIIA than before IA. Some contemporary chemists use the term "B group" for the families of elements in the five columns preceding the zero group; others call these families the "A group," which is the terminology adopted in this book. Series 1, 2, and 3 have become *periods* 1, 2, and 3; series 4 and 5; 6 and 7; 8 and 9; and 10, 11, and 12 have become periods 4, 5, 6, and 7, respectively. With this arrangement, elements with strong similarities are placed in the same vertical column. The A and B subgroups are in separate columns, and their similarities are noted by having the same Roman numeral for both.

The elements having atomic numbers between those of Ba (56) and Hf (72) are so similar chemically that their separation is extremely difficult. The same is true for those having atomic numbers greater than that of Ra (88). The series of chemically almost indistinguishable elements beginning with La (57) and ending with Lu (71), called the *lanthanide series*, is thus considered to occupy only one space in the periodic table. The same arrangement is used for the radioactive elements beginning with Ac (89) and ending with Lw (103), called the *actinide series*.

10.7 The Periodic Table in the Twentieth Century

The periodic table developed in the nineteenth century contained three inconsistencies in the order of increasing atomic weights.

1 Argon (atomic weight 39.948) precedes potassium (39.102).
2 Cobalt (58.933) precedes nickel (58.71).
3 Tellurium (127.60) precedes iodine (126.904).

But it was obvious from the properties of these elements that

1 argon belongs in the family of noble gases
2 potassium belongs with the alkali metals
3 cobalt resembles rhodium (Rh) and iridium (Ir)
4 nickel resembles palladium (Pd) and platinum (Pt)
5 tellurium belongs in the oxygen family
6 iodine belongs with the halogens

Accordingly, these elements were placed where they belonged chemically even though their atomic weights were out of sequence.

In the twentieth century it was shown (Chap. 8) that atoms consist of a central nucleus composed of protons and neutrons and that this nucleus is surrounded by a cloud of electrons. There are as many electrons in this cloud as there are protons in the nucleus, so that the atom as a whole is electrically neutral. The number of protons (or the number of electrons) in a neutral atom is called the atomic number.

Almost immediately it was noted that listing the elements in order of increasing *atomic number* (instead of atomic weight) straightened out the inconsistencies in the nineteenth-century table.

1 Argon came in the family of noble gases.
2 Potassium came with the alkali metals.
3 Cobalt was in the same column with rhodium and iridium.
4 Nickel was in the same column with palladium and platinum.
5 Tellurium came in the oxygen family.
6 Iodine came in the halogen column.

Clearly the atomic number of an element has more to do with determining its chemical properties than does its atomic weight. Consequently we now state the periodic law thus: When the chemical elements are arranged in order of increasing atomic number, elements with similar sets of properties appear periodically in the list.

Element 104 is the subject of a continuing international argument. Russian scientists working at the Joint Institute of Nuclear Research in Dubna claimed that the isotope $^{260}104$ was detected by physical methods in 1964 and the isotope $^{259}104$ was isolated by chemical methods in 1966. They named the element kurchatovium (Ku). In 1969 scientists at the Lawrence Radiation Laboratory of the University of California announced that they were unable to duplicate the Russian experiments, but that they had succeeded by other means in producing the isotopes $^{257}104$ and $^{259}104$. They named the element rutherfordium (Rf). Since then papers have continued to appear in scientific journals in which the scientists in each laboratory question each other's apparatus and methods. Today the name you give to element 104 depends on where you live! Eventually, an international commission will settle the controversy.

10.8 Classifying the Elements by the Properties of Their Oxides: Basic and Acidic Oxides

The classification of elements into metals, nonmetals, and metalloids on the basis of their physical properties (Sec. 10.2) is reinforced when we consider their chemical properties. Atmospheric oxygen is one of the most available and reactive of the common elements. When we burn a metal in air, we find that the ash produced is a compound of the metal and oxygen. We call such a compound an oxide:

$$2Mg(s) + O_2(g) \longrightarrow 2MgO(s)$$

The oxides of metals are solids, but the oxides of nonmetals are gases, except for hydrogen oxide (water), which is a liquid, and phosphorus oxides, which are solids:

$$2H_2(g) + O_2(g) \longrightarrow H_2O(l) \qquad S(s) + O_2(g) \longrightarrow SO_2(g)$$

Some of the oxides of metalloids are solids and some are gases:

$$C(s) + O_2(g) \longrightarrow CO_2(g)$$

If we shake up solid metal oxides with water, some dissolve readily, but most are only very slightly soluble. If a metal oxide dissolves, it produces a solution that is bitter to taste (*never* taste anything in the laboratory unless you are specifically instructed to do so), slippery when rubbed between thumb

litmus
basic solution

base
basic oxide
basic anhydride
hydration

and finger, and that changes the color of a number of vegetable juices. The juice most often used is that from the *litmus* plant, which turns blue when added to a solution of a metal oxide. We define such a solution as a *basic solution*. A basic solution turns litmus blue. If we boil off the water and analyze the solid substance remaining, we find that it contains the metal, oxygen, and hydrogen. We call this substance a hydroxide. A hydroxide is also called a *base*. We call the oxide that reacts with the water to produce the hydroxide a *basic oxide*. We also call it a *basic anhydride* because it produces a base by the process of *hydration*, by taking on water. The chemical equations for the formation of the base from the metal sodium, Na, are

$$4Na(s) + O_2(g) \longrightarrow 2Na_2O(s) \qquad \text{sodium oxide (a basic oxide or basic anhydride)}$$

$$Na_2O(s) + H_2O(l) \longrightarrow 2NaOH(aq) \qquad \text{sodium hydroxide (a base)}$$

aqueous

The symbol (*aq*) signifies that the substance whose formula precedes it is present in *aqueous* (water) solution.

acidic solution
acid

acidic oxide
acidic anhydride

If we shake up a nonmetal oxide with water, it forms a solution that is sour in taste, produces hydrogen gas when added to a piece of magnesium, and changes the color of litmus extract to red. We call such a solution an *acidic solution*, and we say the solute is an *acid*. An acid is a solution that turns litmus red. We call the oxide that reacts with the water to produce the acid an *acidic oxide*; we also call it an *acidic anhydride*, because it produces an acid by hydration. The chemical equations for the formation of an acid from the nonmetal sulfur, S, are

$$S(s) + O_2(g) \longrightarrow SO_2(g) \qquad \text{sulfur dioxide (an acidic oxide or acidic anhydride)}$$

$$SO_2(g) + H_2O(l) \longrightarrow H_2SO_3(aq) \qquad \text{sulfurous acid}$$

If sulfur dioxide and air are passed through a red-hot gauze made of platinum, another oxide of sulfur is produced. It, too, is an acidic anhydride, as shown by the following equations:

$$2SO_2(g) + O_2(g) \xrightarrow{Pt} 2SO_3(g) \qquad \text{sulfur trioxide (an acidic oxide or acidic anhydride)}$$

$$SO_3(g) + H_2O(l) \longrightarrow H_2SO_4(aq) \qquad \text{sulfuric acid}$$

The symbol "Pt" over the arrow means "in the presence of platinum." In Sec. 7.9 we noted that these oxides and acids are present in acid rains from polluted air.

Some of the oxides of metalloids are soluble in water, and some are not.

If such an oxide is soluble, the solution turns out to be neither strongly acid nor strongly basic:

$$CO_2(g) + H_2O(l) \longrightarrow H_2CO_3(aq)$$

H_2CO_3, called carbonic acid, is present in all "carbonated" beverages. Obviously, it is a very weak acid or we wouldn't drink it! Thus metalloids show *physical and chemical* properties that are intermediate between those of metals and nonmetals.

In terms of chemical properties:

1 Metals are elements whose oxides are basic.
2 Nonmetals are elements whose oxides are acidic.
3 Metalloids are elements whose oxides are very nearly neutral.

10.9 The Process of Neutralization

If we drop a piece of magnesium metal and a strip of litmus paper into a test tube and add an acid, we observe that bubbles of gas form on the surface of the metal and rise up through the solution. We note also that the litmus is red. If we slowly add a base to the acid, the evolution of gas gradually slows down and eventually stops. When we have added just enough base to stop the evolution of gas, the litmus paper has become lavender (a color intermediate between its red color in acid and its blue color in base). Evidently, the base has somehow destroyed the acid. If we put a piece of litmus paper in a test tube containing a base and gradually add acid, the blue color of the litmus becomes lavender, and the solution no longer feels slippery. The acid has somehow destroyed the base. To this interaction between acids and bases we give the name *neutralization*. A *neutral solution* is one which is neither acidic nor basic.

neutralization
neutral solution

suspension

When we suffer from "sour stomach" (excess acid in the stomach), we take milk of magnesia, a *suspension* of magnesium hydroxide, $Mg(OH)_2$, in water. This reacts with the excess hydrochloric acid of the stomach to neutralize it, relieving our discomfort.

$$Mg(OH)_2(aq) + 2HCl(aq) \longrightarrow MgCl_2(aq) + 2H_2O$$

If we gradually add sodium hydroxide solution to a solution of sulfuric acid until a piece of litmus paper in it turns lavender, this neutral solution tastes salty. From similar experiments we note that any acid added to any base produces a salty substance and water. We can summarize the process of neutralization in the statement

$$\text{Acid} + \text{base} \longrightarrow \text{salt} + \text{water}$$

With sulfuric acid, H_2SO_4, and sodium hydroxide, NaOH, the salt produced is sodium sulfate, Na_2SO_4:

$$H_2SO_4(aq) + 2NaOH(aq) \longrightarrow Na_2SO_4(aq) + 2H_2O$$

salt

We may define a *salt* as the product, other than water, of neutralizing an acid by a base. If the salt solution is evaporated carefully, the solid salt is obtained.

10.10 Reactions between Anhydrides

The oxides of many elements are not soluble in water, so that classifying them by the interaction of their oxides with water is of limited usefulness. But the concept of the acidic or basic nature of the oxides is very useful to chemists studying metallurgy and ceramics in furnaces at high temperatures. Just as basic solutions of metallic oxides in water can be neutralized by acidic solutions of nonmetallic oxides to form salts and water, so an anhydrous "neutralization" of a basic metallic oxide by an acidic nonmetallic oxide can take place in a furnace at high temperature. For example, large amounts of sand (silicon dioxide, SiO_2) are present as an impurity in iron ores. This impurity can be removed by making it combine with calcium oxide, CaO, added

flux

to the blast furnace as a *flux* to form the salt calcium silicate, $CaSiO_3$. This has a comparatively low melting point and can be run off as a molten, glassy

slag

slag. The strongly basic oxide CaO has a greater tendency to unite with the impurity SiO_2 than do the less basic iron oxides. When the CaO combines with the SiO_2 in the iron ore, the iron oxides are freed from the sandy impurities.

Neutralization of a basic anhydride by an acidic anhydride may be illustrated as follows:

Basic anhydride		acidic anhydride		salt	
CaO(s)	+	$SiO_2(s)$	$\longrightarrow$	$CaSiO_3(l)$	calcium silicate
CaO(s)	+	$SO_2(g)$	$\longrightarrow$	$CaSO_3(s)$	calcium sulfite
FeO(s)	+	$SO_3(g)$	$\longrightarrow$	$FeSO_4(s)$	iron sulfate
$Na_2O(s)$	+	$CO_2(g)$	$\longrightarrow$	$Na_2CO_3(s)$	sodium carbonate

10.11 Group IA Elements: The Alkali Metals

Typical reactions of the elements in group IA—the alkali metals Li, Na, K, Rb, and Cs—are given below. (Francium, Fr, is a manmade element that is very radioactive, and its chemical reactions have not been studied extensively.) In each equation, the symbol M′ stands for any alkali metal.

$$4M' + O_2 \longrightarrow 2M_2'O \qquad\qquad (10.1)$$
$$2M' + 2H_2O \longrightarrow 2M'OH + H_2 \qquad\qquad (10.2)$$
$$2M' + H_2 \longrightarrow 2M'H \qquad\qquad (10.3)$$
$$2M' + X_2' \longrightarrow 2M'X' \qquad \text{where X′ is any halogen} \qquad (10.4)$$
$$M_2'O + H_2O \longrightarrow 2M'OH \qquad\qquad (10.5)$$

reactivity

metallicity

In all these reactions, Li is the least reactive and Cs the most; *reactivity* (intensity of reaction) increases downward in the column in Table 10.3. For instance, when small pieces of the alkali metals are placed in cold water, Li barely fizzes in the production of gaseous hydrogen, Na effervesces moderately, K reacts so vigorously that the heat produced ignites the hydrogen, Rb reacts violently, and Cs explosively. This characteristically metallic behavior of displacing hydrogen from H_2O increases in activity downward in the column. Another way of saying this is that the *metallicity* of the elements in group IA *increases downward*. Lithium is the least active alkali metal and cesium the most.

10.12 Group IIA Elements: The Alkaline Earth Metals

Typical reactions of the elements in group IIA—the alkaline earth metals Be, Mg, Ca, Sr, Ba, Ra—are given below. M″ stands for any alkaline earth metal.

$$2M'' + O_2 \longrightarrow 2M''O \qquad\qquad (10.6)$$
$$M'' + 2H_2O \longrightarrow M''(OH)_2 + H_2 \qquad\qquad (10.7)$$
$$M'' + H_2 \longrightarrow M''H_2 \qquad \text{except Be and Mg} \qquad (10.8)$$
$$M'' + X_2' \longrightarrow M''X_2' \qquad\qquad (10.9)$$
$$M''O + H_2O \longrightarrow M''(OH)_2 \qquad \text{except Be} \qquad (10.10)$$

Again we find that metallicity (the intensity of metallic reactivity) increases down the column in Table 10.3. BeO does not react with water, MgO reacts slightly to form a weak base, $Ca(OH)_2$ is a moderately strong hydroxide, and $Ba(OH)_2$ quite strong. Because of its radioactivity, the basicity of $Ra(OH)_2$ is not well known. Beryllium is the least metallic of the alkaline earth metals and barium the most (excluding radium).

10.13 Group VIIA Elements: The Halogens

Typical reactions of the elements in group VIIA—the halogens F, Cl, Br, I (At is too radioactive to be handled with ease)—are

$$2M' + X'_2 \longrightarrow 2M'X' \tag{10.11}$$

where M' is any alkali metal and X' is any halogen;

$$M'' + X'_2 \longrightarrow M''X'_2 \tag{10.12}$$

where M'' is any alkaline earth metal;

$$H_2 + X'_2 \longrightarrow 2HX' \tag{10.13}$$
$$2X'_2 + 2H_2O \longrightarrow 4HX' + O_2 \tag{10.14}$$

halide

nonmetallicity

In this family of nonmetals we find that the intensity of reactivity increases as we go up the column. Fluorine is the most active and iodine the least. Thus when a halogen combines with hydrogen [Eq. (10.13)], the reaction is most violent with F_2 and least violent with I_2. When a halogen reacts with water to form the hydrogen *halide* and to liberate oxygen [Eq. (10.14)], fluorine liberates oxygen most rapidly and iodine least so. Another way of saying this is that the *nonmetallicity* of the halogens *increases upward*. We note that this statement is roughly equivalent to saying that the metallicity of the halogens *increases downward*. This brings out the similarity with groups IA and IIA, in which metallicity increases downward. It is interesting to note that the appearance of the halogens also becomes more metallic downward; iodine is a crystalline solid at room temperature with a slightly shiny, metallic look.

10.14 Group VIA Elements: The Oxygen Family

Typical reactions of the elements in group VIA—O, S, Se, Te (Po is too radioactive to deal with comfortably)—are

$$2M' + X'' \longrightarrow M'_2X'' \tag{10.15}$$

where M' is any alkali metal and X'' is a group VIA element;

$$M'' + X'' \longrightarrow M''X'' \tag{10.16}$$

where M'' is any alkaline earth metal; and

$$H_2 + X'' \longrightarrow H_2X'' \tag{10.17}$$
$$X'' + O_2 \longrightarrow X''O_2 \tag{10.18}$$

In this family of nonmetals we find that the intensity of reactivity increases upward in the column; oxygen is the most active and tellurium the least. Thus when these elements combine with hydrogen [Eq. (10.17)], water (hydrogen oxide) is formed most vigorously and hydrogen telluride least so. Also, sulfuric acid, H_2SO_4, is stronger than selenic acid, H_2SeO_4. As in group VIIA, nonmetallicity of the group VIA elements *increases upward*, and metallicity *increases downward*. The elements selenium and tellurium have a decidedly metallic appearance.

10.15 The Elements in Groups IIIA, IVA, and VA

The elements of group IIIA—the boron group B, Al, Ga, In, Tl—form oxides and halides thus:

$$4E + 3O_2 \longrightarrow 2E_2O_3 \qquad\qquad (10.19)$$
$$2E + 3X'_2 \longrightarrow 2EX'_3 \qquad\qquad (10.20)$$

Those in group IVA—the carbon group C, Si, Ge, Sn, Pb—form oxides, halides, and hydrides:

$$E + O_2 \longrightarrow EO_2 \qquad\qquad (10.21)$$
$$E + 2X'_2 \longrightarrow EX'_4 \qquad\qquad (10.22)$$
$$E + 2H_2 \longrightarrow EH_4 \qquad\qquad (10.23)$$

Those in group VA—the nitrogen group N, P, As, Sb, Bi—form oxides having the general formula E_2O_5 and chlorides with formula ECl_5, except that N does not form this chloride. Nitric acid, HNO_3, is a stronger acid than phosphoric acid, HPO_3, which in turn is stronger than arsenic acid, $HAsO_3$.

The topmost element in each of these three groups (B, C, and N) has nonmetallic properties; in each case the oxide of the element dissolves in water to produce an acid. The bottom element in each group is decidedly metallic; thallium, lead, and bismuth are typical metals. Again we see that *metallicity increases downward* in each group, and *nonmetallicity increases upward*. Somewhere in each group there is a change from predominantly nonmetallic properties to predominantly metallic ones. The stairstep line through these groups in Table 10.3 marks this transition from metal to nonmetal in each group. The elements adjacent to this line are considered metalloids, because they have some metallic and some nonmetallic properties. Some trends in physical properties in some families are shown in Table 10.4.

TABLE 10.4 *Trends in Some Properties of Elements in a Family*

Specific heats of the alkali metals, cal/g		Densities of elements in group IVA, g/cm³		Melting points of the halogens, °C		Boiling points of the noble gases, °C	
Li	0.837	C	2.25	F	−223	He	−268.9
Na	0.297	Si	2.42	Cl	−101.6	Ne	−245.9
K	0.192	Ge	5.35	Br	−7.3	Ar	−185.7
Rb	0.079	Sn	7.28	I	113.6	Kr	−152.9
Cs	0.048	Sb	11.34	At	?	Xe	−140.0
Fr	?					Rn	−71.0

10.16 General Trends in Properties of Elements from Top to Bottom and from Left to Right in the Periodic Table

We have already noted that metallicity (metallic character) increases in intensity from top to bottom in each group (except in the zero group, where chemical reactivity of any kind is very weak). What about trends in the horizontal periods?

From the observations reported in Secs. 10.11 through 10.15, we see that in period 2 the oxide of Li is strongly basic (Li is strongly metallic), the oxide of Be is insoluble so that Be is not strongly metallic in behavior, the oxide of B dissolves to give a very weak acid, and that of C dissolves to form an acid only very slightly stronger. However, the oxides of N and F dissolve to form strong acids; that is, N and F have strongly nonmetallic characteristics. In the period as a whole we see, then, that *metallicity decreases toward the right*, and that *nonmetallicity increases toward the right*.

In period 3, the oxide of Na dissolves to give a strong base, that of Mg dissolves to give a weak base, the oxides of Al and Si are insoluble (neither base- nor acid-producing), the oxide of P dissolves to give a moderately strong acid, that of S dissolves to give a still stronger acid, and that of Cl to give a very strong acid. Again, in the period as a whole we see that *metallicity decreases toward the right*, and that *nonmetallicity increases toward the right*.

We may sum up a great deal of knowledge about the properties of the elements by noting that *metallicity increases* as we move from *right to left and downward* through the table. *Nonmetallicity increases* as we move from *left to right and upward* through the table. Therefore, the most active metal is in the lower left-hand corner of the table and the most active nonmetal is in the upper right-hand corner (omitting the zero group).

10.17 Main Divisions of the Periodic Table

The elements are classified into three main categories (Table 10.5):

1 the representative elements
2 the short transition series
3 the long transition series

representative
elements

Some of the chemical properties of the *representative elements* have been described very briefly in Secs. 10.8 through 10.15. Most of the environmental chemical processes with which we are concerned in this book are like those we have noted in the behavior of the representative elements.

short transition
series

Many of the metals in the *short transition series* (chromium, iron, nickel, copper, zinc, silver, cadmium, platinum, gold, and mercury) are extensively used in everyday articles of commerce (automobiles, household appliances, electrical equipment, building materials, etc.). These articles contribute to

TABLE 10.5 *Main Subdivisions of the Periodic Table*

Group / Period	IA	IIA	IIIB	IVB	VB	VIB	VIIB	VIII			IB	IIB	IIIA	IVA	VA	VIA	VIIA	0
1	H																	He
2	Li	Be											B	C	N	O	F	Ne
3	Na	Mg											Al	Si	P	S	Cl	Ar
4	K	Ca	Sc	Ti	V	Cr	Mn	Fe	Co	Ni	Cu	Zn	Ga	Ge	As	Se	Br	Kr
5	Rb	Sr	Y	Zr	Nb	Mo	Tc	Ru	Rh	Pd	Ag	Cd	In	Sn	Sb	Te	I	Xe
6	Cs	Ba	See* below	Hf	Ta	W	Re	Os	In	Pt	Au	Hg	Tl	Pb	Bi	Po	At	Rn
7	Fr	Ra	See** below	Ku	Ha													

Representative Elements

Short transition series

Long transition series

	La	Ce	Pr	Nd	Pm	Sm	Eu	Gd	Tb	Dy	Ho	Er	Tm	Yb	Lu
*Lanthanide series	La	Ce	Pr	Nd	Pm	Sm	Eu	Gd	Tb	Dy	Ho	Er	Tm	Yb	Lu
**Actinide series	Ac	Th	Pa	U	Np	Pu	Am	Cm	Bk	Cf	Es	Fm	Md	No	Lw

the problems of recycling solid wastes to avoid environmental clutter and to reduce demands on the ores from which these metals are extracted. Today, the only transition elements whose compounds are polluting our water supplies are iron, cadmium, and mercury. Their chemical properties will be studied in connection with methods for reducing such pollution.

The metals in the lanthanide *long transition series* are very rare and are very little used in our society. Of the actinide metals in the long transition series, we use only thorium, uranium, and plutonium. The environmental problems associated with the use of these elements as nuclear fuels were considered in Chap. 9. The remaining elements in the actinide series are manmade and of primarily theoretical interest.

Summary

When the chemical elements are listed in order of increasing atomic number, elements with similar sets of properties appear periodically in the list. This list becomes more useful if a periodic table is made with elements having similar properties arranged in vertical columns. These columns then contain families or groups of similar elements.

The form of the periodic table currently favored by most chemists is that shown on the color insert in this book. The type formulas for the oxides and hydrides of the representative elements (groups IA through VIIA) are shown at the top of each of the columns in Table 10.3. The oxides and hydrides of the short transition series of the elements have many different formulas. The oxides of the long transition series elements also have quite variable formulas.

The elements in group IA are called the alkali metals. They react violently with oxygen to produce oxides that dissolve in water to yield strongly basic solutions containing hydroxides with the formula EOH. They also react vigorously with chlorine to form chlorides with the formula ECl.

The elements in group VIIA are called the halogens. They react vigorously with hydrogen to form compounds of the type HX', with any alkali metal to form compounds of the type M'X', and with any alkaline earth metal to form compounds of the type $M''X_2'$, where X' stands for any halogen. The halogens are nonmetallic elements whose oxides dissolve in water to form strong acids.

The elements in group VIA are called the oxygen group. They react with hydrogen to form compounds of the type H_2X'', where X'' stands for any element in the group. These elements are mostly nonmetallic, and their oxides form acid solutions.

The elements in group VA are called the nitrogen group. The upper members of the group form oxides that dissolve in water to give strong acids, but the lower members form less strong acids. The upper members are predominantly nonmetallic, but the lower ones have some metallic character.

The oxides of the upper elements in groups IIIA and IVA form mildly acidic solutions in water, whereas the oxides of the lower elements yield mildly basic solutions. The upper elements are predominantly nonmetallic in character, but the lower ones are definitely metallic.

A stairstep line traverses groups IIIA through VIA. Since the elements along this line are neither strongly metallic nor strongly nonmetallic, they are called metalloids.

As we compare the properties of the elements from top to bottom in a given group, we find that metallicity increases downward. Similarly, the elements in a given period are increasingly nonmetallic from left to right. The most metallic of the elements is francium, at the lower left of the periodic table. The most nonmetallic of the elements is fluorine, at the upper right (not counting the family of noble gases, which are very unreactive chemically).

Any acid will react with any base to form a salt and water. This process is called neutralization. Acidity is associated with nonmetallic oxides, basicity with metallic oxides. Acidic anhydrides (oxides of nonmetals) can neutralize basic anhydrides (oxides of metals) when they are melted together. The product is saltlike.

New Terms and Concepts

ACID: A substance that has a sour taste, turns litmus extract red, and effervesces when added to magnesium.

ACIDIC ANHYDRIDE: A substance that produces an acid when added to water.

ACIDIC OXIDE: An oxide that dissolves in water to produce an acid or that fuses with a metallic oxide to form a slag.

ACIDIC SOLUTION: A solution that turns litmus red.

ACTINIDE SERIES: The elements with atomic numbers 89 to 103.

ALKALI METAL: Lithium, sodium, potassium, rubidium, cesium, francium.

ALKALINE EARTH METAL: Beryllium, magnesium, calcium, strontium, barium, radium.

AQUEOUS: Relating to or resembling water.

BASE: A substance that has a bitter taste, turns litmus extract blue, and is slippery to the touch.

BASIC ANHYDRIDE: A substance that produces a base when added to water.

BASIC OXIDE: An oxide that dissolves in water to produce a base or that fuses with a nonmetallic oxide to form a slag.

BASIC SOLUTION: A solution that turns litmus blue.

CHLORIDE: A compound composed of chlorine and another element.

FLUX: A substance used to promote melting of minerals.

FLY ASH: Fine solid particles of noncombustible ash carried out of the bed of a solid fuel by the draft in a furnace.

HALIDE: A compound of a halogen with another element.

HALOGEN: Fluorine, chlorine, bromine, iodine, astatine.

HYDRATION: The combination of water with another substance.

HYDRIDE: A compound composed of hydrogen and another element.

HYDROXIDE: A compound containing the atom of a metal and one or more OH groups.

LANTHANIDE SERIES: The elements with atomic numbers 57 to 71.

LITMUS: A coloring matter from the litmus plant that turns red in acid solutions and blue in basic solutions.

LONG TRANSITION SERIES: Another name for the lanthanide and actinide series.

METALLICITY: Intensity of reactivity as a metal.

METALLOID: An element with properties intermediate between those of typical metals and nonmetals.

NEUTRAL SOLUTION: A solution which is neither basic nor acidic.

NEUTRALIZATION: The reaction of an acid and a base to produce a neutral solution.

NOBLE GAS: Helium, neon, argon, krypton, xenon, radon.

NONMETALLICITY: Intensity of reactivity as a nonmetal.

OXYGEN FAMILY: Oxygen, sulfur, selenium, tellurium, polonium.

PERIOD: A horizontal row in the periodic table.

PERIODIC LAW: When the chemical elements are arranged in order of increasing atomic number, elements with similar sets of properties appear periodically in the list.

RARE GAS: Helium, neon, argon, krypton, xenon, radon.

REACTIVITY: Intensity of reaction.

REPRESENTATIVE ELEMENTS: The light elements of the two short periods (Li through Ar) plus the other members of the families of these elements.

SALT: The compound other than water formed by neutralization.

SHORT TRANSITION SERIES: The elements with atomic numbers 21 to 30, 39 to 48, and 72 to 80.

SLAG: The glassy waste produced during the extraction of a metal from its ore in a furnace.

SUSPENSION: A heterogeneous mixture of a liquid with an insoluble solid or a gas with particles of a solid or liquid.

TRIAD: A group of three elements with closely similar properties.

ZERO GROUP: Helium, neon, argon, krypton, xenon, radon.

Testing Yourself

10.1 What is meant by a group of elements? A period? Give illustrations for your answers.

10.2 How did Mendeleev know where he should leave blank spaces in his periodic table for elements that were as yet undiscovered?

10.3 Give some reasons for considering the elements Li, Na, K, Rb, and Cs to be members of one family.

10.4 $^{90}_{38}Sr$ is very radioactive and thus detrimental to human health. Milk and bones contain more calcium than other tissues of the animal body. Why are babies particularly subject to damage from $^{90}_{38}Sr$ in the environment?

10.5 Name the members of the halogen family. List some of their common characteristics.

10.6 What are the characteristics of metallic elements? Of nonmetallic elements? Of metalloids?

10.7 Write balanced chemical equations to show what products are formed when the following elements are burned in oxygen: aluminum, barium, cesium, selenium, silicon.

10.8 The oxides of barium, cesium, francium, lithium, and radium are soluble in water. Write balanced chemical equations to show the formula of the acid or base produced in each case.

10.9 Would you expect water solutions of the oxides of tellurium and iodine to be acidic, basic, or neutral?

10.10 What changes in chemical properties of the elements does one observe when proceeding from left to right in the periodic table? From top to bottom?

The Wankel engine, because it has many fewer moving parts, makes efficient use of energy and consequently is less polluting. Benzene, shown in the schematic representation, exhibits one common hydrocarbon structure.

11. Energy Production and Its Impact on the Environment

In the preceding ten chapters we have concentrated on changes in matter and on the ways that mental models can help us understand the invisible fundamental processes that cause the changes in appearance that we see. Though we have mentioned the role of energy in some of these changes, we have focused our main attention on the matter involved. In this chapter we shall look at the other side of the coin and consider the importance of changes in energy and the impact these changes have upon the world we live in.

The kind of life a person leads today depends primarily on the amount of energy available to produce the necessities that keep him alive and the luxuries that make his life pleasanter. A person in a primitive culture living close to the land makes few demands for energy. Manual labor supplies him with the food he eats, the clothes he wears, and the shelter he lives in. The decoration of these necessities to make them more esthetically pleasing is largely manmade.

When man moves into cities, more energy is required to meet his needs. Water has to be piped and pumped into his homes. Waste products have to be hauled away. Foods, fibers, and building materials have to be brought to him from fields, forests, brick kilns, and quarries. It takes extra energy to do all this transporting, but so long as preparing food, making clothes, and building houses are largely a family affair, there is no great change in life-style.

But the availability of energy leads to the invention of machines, and the use of machines completely transforms man's life. Now huge amounts of energy are required to heat his homes, factories, and offices; to run the mills that process his foods and make his paper; to run the looms that weave his fabrics; to power the machines that make his railroads, ships, planes, cars, trucks, buses, tractors; to operate the smelters that extract the metals used in these machines; to supply the chemical plants that make the fuels, plastics, rubber, alloys, drugs, and medicines he uses; and to run the power plants that feed electric energy into the network of transmission lines that hold the cities in their web. The flow of energy through a culture is an index of its standard of living.

In this chapter we shall see how energy is produced in our country, how this production pollutes our air, water, and land, and how we can minimize these undesirable side effects.

11.1 Energy and the Gross National Product

gross national
product (GNP)

The annual *gross national product* (*GNP*) of a country is the total value of all the goods processed in the country and all the services rendered to its inhabitants in 1 year. Miners who dig ores out of the earth, lumbermen who cut down forests, farmers who raise crops, mill workers who produce steel, lumber, paper, and flour, laborers and managers who run factories that make cars, washing machines, and clothing, businessmen who distribute goods, servicemen who repair television sets—all are contributing to the GNP. The greater the production and consumption of things, the greater the GNP. Teachers, preachers, doctors, nurses, lawyers, and other professional workers who deal primarily with services to people instead of making things for them contribute to the service sector in the GNP. The greater the value of the services they render, the greater the GNP. The GNP is one way of measuring the standard of living in a given country.

As noted in Fig. 11.1, the annual GNP per person in a country is roughly proportional to the annual consumption of energy per person. As the GNP goes up, so does the consumption of energy. A unit commonly used for the measurement of energy in the world of engineering and technology is the *British thermal unit* (*Btu*). This is the amount of heat required to raise the temperature of one pound of water one degree Fahrenheit: 1 Btu equals 252 cal.

British thermal
unit (Btu)

Each person in India uses about 3 million Btu/yr. Each person in the United States uses 180 million, or 60 times as much. Each American puts as great a drain on the world's resources of energy as 60 Indians. The United States,

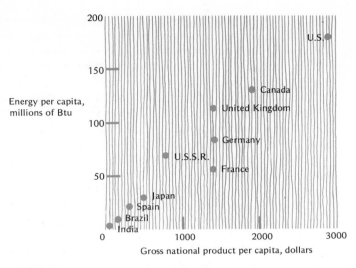

FIGURE 11.1 The relation between annual energy consumption per capita and gross national product per capita

with 6 percent of the world's population, uses 35 percent of the world's energy.

11.2 The Energy Problem in the United States

"A crisis exists right now. For the next three decades we will be in a race for our lives to meet our energy needs." These words of Federal Power Commissioner John A. Carver, Jr., in December of 1969 epitomize the energy problem we face in the United States. Blackouts due to power failures and "brownouts" necessitated by overloads on power plants underscore the commissioner's words. He pointed out that we need to build the equivalent of 670 more Hoover Dams or 1125 more Hanford (Washington) Nuclear Plants in the next 30 years to meet our growing needs.

One part of the energy problem arises from the fact that today we import about a third of the petroleum and petroleum products (gasoline, fuel oil, lubricants) that we consume. In 1990 we shall have to import a third of the natural gas we use. These materials come largely from "underdeveloped" countries. As the people in these nations develop their own technology, they will want to use their resources for their own needs. They will become more and more reluctant to feed our voracious appetites for more and more petroleum.

Another facet of the problem is that in 1970, 76 percent of our energy came from oil and gas, 20 percent from coal, and 4 percent from water power and nuclear power plants. From these rates and from our knowledge of the reserves we have in our 48 contiguous states, it appears that by the year 2000 we shall have used up 90 percent of our oil and 80 percent of our gas. Full exploitation of the deposits in Alaska will probably carry us another 10 years. By then, we must have built enough power stations using other fuels so that we need to depend very little on petroleum and natural gas for energy.

About a fifth of the energy we consume is used in transportation by car, truck, bus, train, and plane. Almost all this energy comes from petroleum derivatives: gasoline, diesel fuel, air-jet fuel. As we noted in Chap. 7, these fuels also produce air pollutants. Later in this chapter we shall discuss ways to minimize this pollution.

As we cut down on our consumption of oil and gas, we will have to increase our use of other energy sources. The amount of undeveloped water power in our country is negligible compared to our total demands. Therefore, we must build many nuclear power plants in the next 30 years. During the same 30 years we must step up our use of coal in power plants for generating electricity. Combustion of coal will considerably increase air pollution unless we develop more effective processes for purifying stack gases. Such processes are discussed in the next section.

11.3   How Can We Reduce Air Pollution by Particulates from Power Plants?

As noted in Sec. 7.13, particulates are the most dangerous of common pollutants. Of the 35 million tons of particles discharged into our air each year, about 9 million tons come from power plants burning coal. Removal from stack gases of particles larger than 0.1 micron is not difficult, and 99.7 percent of the mass of particulates are of this size. But this 99.7 percent of the *mass* contains only 12 percent of the total *number* of particles, and we do not yet know how to catch the other 88 percent.

Dusty gas from coal-fired boilers can be passed through cloth filters in "baghouses" and removed as we remove household dust in the bag of a vacuum cleaner. The bags are shaken periodically to recover the dust and clean the filter. Dust can also be removed from furnace gases in a "cyclone" separator like that shown in Fig. 11.2. The whirling motion of the gas-particulate mixture in such a device throws the particles to the sides of the cone, where they are collected and discharged into a storage bin. The cleaned gas passes from the bottom of the cone to the stack. Water sprayed into the whirling mixture will catch some particles too small to be removed from dry gas. In a wet scrubber, which is similar, dusty gas may be passed through a curtain of falling water as shown in Fig. 11.3. The water washes out the particulates much as rain washes the atmosphere.

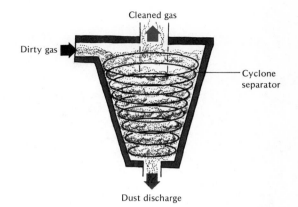

FIGURE 11.2 A cyclone separator

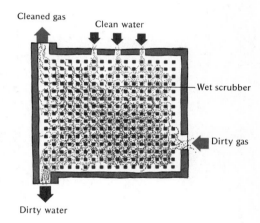

FIGURE 11.3 A wet scrubber

In an electrostatic precipitator like that shown in Fig. 11.4, the stream of dusty gas is passed across electrically charged wires. The particles pick up the electric charge from the wires and then pass metal plates that carry the opposite charge. The particles are attracted to these plates and adhere to them, the clean air passing on to the stack. The plates are vibrated periodically to remove the deposits of particles, which then drop into a storage hopper. Electrostatic precipitators can collect particles smaller than those caught by other cleaning devices, but most of those smaller than 0.1 micron escape.

The fly ash (dust particles) removed from stack gases by particulate separators is used for various purposes. It can be combined with plastics to make a material that can be molded like plastic but can also be worked like wood.

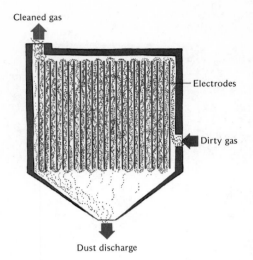

FIGURE 11.4 An electrostatic precipitator

This material is being used to make furniture. Fly ash is also useful for making bricks that are much lighter and easier to handle than bricks made from clay. It is also used as filler in roofing, in insulating materials, and for road surfacing. As markets for fly ash develop, they will help to pay the cost of reducing pollution by particulates.

All the methods described above are also used to control particulates in processes other than steam generation. Baghouses are used to clean gases from kilns (furnaces) and grinders used in the production of portland cement (the binding ingredient in concrete), from iron and steel furnaces, and from glass furnaces. Cyclone separators are used to precipitate dust generated by rock and ore crushing, petroleum refining, and wood-working mills. Wet scrubbers are used in crushing and grinding plants, iron foundry furnaces, and kilns for making lime from limestone. Electrostatic precipitators are used in incinerators, smelters, and steel and paper mills.

11.4 Controlling Sulfur Oxides (SO$_x$) from Power Plants

The SO$_x$ that enters the atmosphere from power plants comes from those fueled by coal or oil containing sulfur. We do not have enough sulfur-free fuels to meet our needs, and so we must learn how to reduce pollution from sulfur-bearing fuels. This may be done by (1) removing the SO$_x$ from the stack gases, (2) removing the sulfur from the fuel before burning it, or (3) burning the fuel by a process that does not produce SO$_x$. All these alternatives are being studied intensively by scientists and engineers. To date, only the first is being used in commercial installations.

Most processes for removing SO_x from stack gases make use of the fact that these oxides are acid anhydrides (Sec. 10.8). When stack gas is brought into intimate contact with the basic anhydride magnesium oxide, MgO, the sulfur oxides are removed:

$$SO_2(g) + MgO(s) \longrightarrow MgSO_3(s) \quad \text{magnesium sulfite (solid)}$$
$$SO_3(g) + MgO(s) \longrightarrow MgSO_4(s) \quad \text{magnesium sulfate (solid)}$$

If powdered dry limestone ($CaCO_3$) is injected into the flue gas, the reactions are

$$SO_2(g) + CaCO_3(s) \longrightarrow CaSO_3(s) + CO_2(g)$$
$$SO_3(g) + CaCO_3(s) \longrightarrow CaSO_4(s) + CO_2(g)$$

A mixture of molten carbonates of the alkali metals can be sprayed into stack gases to remove the sulfur oxides:

$$SO_2(g) + K_2CO_3(l) \longrightarrow K_2SO_3(s) + CO_2(g)$$
$$SO_3(g) + K_2CO_3(l) \longrightarrow K_2SO_4(s) + CO_2(g)$$

Similar reactions take place with other alkali metal carbonates in the melt.

If a water solution of alkali metal carbonates is used instead of a molten mixture, the chemical reactions are the same as those just noted. As the spray is carried along by the hot stack gases, the water is evaporated, and solid alkali metal sulfite and sulfate particles are produced. In all the processes just described, the solids are removed from the stack gases by one of the methods noted in Sec. 11.3.

slurry
A *slurry* (a dilute suspension of solid particles in water) of calcium hydroxide, $Ca(OH)_2$, may be used in a wet scrubber to remove the sulfur oxides:

$$SO_2(g) + Ca(OH)_2 \longrightarrow H_2O(l) + CaSO_3(s)$$
$$SO_3(g) + Ca(OH)_2 \longrightarrow H_2O(l) + CaSO_4(s)$$

The calcium sulfite and sulfate are carried off in the liquid draining from the wet scrubber. In one experimental plant, a spray of seawater was found to be effective in removing SO_2 by a wet scrubber. None of the preceding processes has yet been used in a commercial power plant long enough to have proved itself, but several are being used experimentally to test their effectiveness.

If the sulfur in coal is in the form of metal sulfide minerals (of which iron sulfide, FeS, is the most common), it can be removed by crushing the coal and blowing it through a cyclone separator. The metal sulfides are much denser than the coal and are thrown to the sides of the cone, while the less dense coal particles are carried through by the high-speed blast of air. If the crushed coal is washed through a trough with water at high speed, the dense sulfides settle out, while the lighter coal is carried through.

Sulfur that is combined with carbon and hydrogen in coal is much more difficult to remove. If the coal is heated to a high temperature in the absence of air, the sulfur is driven off as hydrogen sulfide gas, and the carbon is left as a solid called char. Essentially pure carbon, it is a very clean fuel.

Sulfur compounds in fuel oil can be converted to H_2S by heating the oil to a high temperature in the presence of hydrogen gas and a catalyst. It is also possible to remove the sulfur compounds in petroleum by adding a substance that will combine selectively with these compounds to form water-soluble products. These can then be washed out of the oil with water. This process is still in the research stage, and the nature of the special additive is a secret.

So far only one process has been suggested for burning sulfur-bearing coal without producing SO_x. The coal is dissolved in molten iron and is burned at the surface of the melt with a limited supply of oxygen. The hot carbon monoxide gas so produced is delivered to a steam boiler where more oxygen is added to complete the combustion to CO_2. Under these conditions, the sulfur in the coal is retained in the molten iron. It is subsequently removed from the iron by forming a slag. It will take several years to develop this process for commercial use.

11.5 Manmade Fuels

Natural gas is the cleanest fuel we have; it can be burned in steam boilers or gas turbines to drive electric generators without producing any pollution problem. Desulfurized fuel oil is almost as clean, but coal is dirty. Unfortunately, it now appears that we will run out of gas in a few years and out of oil in a few more. Because we have vast supplies of coal, we need to find ways to convert it into clean-burning fuel oil and gas.

If coal is pyrolyzed (heated in a furnace with a very carefully controlled amount of air), it produces hydrogen, carbon monoxide, oil, and ash. The mixed gases and the oil are very clean fuels, free of sulfur. The ash contains all the sulfur originally in the coal. An experimental plant to test this method is now being developed. As the process is perfected, it will be adapted for commercial power plants.

Coal may be completely gasified by another process. If powdered coal is injected into a stream of hot gases consisting of oxygen and steam, the products are methane, hydrogen, and carbon monoxide. Small amounts of tarry mixtures, dust, sulfur compounds, and carbon dioxide are also produced. When these are removed by the processes described for scrubbing stack gases, a very clean fuel gas remains. The heating value of this gas may be enhanced by causing the CO and H_2 to react:

$$CO + 3H_2 \longrightarrow CH_4 + H_2O$$

The water is condensed out by cooling. By this process the heating value of the gas is increased from 500 to more than 900 Btu/ft^3, which is about the heating value of natural gas.

In Colorado, Wyoming, and Montana there are mountainous deposits of shale that contains compounds of carbon and hydrogen. When this rock is pulverized and heated in a furnace with no air, the *carbonaceous* (carbon-bearing) material is driven off as a gas. Some of this can be condensed to a liquid much like crude petroleum. The remainder is a combustible gas that can be used to heat the shale furnace. About 25 to 30 gal crude oil can be recovered from a ton of rock. If the shale is heated with hydrogen gas at 1200°F and under a pressure of 2000 lb/in.^{2}, all the carbonaceous material is converted to gas. This gas is free of sulfur and is a substitute for natural gas. One of the most difficult problems in producing oil from shale is disposing of the processed rock. It becomes more porous by heating and increases in volume. It will have to be disposed of by filling in whole valleys in the mountains where the oil shale is found. This prospect makes the ruin created by strip mining seem insignificant. The problem of waste disposal may be so acute that the carbonaceous material will have to be extracted by heating the oil shale underground. This could be done by boring holes through the deposits and using combustion in these channels to produce enough heat to drive the oil out of the rest of the deposit. It seems likely that crude oil will be produced from coal long before significant quantities of oil are extracted from shale. At the end of World War II, Germany was making 40 percent of her oil and gasoline from coal.

Along the Athabasca River 300 mi north of Edmonton in Alberta, Canada, there are huge deposits of sand carrying a carbonaceous material that can be pyrolyzed to produce a crude oil readily refinable into gasoline, diesel oil, and other commonly used petroleum products. Experts estimate that 25 years from now the United States will be importing 2.5 to 3 million barrels/day of crude oil from the Athabasca tar sands.

11.6   Reducing Pollutants from Internal Combustion Engines

As noted earlier, about a fifth of the energy we consume in our country is used for transportation by car, truck, bus, train, plane, and boat. With the exception of a few electric railway trains and fewer electric trucks and buses, all this transport is driven by internal combustion engines. Because so much transportation of people and goods takes place in cities, air pollution is particularly acute there. In open country, natural processes remove most air pollutants (except very fine particulates) about as quickly as they are produced. Engines produce negligible amounts of SO_x, but large amounts of NO_x, hydrocarbons, carbon monoxide, and particulates.

The exhaust gases from our 100 million cars are the prime offenders. The gasoline we use is composed of hydrocarbons with 4 to 13 carbon and 10 to 18 hydrogen atoms per molecule. When a hydrocarbon is burned with plenty of air, it produces CO_2 and H_2O. For the complete burning of octane, a typical component of gasoline, the equation would be

$$2C_8H_{18} + 25O_2 \longrightarrow 16CO_2 + 18H_2O$$

In the automobile engine the gasoline is burned in a limited supply of air, so that considerable amounts of unburned hydrocarbons and carbon monoxide are produced. Because the temperature of combustion is high, some of the nitrogen of air unites with the oxygen to form NO_x. These undesirable products of combustion come out the tailpipe of the car along with the harmless CO_2 and H_2O. Exhaust gases have to be treated to remove the pollutants.

One way of purifying exhaust gases is to pass them through a catalyst that facilitates the conversion of NO_x to nitrogen and water. Passage through a second catalyst burns carbon monoxide to CO_2 and hydrocarbons to CO_2 and H_2O. Automobile manufacturers and petroleum refiners are working on these "afterburners" or "converters," which will probably be placed between the exhaust manifold and the muffler in most cars. Similar devices are being developed to purify exhaust gases from diesel engines. Many different materials are being tried out as catalysts—precious metals like platinum, and oxides of more common metals like iron, nickel, zinc, vanadium, and cobalt. The exact composition of the catalysts and the manner in which they are fabricated are trade secrets until the devices are patented. Most of them are much more effective with nonleaded gasoline, so that their use on many cars and trucks will encourage the production of lead-free fuels.

An automobile engine that burns a gas as fuel produces fewer pollutants than one that burns gasoline. Natural gas—which is mostly methane, CH_4, and ethane, C_2H_6—is an excellent fuel for this purpose. Natural gas under a pressure of about 2000 lb/in.^{2} at ordinary temperatures condenses to a liquid. This can be conveniently stored under pressure in steel cylinders, and is called *liquefied natural gas*, LNG. Propane, C_3H_8, and butane, C_4H_{10}, are contained in natural gas and are produced by the refining of petroleum to make gasoline. These two compounds are easily liquefied at a pressure much less than that required to produce LNG. A mixture of these liquids, called *liquefied petroleum gases*, or LPG, can also be stored in steel cylinders. It is an excellent fuel for the clean running of gasoline engines. When either LNG or LPG is released from its container, it changes to a gas that can be fed directly into the conventional auto engine.

liquefied natural gas (LNG)

liquefied petroleum gases (LPG)

In order to promote the use of gas as engine fuel, the United States government in 1972 ran 2000 of its vehicles on LNG. The state of California used LNG in several hundred of its cars and pickup trucks. Fifteen hundred Chicago buses were powered by LPG. Many fleets of taxis and trucks in London, New York, and Tokyo are using LNG or LPG. In 1972 about 250,000 vehicles in the United States and 200,000 autos in Japan used LPG. But considerable numbers of vehicles have to use these fuels in a given area to justify establishing fuel stations that can deliver liquefied gas to a vehicle's fuel tank under pressure.

11.7 Alternatives to Present Engines

Fuel burned to produce a hot gas directed against the vanes of a wheel like a windmill gives rotary power with much less pollution than fuel burned in the conventional engine used today. This windmill-like engine is called a gas turbine. Because automobile manufacturers estimate that it will cost $7 billion to $8 billion to change over from internal combustion engines to gas turbines, there is little likelihood that pollution will be reduced by such a change in the immediate future. But we should continue to finance research in engine design in the hope that a simple engine can be developed with changeover far less costly than present estimates.

The rate at which an internal combustion engine emits pollutants depends considerably on whether it is accelerating rapidly, running at constant speed, or decelerating. Minimum pollution is produced when the engine is running at a constant speed. To reduce pollution arising from acceleration or deceleration the engine may be coupled to a flywheel that will absorb energy when the car is decelerating and release it when accelerating. The engine can then be run at that constant speed which produces least pollution. Preliminary studies by automotive engineers indicate that a flywheel supplement to the present engine would reduce hydrocarbon pollution by 90 percent, CO pollution by 95 percent, and NO_x pollution by 75 percent. The mechanical problems of transmitting energy from the engine to such a flywheel and from it to the wheels are formidable, but they are yielding to continuing research.

The German Wankel engine produces circular motion by explosions of an air-fuel mix which propel an off-center rotor on a shaft enclosed in a combustion chamber. In the standard reciprocating engine the exploding fuel pushes pistons up and down in separate combustion chambers. This up-and-down motion then has to be changed to rotary motion by a crankshaft. While the Wankel engine produces as many pollutants as the conventional engine, it has fewer than half as many moving parts. Consequently, it costs much less to build and to maintain in good working condition, features that minimize pollution. In 1972 Japan's Toyo Kogyo Company manufactured about 4000 Wankel-powered cars per month. General Motors invested millions of dollars in rights to produce and sell the Wankel engine. It is likely that within a decade these engines will be used on many small cars built in the United States.

For years steam engines have been studied as power plants for automobiles.

They emit far less pollutants than internal combustion engines do: 70 percent less hydrocarbons, 90 percent less CO, and 75 percent less NO_x. The drawbacks are that the traditional steam engine used in cars is more complicated than a gasoline engine, twice as heavy, costs twice as much to construct, and burns 50 percent more fuel. Still, many scientists and engineers are trying to improve the steam engine to make it competitive. The new Hinckley-Beloit-Hornbostel steam engine, with a driving rotor somewhat like that in the Wankel engine, is lighter and more compact than the conventional steam engine, and it is being considered seriously by engine makers.

Why can't we use pollution-free electric motors to drive cars and trucks? We can—when we develop more efficient storage batteries and devices to convert energy from fuels directly into electricity, thus cutting out the costly, wasteful, and pollution-prone process of burning fuels to drive electric generators. With our present knowledge we can build electric cars that move quietly and efficiently. But they can go only a few miles, and at low speeds, before their batteries have to be recharged. Intensive research is now under way to develop new kinds of storage batteries and new kinds of electrochemical cells that will use fuels for the direct production of electricity. To understand these developments, we need to know much more about the characteristics of solutions of various chemicals. These will be studied in Chap. 12.

11.8 Generating Electric Energy from Hot Springs and Geysers

In 1847 a hunter in northern California stumbled onto a region where puffs of steam were escaping from the ground, and having a vivid imagination, he named the area "the gates of hell." Today, "The Geysers" power project has harnessed the steam issuing from the gates to generate enough electricity for a city of 200,000 people. To be sure, the gates have had to be opened up a bit by drilling wells to get at the steam. The most recent well is 9029 ft deep and releases enough steam to generate electricity for 10,000 people. By 1974 "The Geysers" project will be generating enough electricity to meet the needs of another 100,000 people.

geothermal
magma
earth's mantle

Where does this *geothermal* steam come from? Geological explorations show that in some places the molten rock (*magma*) core of the earth has intruded upward into the solid rocky *mantle* above it and heats this rock to a high temperature. As underground water seeps into this layer of hot rock, it is heated to temperatures of 350 to 700°F. Because it is under high pressure, the water cannot boil but remains in the liquid state. This layer of water-filled rock is called a reservoir. If the reservoir is covered by a layer of rock (called caprock) that is impervious to the flow of water, the hot water is trapped in the reservoir. A well drilled down through the top layer of rock

earth's crust

(the *crust*) and through the caprock into the rock of the reservoir permits

The Geysers power project in northern California

the hot water to flow upward. As the pressure on this water decreases, it boils rapidly to produce steam that is released through the top of the well. The formation of a natural opening through the crust and caprock and into the reservoir enables the hot water to leak to the surface as a hot spring or geyser. The production of geothermal steam is indicated schematically in Fig. 11.5.

Geothermal steam usually carries considerable amounts of liquid water in droplet form. These droplets contain various dissolved substances including hydrogen sulfide, silica, salt, fluorides, and compounds of boron and arsenic. These impurities corrode and encrust the metal pipes used to deliver steam from the wells to the turbines. The hydrogen sulfide may be transformed into sulfuric acid in the air or in the waste water from the wells and turbines:

$$H_2S + 2O_2 \longrightarrow H_2SO_4$$

Fluorides and compounds of boron and arsenic are poisonous if present in drinking water in uncontrolled amounts. In some fields, the waste water is injected back into the ground. It is important that potential environmental pollution be evaluated before any geothermal steam is used on a commercial scale. Though the steam is free, the expense of drilling wells, piping the steam to the turbo-generators, and disposing of waste water brings the cost close to that of electricity produced from fossil fuels.

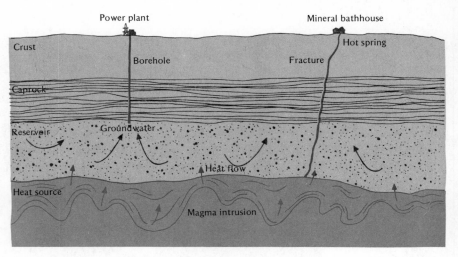

FIGURE 11.5 The generation of geothermal steam. (Figure adapted from "Power from the Earth," *Environment Magazine*, December 1971, p. 21.)

Geothermal steam and hot water are used in Iceland to heat 90 percent of the homes in the capital city of Reykjavik. The annual thermal output of Iceland's 30 volcanoes and 700 hot springs is equivalent to the burning of 7 million tons/yr of oil. Since Iceland has no fossil fuels, geothermal energy is of great importance there. Considerable amounts of electric energy are generated from geothermal steam at several places in the world today. The world production of geothermal power is summarized in Table 11.1. The output of electric power from a generating plant is conveniently measured in kilowatts (kW). One kilowatt equals approximately 1 Btu/sec. Large quantities of electric power are conveniently measured in megawatts (MW); 1 MW is 1,000,000 *watts* or 1000 kW.

watt

Supplies of geothermal steam are not inexhaustible in a given area, but by keeping a record of the decrease in flow as a given field is developed, the probable reserves can be estimated. Conservative estimates indicate that a geothermal field is likely to have a lifetime of 50 to 100 yr. A scientist of the U.S. Geological Survey has stated that if all the known hot springs and all the geothermal sedimentary basins in the world are taken into account, they provide an almost limitless source of electric power.

11.9 Energy from the Sun

When we compare the amount of energy the sun pours onto the earth with the amounts we produce from falling water, burning fuels, and splitting atoms,

224   *Energy Production and Its Impact on the Environment*

TABLE 11.1 *World Geothermal Power Production, 1971*

Country	Field	Electric capacity, MW	
		Operating	*Under construction*
Italy	Larderello	358.6	
	Mt. Amiata	25.5	
U.S.A.	The Geysers	192.0	110.0
New Zealand	Wairakei	160.0	
	Kawerau	10.0	
Japan	Matsukawa	20.0	
	Otake	13.0	
Mexico	Pathe	3.5	
	Cerro Prieto		75.0
U.S.S.R.	Pauzheika	5.0	
	Paratunka	0.7	
Iceland	Namafjall	2.5	
		790.8	185.0

our efforts seem puny indeed. If we gathered together all known reserves of wood, coal, oil, and gas in the world and built one big bonfire, the heat it produced would be equal to 3 days of sunshine around the world. Or, if we compute the amount of solar energy falling on the United States in a year, we find that it is more than 2000 times our present annual consumption of energy. If we could collect (ay, there's the rub!) the solar energy falling only on the Mojave Desert in a year with an efficiency of only 1 percent, we'd have $1\frac{1}{2}$ times the amount of energy we now use. At present, all these "if's" are only dreams. But as the economic and environmental costs of energy from fossil fuels and nuclear reactors rise, there will be greater incentives to make the dreams come true.

About a third of all the energy we consume in the United States is used to heat our homes, schools, factories, churches, offices, and other buildings. We could considerably reduce air pollution if we did our heating with solar energy. The simplest device for doing this is the greenhouse. It heats up because sunlight gets to the interior through the glass more easily than heat rays can escape to the outside. In a sunny climate where the temperature is moderate, water circulated through coils of metal pipe under a glass roof will absorb enough of the sun's energy to heat a house and provide it with hot water. In Japan there are 300,000 solar water heaters; in Australia and other countries where sunshine is plentiful and fuel scarce, there are many more. Many homes in the southern part of the United States are heated this way in winter.

They can also be cooled in summer by using the heat of the hot water to run a refrigerating unit. At present, such refrigerators cost more than those driven by electricity, but as the technology of building them is improved and the cost of electricity rises, they will become more widely used.

The experimental sun-heated house shown in Fig. 11.6 was built by the Massachusetts Institute of Technology in nearby Lexington, where the winters are cold and there are frequent cloudy days. The half of the roof facing south is made of glass with an area of 640 ft². Water circulating through copper tubing under the glass carries the sun's heat to a 1500-gal storage tank in the basement. This hot water is used to heat air that is circulated through the house. Enough heat can be stored in the hot water tank to carry the house through three successive cloudy days. A small oil-burning furnace supplements the solar heater, but 75 percent of the heat needed in the house comes from the sun. Because a solar heater installation costs about 6 times as much as a conventional furnace, only rising costs of fuels and increasing demands to clean up air pollution will make it competitive in northern climates.

For almost 100 years men have dreamed of harnessing the sun's energy to generate power. The solar-powered steam engine shown in Fig. 11.7 was displayed at the Paris Exhibition of 1878. Sunlight falling on a large conical mirror is focused on a steam boiler mounted on a shaft through the center of the cone. The steam is conducted to an engine behind the cone.

Today the solar engine is still a laboratory curiosity. Although it is technically possible to design and manufacture a solar boiler to provide steam for driving an engine, the cost of the equipment is so high and the sun is so undependable that such a device cannot compete with boilers fired by fossil

FIGURE 11.6 Solar-heated house in Lexington, Massachusetts. (Figure © 1958 by The New York Times Company. Redrawn with permission.)

FIGURE 11.7 The solar engine at the Paris Exhibition of 1878. (Photograph courtesy of Peter Glaser, "Solar Energy, Prospects for Its Large Scale Use," *The Science Teacher*, March 1972.)

fuels. How can we store the energy produced when the sun shines and use it at night and on cloudy days? When we find an answer that is both technologically workable and economically feasible, solar energy may become an important substitute for fossil fuels.

photocell

It is possible to generate electricity directly from sunlight falling on *photocells*, but the chemical principles by which this can be done are too complicated to consider in this book. Furthermore, the cost of making such cells is so high as to prohibit their use except as small power generators in remote and inaccessible places (such as weather stations on mountaintops) and in spaceships.

Building a solar power station as a satellite in space is an intriguing possibility. Such a station placed in orbit about 22,300 mi above the equator would remain in a fixed position relative to any spot on the earth's surface, since it would orbit the earth every 24 hr. Such a satellite would receive the sun's rays 24 hr a day except for short periods near the equinoxes. It would consist of two main parts: (1) a large array of photocells to absorb solar energy and produce electricity, and (2) a device to convert this electricity into *microwaves*

microwave

(very short radio waves). These waves could be focused on a spot on the earth's surface and transformed into electric current (Fig. 11.8). Microwaves can penetrate the earth's atmosphere with little loss of energy. At the edges of a receiving antenna on the earth, the intensity of radiation would be about one-tenth that of direct sunlight; at the center of the antenna, it would be about equal to that of sunlight. Such a space station might well be a project for the United Nations. It would provide the nations of the world with a cooperative venture that might prove to be of inestimable value to all mankind.

11.10   What Shall We Do?

Clearly, our economy of abundance and the affluent society that grows out of it depend on large amounts of energy. Equally clearly, the production and distribution of energy have undesirable side effects: our air is polluted with noxious fumes, our waters are warmed with waste heat that reduces the production of fish and other aquatic foods, our farmlands are crisscrossed with power lines. If we are to preserve those amenities of life that depend on ready energy, we must learn to produce it with less insult to the environment.

Our first task is to create power plants that transform fossil fuels into clean electricity without dirtying the air and heating our rivers. We shall consider the problems of waste heat disposal in Chap. 13, when we study water pollu-

FIGURE 11.8   A satellite space station for catching and transmitting solar energy to the earth. (Figure adapted from Peter Glaser, "Solar Energy, Prospects for Its Large Scale Use." *The Science Teacher*, March 1972.)

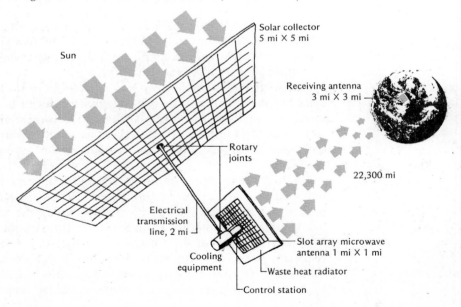

tion. The technology for reducing pollution from stationary power plants is developing rapidly enough that we can expect to eliminate all but a minimum of pollution. Power plants will emit cleaner air and cooler water if we have the determination to demand these things through effective social and political channels.

Our second task is more difficult. We must curb air pollution from cars, trucks, and buses. This problem should be attacked on two fronts: (1) we must reduce our dependence on the automobile for urban transport, and (2) we must reduce pollution from the cars, trucks, and buses we use. This is primarily a political and a social problem. We have now the basic technology to replace myriads of cars on freeways by rapid transit much less insulting to the environment. We can clean up the exhaust gases from our cars if we are willing to pay the cost. We know how to build cleaner engines, but changing over to produce them will be expensive. Will we pay the price for clean air? The answer is up to all of us. Will we elect government officials who will strive for these ends? Will we support them when the going gets tough?

Our third task is to continue to develop methods to minimize the hazards of nuclear reactors in power plants. The technology of generating electricity safely by nuclear power plants is not yet fully developed. There are risks in building more nuclear power plants. We must support research on the safe production of nuclear fuels, the safe operation of nuclear plants, and the safe disposal of nuclear wastes. We must learn these techniques in the next decade or so, lest we be forced to make the hard choice between a healthful environment and a lowered standard of living. If we develop atomic energy as we should, we need not be forced to such a choice. Again, the problem is social and political. Will we, the electorate, support research and development for generating nuclear power?

Our fourth task is to develop the use of solar energy. Fossil and nuclear fuels will see us through the next century, but if we are to provide beyond that, we must plan ahead. Clearly the utilization of solar energy is the greatest technological challenge of the next 100 years. It will demand a dedication like that required to put a man on the moon in the 1960s. Do we want solar energy? Will we pay the price?

Today we are behaving like Goldilocks in the home of the three bears. The first bowl of porridge was too hot. The second was too cold. The third was just right, so she gobbled it up. Atomic energy is too hot to handle and solar energy is too cold. So we are rapidly gobbling up the fossil fuels—which we find are just right.

Summary

The standard of living in a society is proportional to the annual gross national product per person. The annual GNP per person is proportional to the amount of energy used per person. A high standard of living requires a high level of energy production.

We are facing an energy crisis in the United States because our demand for oil and natural gas exceeds the rate at which we can extract them from our own deposits or import them from abroad. In the next three decades we must increase our use of coal and nuclear fuels to produce energy.

To avoid high levels of air pollution from burning coal, we must develop new technologies for removing particulates and sulfur oxides from stack gases or prevent them from being formed during combustion. The production of fuel gas and oil from coal is one important step in this direction. Some fuel oil may be produced from oil shales and oil sands, but this is not likely to be a major development. Considerable numbers of nuclear power plants will have to be built. Intensive research on minimizing the hazards of using nuclear fuels must be continued.

In the years immediately ahead, we must reduce pollutants from internal combustion engines. Methods are now being developed to accomplish this with existing types of engines, and cleaner engines are being developed.

In the next few decades considerable resources of geothermal steam may be tapped for production of energy, but for the long run we pin our hopes on learning how to transform sunshine into other forms of energy cheaply and efficiently.

New Terms and Concepts

BRITISH THERMAL UNIT (BTU): The amount of heat required to raise the temperature of one pound of water one degree Fahrenheit.

CARBONACEOUS: Carbon-bearing.

EARTH'S CRUST: The outermost solid layer of the earth.

EARTH'S MANTLE: The layer of the earth between the crust and the core.

GEOTHERMAL: Pertaining to the internal heat of the earth.

GROSS NATIONAL PRODUCT (GNP): The total market value of all the goods and services produced by a nation in a specified period.

LIQUEFIED NATURAL GAS (LNG): Natural gas that has been cooled and compressed until it becomes a liquid.

LIQUEFIED PETROLEUM GASES (LPG): Petroleum refinery gases which have been cooled and compressed until they become a liquid.

MAGMA: The molten matter under the earth's crust from which igneous rock is formed by cooling.

MICROWAVE: Radiation with wavelengths between infrared and shortwave radio.

PHOTOCELL: An electronic device having an electrical output that varies in response to incident radiation, especially to visible light.

SLURRY: A suspension of solid particles in a liquid.

WATT: A unit of electric power; an electric toaster uses about 600 W.

Testing Yourself

11.1 What evidence have you observed of the energy crisis in the United States?

11.2 A man on a treadmill can put out energy at a continuous rate of about 75-W power. That is, he can generate enough electricity to run one medium-sized light bulb. The production of power in the United States today is about 1 kW per person. How many men would you have to have working for you full-time in order to provide you with your kilowatt?

11.3 At the "four corners" in southwest Indian country, where Utah, New Mexico, Arizona, and Colorado meet, there is a large coal-burning steam-electric plant generating power for Los Angeles, Las Vegas, Phoenix, and Albuquerque. Large amounts of coal are strip-mined, large amounts of stack effluents are poured out, and the Colorado River is tapped for cooling water. Local Indian tribes are protesting this development; their battle cry is, "They get the energy, we get the pollution." How shall we deal with this issue?

11.4 What by-products from pollution control may help pay for particulate recovery? For SO_x recovery?

11.5 Why should we remove sulfur from coal and oil before burning them? What methods of sulfur removal seem best to you?

11.6 Compare the good and bad features of manmade fuels. For what process would you prefer to see government money spent in development?

11.7 As pollution control devices are added to cars, the number of miles we can get from each gallon of gasoline decreases. This means we shall have to burn more gasoline to go the same distance and thus produce more pollutants to be removed. Does this make pollution control seem self-defeating to you? How might the situation be improved by redesigning cars?

11.8 Keeping in mind the problem stated in 11.7, what alternatives do you think we should employ to reduce pollution by internal combustion engines in large cities?

11.9 Noting that about one-third of the energy we consume in the United States is used for space heating, would you encourage the development of solar heating for this purpose? Would you encourage the development of other ways of using solar energy? Why or why not?

In the photomicrograph we see crystals of the salt, potassium chromium sulfate. The drawing illustrates the hydration of ions in an aqueous medium. (Figure redrawn from Theodore L. Brown, General Chemistry, *Charles E. Merrill Books, Inc., Columbus, Ohio, 1968.)*

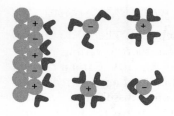

12. The Properties of Water Solutions

As we have repeatedly seen, one of the paradoxes of a modern industrial culture is that plenty promotes pollution. We cover acres of land with outworn automobiles and discarded refrigerators. Our cars and factories fill the air with noxious gases and damaging dust and ash. Perhaps most serious of all, we taint the waters we must use and drink with the wastes from our cleansing processes, our food, and our own bodies.

But fortunately we are learning—and, it is hoped, learning in time to avert disaster—that these damaging processes are at least in part reversible. We can reuse old metal, paper, and glass, and in a time of growing shortages it grows more and more attractive to do so. We can reduce the pollutants in our atmosphere. And we can do a good deal more than we have been doing to purify our waters to make them fit to be used again and again and again.

In earlier chapters in this book we have seen what chemistry and an understanding of chemical processes can do to help in the fight against solid wastes and tainted atmosphere. In the chapters that follow we shall turn to a discussion of the problems of water pollution. But before we do so, we must first learn something about the properties of water and what happens when other substances are put into it. We turn, then, to the nature of *solutions in water*.

12.1 The Importance of Solutions

The air we breathe and the water we drink are solutions. The food we eat is first digested and then dissolved in the blood for transport to various parts of the body where it is used to build tissues or produce energy. Carbon dioxide produced along with the energy is dissolved in the blood and carried to the lungs, where it is exhaled. Carbon dioxide in the air is dissolved in water in the leaves of green plants, which use it to build their tissues and produce oxygen. Oxygen dissolved in water is used by aquatic organisms for respiration. The CO_2 produced by this respiration is dissolved in the water and used by green water plants to build their tissues and liberate dissolved oxygen. The lives of all creatures are intimately bound up with processes involving solutions.

Many of the processes that influence the nature of our environment involve substances in solution. The rain dissolves gaseous impurities from the atmosphere and helps freshen the air we breathe. The quality of the water we drink from wells depends on the chemical properties of the substances dissolved from the rocks through which the underground water percolates. Sewage, industrial wastes, and runoff from feedlots and heavily fertilized farmlands pollute our rivers with hundreds of dissolved substances that must be removed if surface water is to be used for human consumption. Without some understanding of the properties of solution we cannot understand these and related environmental problems.

12.2 The Nature of Solutions

A solution is a homogeneous mixture of two or more substances. All gases are completely miscible with one another; i.e., they form solutions when mixed. Air is a mixture of many gases (Sec. 7.6). Some liquids—like grain alcohol and water—are completely miscible in all proportions, while many dissolve in one another only slightly or not at all. We describe the latter as *immiscible*; thus oil and water are immiscible liquids. Some gases dissolve in some liquids: "carbonated water" is a solution of carbon dioxide in water. Many solids (like sugar and salt) can be dissolved in liquids of one kind or another, but many others (like sand and rocks) are insoluble. Water dissolves a far greater number of natural substances than any other liquid encountered in nature or in the laboratory.

immiscible

concentrated
solution
dilute solution

In a *concentrated solution* there is a large amount of solute for a given amount of solvent; in a *dilute solution* there is little. If we put a few grains of table salt (NaCl) in a test tube half-full of water and shake it, the solid soon dissolves (disappears) to form a solution. If we add more solid in small amounts, and shake the test tube between additions, there comes a time when the last solid we add does not dissolve, no matter how hard we shake the tube. When a solution is in contact with undissolved solid solute and some of the solute remains undissolved, we say that the solution is saturated with respect to that solute. We may prepare a *saturated solution* of a liquid or a gas in the same way. A convenient operational definition of a saturated solution is: A solution is saturated with a given solute when prolonged contact of the solution with an excess of the solute produces no further dissolving of the solute. Because NaCl is quite soluble, it takes a considerable amount of it to saturate a small amount of solution; the saturated solution is a concentrated solution. On the other hand, because silver chloride, AgCl, is very slightly soluble, only the tiniest bit is required to make a saturated solution; the saturated solution is a dilute solution. A saturated solution of salt contains about 35 g NaCl per 100 ml water; a saturated solution of silver chloride contains 0.00009 g AgCl per 100 ml water.

saturated solution

12.3 Suspensions and Colloidal Systems

It is difficult to distinguish between solutions and suspensions. The fact that cream rises to the top of milk as it comes from the cow leads us to conclude that untreated milk is a suspension, a heterogeneous mixture. Through a low-power microscope we can see globules of liquid fat suspended in the water present in milk. These globules rise and coalesce to form cream. If milk is "homogenized," cream no longer rises to the top, and we conclude that the liquid is homogeneous. But if we examine homogenized milk with a high-power microscope, we see that the globules of fat are still present but much smaller. Our method of examining milk obviously determines whether we classify it as homogeneous (a solution) or heterogeneous (a suspension).

The same is true of mixtures of clay and water. If we shake finely ground clay in water, some—but not all—of it settles out of the liquid when the mixture stands for a while. The cloudy liquid from which the clay does not settle out looks homogeneous to the naked eye, but through a high-power microscope we can see individual particles of clay suspended in the water. Is the cloudy mixture of clay and water homogeneous or heterogeneous? Again our answer depends on the method we use in examining the mixture.

In a mixture of sugar in water, we are unable to see any particles, no matter how powerful our microscope. We conclude, therefore, that the particles of solute are individual molecules far too small to be seen. Obviously there is no clear line of demarcation between suspensions and solutions. So we say a mixture is a suspension if a solid or a liquid component is present in particles large enough to be seen with our highest powered optical microscopes; such particles have a diameter of about 10^{-4} cm. And we say a mixture is a solution

colloid

if the particles are about the size of ordinary molecules; such particles have a diameter of about 10^{-7} cm. Mixtures with solute particles between these limits are called colloidal suspensions, colloidal solutions, or simply *colloids*.

12.4 Electrolytes and Nonelectrolytes

electrical
conductivity

electrolyte

nonelectrolyte

Because so many substances are soluble in water, it is convenient to divide them into categories. One of the most useful classifications is based on *electrical conductivity*. Figure 12.1 shows a simple apparatus for testing conductivity. When the apparatus is connected to an automobile battery and a piece of metal is held against the two electrodes, the electric circuit is completed, and the lamp glows. If the electrodes are placed in a solution in the beaker and the lamp glows, we know electricity is being conducted from one electrode to the other through the solution. We call such a conductive solution an *electrolyte*. We also give this name to a solute that forms a conductive solution. Solutions (and the solutes in them) that do not conduct electricity are called *nonelectrolytes*. There are a few hundred electrolytes in the environment, but there are many thousands of nonelectrolytes. Electrolytes are found in a few salt mines and the beds of dried up lakes, but occur in nature mostly in seawater and to a lesser extent in freshwater. Tapwater almost always contains enough electrolytes to make it conductive. You must handle

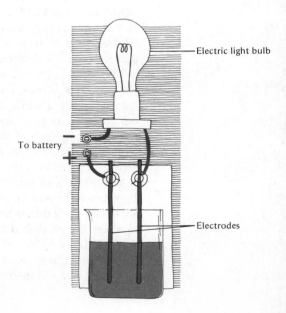

Electric light bulb

To battery

Electrodes

FIGURE 12.1 Apparatus for testing the electrical conductivity of substances

electrical equipment with extreme care when your body is in contact with tapwater. Most of the substances in living organisms are nonelectrolytes, such as starches, sugars, alcohols, fats, cellulose, and hundreds of other carbon compounds. Many common manufactured materials like plastics, paints, dyes, adhesives, and petroleum products are also nonelectrolytes.

12.5 The Freezing and Boiling Points of Solutions of Nonelectrolytes

When you add antifreeze to the water in your car radiator, you know that the solution it forms will freeze at a lower temperature than water. More antifreeze lowers the freezing point of the solution. To correlate the lowering of the freezing point of a solution with the amount of solute added, we use an expression of concentration called *molality*. A solution containing 1 mole of solute in 1000 g solvent is defined as *1 molal*. A solution containing $\frac{1}{10}$ mole of solute in 1000 g solvent is called 0.1 molal, often abbreviated to 0.1 *m*.

molality

From hundreds of experiments with the freezing points of various solvents and solutions we learn that solutes which are nonelectrolytes display a simple pattern in their effect on freezing points of solvents. When water is the solvent, a 0.1 *m* solution of any nonelectrolyte freezes at −0.186°C, a 0.2 *m* solution freezes at −0.372° (2 × −0.186°C), a 0.3 *m* solution at −0.558° (3 × −0.186°C), etc. From these data it appears that a 1.0 *m* solution would freeze at −1.86°C. Therefore, we say that the *molal freezing point lowering* of water is 1.86°C; we give this quantity the symbol K_f.

molal freezing
point lowering
nonvolatile solute

When a *nonvolatile solute* (one with very little tendency to evaporate) is dissolved in a pure liquid, the boiling point of the solution is higher than that of the solvent. Elevation of the boiling point depends on molality much as does depression of the freezing point. With nonvolatile nonelectrolytes as solutes, the *molal boiling point elevation*, K_b, for water is 0.512°C. The boiling point of a 0.1 *m* solution of sugar in water is 100.05°C.

molal boiling
point elevation

12.6 The Kinetic-Molecular Concept Extended to Solutions

In Chap. 6 we interpreted the behavior of gases, liquids, and solids in terms of molecules moving at a considerable range of speeds but with their average speed proportional to their temperature. When the temperature of a liquid is lowered continuously, its molecules move past one another at slower and slower average speeds until forces of attraction cause them to occupy relatively fixed points in space. The free-flowing liquid then becomes a crystalline solid, with the molecules oscillating about positions in a geometrically regular three-dimensional pattern. When a solution is cooled, the molecules of solute impede the molecules of solvent from forming their geometrically patterned crystals at the same temperature as that for their formation in pure solvent.

The temperature has to be lowered further for the solvent to freeze—that is, to form crystals of solid solvent—in the presence of the solute.

When a liquid is heated, its molecules move faster and faster and have a greater tendency to escape from the liquid into the gaseous state. That is, the vapor pressure of the liquid increases with temperature. The presence of a nonvolatile solute in a solution impedes this tendency to escape into the gaseous state; the vapor pressure of the solvent in a solution is less than in the pure solvent at the same temperature. Therefore the solution must be heated to a higher temperature for the vapor pressure of the solvent to attain that of the atmosphere. The boiling point of the solution is higher than that of the pure solvent.

12.7   Characteristics of Electrolytes

When we use the simple apparatus shown in Fig. 12.1 to study the electrical conductance (conductivity) of various substances, some interesting facts emerge. Most solid substances other than metals are nonconductors, but some become conductors when they are melted. Pure liquids at room temperature (except metallic mercury) are nonconductors (water, alcohol, benzene). Solutions of some substances in some solvents conduct electricity very well (salt in water), others conduct it poorly (ammonia in water), still others not at all (sugar in water). In most solutions that conduct electricity, water is the solvent, but solutions of many substances in water are nonconductors. Most substances dissolved in carbon tetrachloride, benzene, and many other liquid compounds containing carbon do not yield conducting solutions. This is true even of solutes that become conductors in water. From this we conclude that the conductance of a solution is determined by the properties of both the solute and the solvent.

strong electrolyte
weak electrolyte

When we study a list of substances that are electrolytes, we are struck by the fact that it includes acids, bases, and salts. Those that give solutions with high conductivity we call *strong electrolytes*; those with low conductivity, *weak electrolytes*. Salts, which in the solid state do not conduct electricity, are good conductors when melted.

12.8   The Freezing Points of Solutions of Electrolytes

We noted in Sec. 12.5 that when 0.1 mole (6×10^{22} molecules) of any non-electrolyte is dissolved in 1000 g water, the freezing point is lowered to $-0.186°C$ and that the presence of 0.2 mole (12×10^{22} molecules) lowers the freezing point twice as much. Apparently, the lowering of the freezing point is proportional to the number of particles of solute present.

But when 0.1 mole of salt, NaCl, which is an electrolyte, is dissolved in

1000 g water, the freezing point is lowered almost *twice* as much as when 0.1 mole of nonelectrolyte is present. It seems likely that somehow or other the 0.1 mole of NaCl has become 0.2 mole of particles in solution. Similar results are obtained with 0.1 molal solutions of the electrolytes HCl, HNO_3, NaOH, NH_4Cl, and $CuSO_4$ (Table 12.1).

Even more pronounced is the lowering of the freezing point when 0.1 mole of H_2SO_4 or 0.1 mole of K_2SO_4 is dissolved in 1000 g water. With these electrolytes, the lowering is nearly 3 times as much as when 0.1 mole of a nonelectrolyte is present (Table 12.1). How can this be explained?

12.9 The Arrhenius Theory for Solutions of Electrolytes

How can it be that 0.1 mole of each of the first six electrolytes behaves like 0.2 mole when it lowers the freezing point—that 0.1 mole of H_2SO_4 and 0.1 mole of K_2SO_4 can have the effect of 0.3 mole of nonelectrolyte? As a graduate student, the Swedish chemist and physicist Svante Arrhenius was puzzled by this situation. On the basis of his doctoral research he proposed in 1884 a "Theory of Electrolytic Dissociation," which states that when an electrolyte dissolves,

1 the solute breaks up to give more than 1 mole of particles per mole of solute
2 these new particles must carry electric charges and be able to move through the solution and thus conduct electricity.

TABLE 12.1 *The Freezing Point Lowering of 0.1 m Solutions of Some Nonelectrolytes and Electrolytes*

Compound	Formula	Approximate freezing point lowering	Apparent number of particles present	Moles of particles per mole of solute
Nonelectrolytes				
Alcohol	C_2H_5OH	0.186	0.1 mole	1
Sugar	$C_{12}H_{22}O_{11}$	0.186	0.1 mole	1
Electrolytes				
Sodium chloride	NaCl	2(0.186)	0.2 mole	2
Hydrochloric acid	HCl	2(0.186)	0.2 mole	2
Nitric acid	HNO_3	2(0.186)	0.2 mole	2
Sodium hydroxide	NaOH	2(0.186)	0.2 mole	2
Ammonium chloride	NH_4Cl	2(0.186)	0.2 mole	2
Copper sulfate	$CuSO_4$	2(0.186)	0.2 mole	2
Sulfuric acid	H_2SO_4	3(0.186)	0.3 mole	3
Potassium sulfate	K_2SO_4	3(0.186)	0.3 mole	3

Arrhenius' proposal was considered so radical that he barely passed his qualifications for the doctor's degree and had a very difficult time obtaining an appointment in the faculty of a university. Yet within a few years the *Arrhenius theory* proved to be the one which most adequately explained many of the properties of electrolytic solutions and became almost universally accepted.

Arrhenius theory

When a water solution of HCl conducts an electric current, hydrogen is evolved at the negative electrode. This indicates that there are hydrogen particles in the solution carrying a positive charge and moving toward the negative electrode. Arrhenius called these hydrogen ions. Since the solution of HCl is electrically neutral, there must be negative particles present and carrying as much charge as the positive hydrogen ions. Arrhenius called these chloride ions. He postulated that HCl *dissociates* in water solution thus:

dissociation

$$HCl \longrightarrow H^+ + Cl^- \tag{12.1}$$

Though he had no evidence for saying that one hydrogen ion carries one positive charge, he made the simplest assumption—a choice scientists always make if there is no evidence calling for a more complicated one.

Sodium chloride also dissociates into two ions (Table 12.1). Since the chloride ion is Cl^-, the sodium ion must be Na^+.

$$NaCl \longrightarrow Na^+ + Cl^- \tag{12.2}$$

When a water solution of nitric acid conducts an electric current, hydrogen is evolved at the negative electrode, indicating the presence of H^+. Since nitric acid dissociates into two ions (Table 12.1) and its solution is electrically neutral, the other ions present must have a single negative charge:

$$HNO_3 \longrightarrow H^+ + NO_3^- \tag{12.3}$$

Similar arguments lead to the conclusion that NaOH and NH_4Cl dissociate:

$$NaOH \longrightarrow Na^+ + OH^- \tag{12.4}$$
$$NH_4Cl \longrightarrow NH_4^+ + Cl^- \tag{12.5}$$

When a water solution of sulfuric acid conducts an electric current, hydrogen is evolved at the negative electrode, indicating the presence of H^+. Sulfuric acid dissociates into three ions (Table 12.1); the equation is

$$H_2SO_4 \longrightarrow 2H^+ + SO_4^{2-} \tag{12.6}$$

Since there are two positive hydrogen ions for each negative sulfate ion, the sulfate ion must have a charge of 2− to preserve electrical neutrality. When K_2SO_4 dissolves in water, it yields three ions (Table 12.1); the equation is

$$K_2SO_4 \longrightarrow 2K^+ + SO_4^{2-} \tag{12.7}$$

Since there are two positive potassium ions for each negative sulfate ion, each potassium ion must bear a single charge of $1+$.

Copper sulfate in solution has two moles of ions per mole of solute (Table 12.1). The equation for its dissociation is

$$CuSO_4 \longrightarrow Cu^{2+} + SO_4^{2-} \tag{12.8}$$

There is one copper ion for each sulfate ion, and so to preserve electrical neutrality each copper ion must carry a charge of $2+$.

12.10 The Arrhenius Theory of Strong and Weak Electrolytes

The electrolytes whose behavior Arrhenius interpreted by postulating dissociation into ions as shown in Eqs. (12.1) through (12.8) were very good conductors of electricity in solution, and so he called them *strong* electrolytes. Arrhenius also found a number of solutes that formed only weakly conductive solutions in water, and he called these *weak* electrolytes. He postulated that weak electrolytes were only partially ionized (dissociated into ions).

A common weak electrolyte is acetic acid, which gives vinegar its sour taste. When acetic acid solution conducts electricity, hydrogen is liberated at the negative electrode; so there must be H^+ ions present. To indicate that acetic acid is a weak acid (weak electrolyte), we may write the equation for its ionization with a very small arrow pointing to the right to indicate that only a little of the acid is ionized:

$$HC_2H_3O_2 \rightarrow H^+ + C_2H_3O_2^- \tag{12.9}$$

Measurements of the freezing points of very dilute solutions of acetic acid indicate that only 2 moles of ions are formed from 1 mole of solute, and so we conclude that when one H^+ is liberated from the molecule, all the remaining atoms together form one ion with a single negative charge, $C_2H_3O_2^-$. We call this the acetate ion.

A common weak base is a solution of ammonia in water, $NH_3(aq)$. It is a weak electrolyte with the same effect on litmus as dilute NaOH solution, and so Arrhenius postulated that NH_3 dissolves in water to yield OH^-:

$$NH_3(g) + H_2O(l) \rightarrow OH^-(aq) + NH_4^+(aq) \tag{12.10}$$

Again we use the small arrow to indicate only partial ionization of the weak electrolyte. Measurements of the freezing points of ammonia solutions indicate that only 2 moles of ions are formed from 1 mole of ammonia. Therefore, we conclude that when OH^- is formed, the remainder of the atom must be combined in a single ion with one positive charge, NH_4^+. An aqueous solution of ammonia is sometimes called ammonium hydroxide. However, no one has been able to find evidence for the existence of any NH_4OH molecules in the solution, so that the name "ammonium hydroxide" for a solution of NH_3 in water is misleading—although much used.

From the data available to him, Arrhenius postulated that all acids contain hydrogen atoms that can be released to yield hydrogen ions (H^+) in solution, and that all bases contain OH groups that can be released to yield hydroxide ions (OH^-) in solution. When an acid is mixed with a base, the H^+ from the acid unites with the OH^- from the base to produce H_2O, and if all the water is boiled off, the remaining ions combine to form salts. Thus the neutralization of hydrochloric acid by sodium hydroxide may be symbolized as follows:

$$H^+ + Cl^- + Na^+ + OH^- \longrightarrow H_2O + Na^+ + Cl^- \qquad (12.11)$$

If all the water is boiled off, solid salt remains:

$$Na^+ + Cl^- \longrightarrow NaCl(s)$$

Any neutralization in a water solution is essentially the union of hydrogen ions and hydroxide ions to produce water molecules. Any acid, then, will neutralize any base. The salt is just what is left over from this production of water. The characteristics of some common acids, bases, and salts are given in Table 12.2.

TABLE 12.2 *Some Common Acids, Bases (Hydroxides), and Salts*

ACIDS (*furnish H^+ ions in water solution*)

Name	Formula	Negative ions		Strength
Acetic	$HC_2H_3O_2$	Acetate	$C_2H_3O_2^-$	Weak
Carbonic	H_2CO_3	Carbonate	CO_3^{2-}	Weak
Hydrobromic	HBr	Bromide	Br^-	Strong
Hydrochloric	HCl	Chloride	Cl^-	Strong
Nitric	HNO_3	Nitrate	NO_3^-	Strong
Phosphoric	H_3PO_4	Phosphate	PO_4^{3-}	Moderate
Sulfuric	H_2SO_4	Sulfate	SO_4^{2-}	Strong
Sulfurous	H_2SO_3	Sulfite	SO_3^{2-}	Moderate

BASES (*furnish OH^- ions in water solution*)

Name	Formula	Positive ions		Strength
Sodium hydroxide (All other group IA hydroxides)	NaOH	Sodium	Na^+	Strong
	M'OH*		$(M')^+$	Strong

TABLE 12.2 *(continued)*

BASES (*furnish OH⁻ ions in water solution*)

Name	Formula	Positive ions		Strength
Calcium hydroxide	$Ca(OH)_2$	Calcium	Ca^{2+}	Moderate
(All other group IIA hydroxides)	$M''(OH)_2$		$(M'')^{2+}$	Moderate
Aluminum hydroxide	$Al(OH)_3$	Aluminum	Al^{3+}	Weak
(All other group IIIA hydroxides)	$M'''(OH)_3$		$(M''')^{3+}$	Weak
Ammonia	NH_3	Ammonium	NH_4^+	Weak

SALTS (*composed of positive metal or ammonium ions and negative nonmetal ions with relative numbers of each to give equal total positive and negative charges, thus making the salt electrically neutral; all true salts are highly ionized in water solution*)

Name	Formula	Positive ions	Negative ions
Ammonium nitrate	NH_4NO_3	NH_4^+	NO_3^-
Sodium chloride	$NaCl$	Na^+	Cl^-
(All other group IA halides)	$M'X\dagger$	$(M')^+$	X^-
Magnesium chloride	$MgCl_2$	Mg^{2+}	$2Cl^-$
(All other group IIA halides)	$M''X_2$	$(M'')^{2+}$	$2X^-$
Sodium sulfite	Na_2SO_3	$2Na^+$	SO_3^{2-}
Barium sulfate	$BaSO_4$	Ba^{2+}	SO_4^{2-}
Calcium phosphate	$Ca_3(PO_4)_2$	$3Ca^{2+}$	$2PO_4^{3-}$
Aluminum sulfate	$Al_2(SO_4)_3$	$2Al^{3+}$	$3SO_4^{2-}$

*M' stands for any alkali metal, M'' stands for any alkaline earth metal, M''' stands for any group IIIA metal.
†X stands for any halogen.

12.11 Net Ionic Equations

The chemical equation for the reaction between hydrochloric acid and sodium hydroxide can be written in various ways.

In words:

Hydrochloric acid solution + sodium hydroxide solution $\longrightarrow$
water + sodium chloride solution

With molecular formulas:

$$HCl + NaOH \longrightarrow H_2O + NaCl \tag{12.12}$$

With formulas showing the ions and molecules present:

$$H^+ + Cl^- + Na^+ + OH^- \longrightarrow H_2O + Na^+ + Cl^- \tag{12.13}$$

With the formulas of only those species that react (canceling Na^+ and Cl^-, which appear as both reactants and product):

$$H^+ + OH^- \longrightarrow H_2O \tag{12.14}$$

This is often called a "net equation." The net equation shows that the presence of Cl^- and Na^+ is of no consequence chemically. They are present before and after the reaction and do not participate in it. They may conveniently be thought of as "spectator ions"; they are not involved in the chemical game. The net equation for the reaction is that for the neutralization of any strong acid by any strong hydroxide; its simplicity underscores the usefulness of considering all neutralizations in water solution as variations on this theme.

12.12 The Brønsted-Lowry Theory of Acids and Bases

ionization

Although Arrhenius' model for *ionization* is very useful in explaining the electrical conductivity of solutions of acids, bases, and salts, it does not take into account the role of water in producing conducting solutions. Using the apparatus in Fig. 12.1, we find that when HCl gas is dissolved in benzene, C_6H_6, the solution is nonconductive; when HCl is dissolved in water, the solution is an excellent conductor. Pure liquid acetic acid, $HC_2H_3O_2$, is nonconductive; pure water is nonconductive; when acetic acid and water are mixed, the solution is conductive. How can we explain these facts?

In 1923, J. N. Brønsted in Denmark and T. M. Lowry in England independently proposed the same explanation. They postulated that an *acid* is a compound which *donates* protons to another compound, and that a *base* is a compound which *accepts* protons from another compound. This concept

Brønsted-Lowry
theory

has become known as the *Brønsted-Lowry theory*. When HCl is dissolved in benzene, no exchange of protons takes place:

$$HCl + C_6H_6 \longrightarrow \text{no reaction}$$

When HCl is dissolved in water, proton exchange does take place:

$$HCl + H_2O \longrightarrow H_3O^+ + Cl^- \tag{12.15}$$

hydronium ion

A molecule of *acid* (HCl) has donated a proton (H^+) to a molecule of *base* (H_2O); the products are a *hydronium ion* (H_3O^+, a water molecule with a proton, H^+, attached) and a chloride ion (Cl^-, a chlorine atom with an extra electron). The hydronium ions and chloride ions present make the solution electrically conductive. When acetic acid is dissolved in water, proton exchange takes place:

$$HC_2H_3O_2 + H_2O \rightarrow H_3O^+ + C_2H_3O_2^- \qquad (12.16)$$

Molecules of $HC_2H_3O_2$ *cannot* conduct electricity through pure acetic acid; *molecules* of H_2O *cannot* conduct electricity through pure water; hydronium *ions* and acetate *ions* ($C_2H_3O_2^-$) *can* conduct electricity through a solution of acetic acid in water.

Pure liquid ammonia, NH_3, does not conduct electricity, but a solution of ammonia in water does; the water solution is definitely basic (turns red litmus blue). How can we account for these facts? We believe that when NH_3 dissolves in water, an exchange of protons takes place:

$$NH_3 + H_2O \rightarrow NH_4^+ + OH^- \qquad (12.17)$$

A molecule of the base (NH_3) accepts a proton from a molecule of the acid (H_2O); the products are an ammonium ion (NH_4^+) and a hydroxide ion (OH^-). The presence of these ions makes the solution electrically conductive.

It is interesting to note that water behaves like an acid (proton donor) in Eq. (12.17) and like a base (proton acceptor) in Eqs. (12.15) and (12.16). Apparently, water is neither very acidic nor very basic; in the presence of a more acidic compound, water accepts protons (behaves like a base); in the presence of a more basic compound, water donates protons (behaves like an acid). The acidic or basic nature of many compounds whose behavior cannot be clearly accounted for by the Arrhenius theory is easily interpreted in terms of the Brønsted-Lowry concept of proton exchange.

12.13 The Stepwise Ionization of Acids with More Than One Hydrogen

polyprotic acid

If the molecule of an acid contains more than one hydrogen atom that can be released as a hydrogen ion (proton), we call the acid a *polyprotic acid*. Phosphoric acid, H_3PO_4, is a common polyprotic acid. If one adds a solution containing 1 mole of sodium hydroxide, NaOH, to a solution containing 1 mole of phosphoric acid and evaporates off the water, he finds that the solid salt produced is NaH_2PO_4, called monosodium dihydrogen phosphate. If 2 moles of dissolved NaOH are added to 1 mole of dissolved H_3PO_4, the solid salt produced has the formula Na_2HPO_4, called disodium monohydrogen phosphate. When 3 moles of NaOH react with 1 mole of phosphoric acid, the salt formed is Na_3PO_4, called trisodium phosphate. The chemical equations for these reactions are

$$NaOH + H_3PO_4 \longrightarrow H_2O + NaH_2PO_4 \qquad (12.18)$$
$$2NaOH + H_3PO_4 \longrightarrow 2H_2O + Na_2HPO_4 \qquad (12.19)$$
$$3NaOH + H_3PO_4 \longrightarrow 3H_2O + Na_3PO_4 \qquad (12.20)$$

These equations indicate that phosphoric acid ionizes in steps:

$$H_3PO_4 \longrightarrow H^+ + H_2PO_4^- \tag{12.21}$$
$$H_2PO_4^- \longrightarrow H^+ + HPO_4^{2-} \tag{12.22}$$
$$HPO_4^{2-} \longrightarrow H^+ + PO_4^{3-} \tag{12.23}$$

The other frequently encountered polyprotic acids are sulfuric, H_2SO_4, and carbonic, H_2CO_3.

$$H_2SO_4 \longrightarrow H^+ + HSO_4^- \tag{12.24}$$
$$HSO_4^- \longrightarrow H^+ + SO_4^{2-} \tag{12.25}$$
$$H_2CO_3 \longrightarrow H^+ + HCO_3^- \tag{12.26}$$
$$HCO_3^- \longrightarrow H^+ + CO_3^{2-} \tag{12.27}$$

The sodium salts of sulfuric acid are sodium hydrogen sulfate, $NaHSO_4$ (often called sodium acid sulfate or sodium bisulfate), and sodium sulfate, Na_2SO_4. The sodium salts of carbonic acid are sodium hydrogen carbonate, $NaHCO_3$ (often called sodium bicarbonate or bicarbonate of soda), and sodium carbonate, Na_2CO_3 (often called washing soda).

Using Brønsted's model for acids and bases, we can show how the acid H_3PO_4 donates protons to the base H_2O in three successive steps:

$$H_3PO_4 + H_2O \longrightarrow H_3O^+ + H_2PO_4^- \tag{12.28}$$
$$H_2PO_4^- + H_2O \longrightarrow H_3O^+ + HPO_4^{2-} \tag{12.29}$$
$$HPO_4^{2-} + H_2O \longrightarrow H_3O^+ + PO_4^{3-} \tag{12.30}$$

It is fairly easy for a water molecule to attract an H^+ from the neutral molecule H_3PO_4 in Eq. (12.28). It is more difficult for a water molecule to attract an H^+ from the negatively charged $H_2PO_4^-$ ion. It is still more difficult for a water molecule to attract an H^+ from the doubly negatively charged HPO_4^{2-} ion. It is not surprising, then, that the first ionization of phosphoric acid is moderately strong (takes place to a considerable extent), the second is weak (takes place to only a small extent), and the third very weak (takes place to a very small extent). With all polyprotic acids, the first ionization is always stronger than the second, and so on.

Summary

A solution is a homogeneous mixture of two or more substances. With the exception of petroleum and its products, the liquid solutions we encounter in the environment contain water, which is generally considered the solvent, and one or more other substances, called solutes. In solution, the solute particles (molecules or ions) are invisible, being roughly 10^{-7} cm in cross section. If they range from 10^{-6} to 10^{-5} cm in size, the mixture is called a colloidal system. If the particles are 10^{-4} cm or larger, we call the mixture a suspension, because we can see the particles as they slowly settle out.

Electrolytes are substances that dissolve in water to give a solution that conducts electricity. The conducting solution is also called an electrolyte. A nonelectrolyte is a substance whose water solution does not conduct electricity.

Solutes that are nonelectrolytes lower the freezing point of water 1.86°C per mole of solute in 1000 g water. Nonvolatile nonelectrolytes dissolved in water raise the boiling point 0.512°C per mole of solute in 1000 g water. The freezing point lowering and boiling point elevation produced by solutes that are electrolytes are greater than 1.86 and 0.512°C (per mole in 1000 g), respectively. This is because electrolytes dissolved in water dissociate into ions (ionize), yielding more than 1 mole of ions for 1 mole of solute. Strong electrolytes are highly ionized (dissociated), and so they conduct electricity strongly when dissolved in water. Weak electrolytes are only slightly ionized, and they conduct electricity weakly when dissolved in water. In any solution of any electrolyte, the total charge on the positive ions present must equal the total charge on the negative ions.

Electrolytes are conveniently classified as acids, bases, and salts. In Arrhenius' model, an acid is a substance that will yield H^+ ions when dissolved in water, and a base is a substance that will yield OH^- ions when dissolved in water. A salt is the solid product formed when an acid reacts with a base and the water present is driven off.

Brønsted and Lowry modified Arrhenius' model to include the role of water in ionization. In their theory, an acid is a substance that can donate protons to another substance, and a base is a substance that can accept protons from another substance. In water solutions, acids donate protons to water molecules to form hydronium ions, H_3O^+. In water solutions, bases accept protons from water molecules to form hydroxide ions, OH^-. Neutralization is the union of hydronium ions with hydroxide ions to produce water:

$$H_3O^+ + OH^- \longrightarrow 2H_2O$$

Acids that can furnish more than one ionizable hydrogen give up H^+ ions step by step. The first step is always more extensive than the second, and so on.

New Terms and Concepts

ARRHENIUS' THEORY: When electrolytes dissolve in water, they dissociate to form solutions containing positive and negative ions.

BRØNSTED-LOWRY THEORY: An acid is a substance that donates protons to another substance, and a base is a substance that accepts protons from another substance.

COLLOID: When a substance is dispersed in a gas or a liquid and the particles are larger than molecules but too small to settle out, it is called a colloid. The dispersion is also called a colloid.

CONCENTRATED SOLUTION: A solution that contains a large amount of solute in a small volume of solution.

DILUTE SOLUTION: A solution that contains a small amount of solute in a large volume of solution.

DISSOCIATION: The process in which a molecule breaks up into atoms or ions.

ELECTRICAL CONDUCTIVITY: The ability to transmit an electric current.

ELECTROLYTE: A substance that dissolves to produce a solution that conducts electricity; or, a solution that conducts electricity.

HYDRONIUM ION: The hydrated hydrogen ion, H_3O^+.

IMMISCIBLE: Two liquids which are insoluble in one another are said to be immiscible.

IONIZATION: The process in which an atom or a molecule breaks up into two or more electrically charged fragments.

MOLAL BOILING POINT ELEVATION: The difference between the boiling point of a pure solvent and the boiling point of a solution containing 1 mole of solute per 1000 g of the pure solvent.

MOLAL FREEZING POINT LOWERING: The difference between the freezing point of a pure solvent and the freezing point of a solution containing 1 mole of solute per 1000 g of the pure solvent.

MOLALITY: The number of moles of solute per 1000 g of solvent present in a solution.

NONELECTROLYTE: A substance that dissolves to produce a solution that does not conduct electricity; or, a solution that does not conduct electricity.

NONVOLATILE SOLUTE: A solute with very little tendency to pass from the liquid or solid state into the gaseous state.

POLYPROTIC ACID: An acid whose molecules contain more than one ionizable hydrogen and that yields them in successive stages of dissociation in solution.

SATURATED SOLUTION: A solution is saturated if the presence of undissolved solute produces no further dissolving.

STRONG ELECTROLYTE: An electrolyte that forms a solution that is a good conductor of electricity; or, a solution that is a good conductor.

WEAK ELECTROLYTE: An electrolyte that forms a solution that is a poor conductor of electricity; or, a solution that is a poor conductor.

Testing Yourself

12.1   If 20 g alcohol (C_2H_6O) is dissolved in 500 g water, what is the molality of the solution?

12.2   What will be the freezing point of the solution in problem 12.1?

12.3   Maple syrup is principally a solution of sucrose ($C_{12}H_{22}O_{11}$) in water; the amount of flavoring and coloring matter present is very small. If maple syrup freezes at $-0.50°C$, what is its molality? What mass of sucrose is present in a batch of syrup that contains 1000 g water?

12.4   So-called "permanent" antifreeze for automobile radiators is largely composed of ethylene glycol, $C_2H_6O_2$. If an antifreeze solution is made by mixing 95% water and 5% glycol (percentages based on weight), at what temperature will it freeze?

12.5   The freezing point of 0.10 m C_2H_5OH in water is $-0.186°C$. The freezing point of 0.10 m HCl in water is $-0.372°C$. The freezing point of 0.10 m $HC_2H_3O_2$ in water is $-0.193°C$. How can these differences be explained?

12.6   In what way is the dissolving of a solid analogous to its melting? (Consider what happens to the arrangement of molecules when a molecular solid dissolves.)

12.7   Is the temperature of a paved road in the winter increased by spreading salt on it? How does the spreading of salt cause the ice on a cold highway to melt?

12.8   Some fish that inhabit Antarctic waters have blood that freezes at $-2°C$. Blood is composed of red and white corpuscles and platelets suspended in plasma. Plasma is a water solution containing many different solutes. How do you suppose the blood of these unusual Antarctic fishes differs from that of most animals, whose blood freezes at about $0°C$?

12.9   List the following solutions in order of decreasing freezing point: 2 m potassium chloride, 1 m calcium chloride, 1 m alcohol, 0.5 m sugar, 0.5 m sodium chloride.

12.10 Complete and balance each of the following equations. All substances are in solution unless otherwise noted.

$KOH + HC_2H_3O_2 \longrightarrow ?$
$Ca(OH)_2 + HBr \longrightarrow ?$
$(NH_4)_2CO_3 + Sr(NO_3)_2 \longrightarrow SrCO_3(s) + ?$
$Li_2CO_3 + H_2SO_4 \longrightarrow CO_2(g) + H_2O + ?$
$Mg(OH)_2 + SO_2(g) \longrightarrow ?$

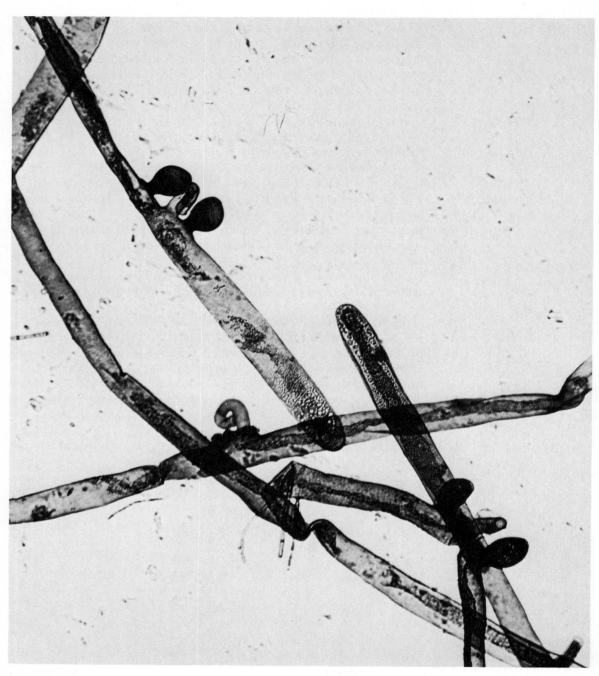

Increased pollution by phosphorus compounds accelerates the natural aging of lakes by the excessive growth of aqueous plants such as green algae, seen in the photomicrograph. A common phosphorus compound is symbolized in the drawing.

13. How Can We Keep Our Water Pure?

Of the three kinds of pollution that plague us—solid waste disposal, air pollution, and water pollution—the last has been with us for the longest time. It arises from the activities of the most people, and it is the most difficult to control. Many of us throw away trash, drive cars that pollute the air, use gadgets that consume electricity. But all of us—from families living on their farms to the millions packed into apartments in our cities—personally produce waste products that must be kept out of our drinking water.

We are finding new ways of recycling solid wastes so that our vast accumulations of trash may be turned into reusable materials. We are building new plants to make use of well-known technologies for purifying stack gases and recovering valuable sulfur compounds and useful fly ash. We are designing absorbers to take the poisons out of the exhaust gases from internal combustion engines. But keeping pollutants out of our water is a much more widespread problem, and the likelihood of turning pollutants into profits is remote.

Strong political pressures will have to be brought to bear to force industries and municipalities to remove pollutants from their effluents effectively. The citizen-voters must have a sound basic understanding of the problems involved in minimizing water pollution if they are to be effective in making their government respond to the challenge. Chapter 13 is designed to provide this basic understanding.

13.1 What Is Our Problem?

Our water problem is twofold:

1 How can we get the quality of water we need?
2 How can we balance its use among domestic needs, agriculture, and industry?

The uneven distribution of our population accentuates these problems. Along the Atlantic and Gulf coasts and around the Great Lakes we have a dense population, heavy industrialization, limited agriculture, and good supplies of water. There our chief effort must be directed to keeping these supplies free of pollution so that water can be used over and over again. In the arid Southwest and in southern California we have major centers of population, some industrialization, extensive agriculture, and strictly limited supplies of water. There the problem is principally one of allocating water among cities, farms, and industries. Water resource management is largely a political, economic, and social problem. Minimizing pollution is principally a scientific and technological problem, but it has considerable social and economic impact. Efforts to improve water quality must have strong political and social backing. In this book we are concerned mainly with scientific and technological problems, and only occasionally with social factors.

The Council on Environmental Quality, an agency in the executive branch of the United States government, in its third annual report (August 1972) indicated that the quality of air in our country had improved in recent years but that pollution of our water supplies had worsened. Our average total water *supply* (the average discharge of all streams) is 1.2 trillion gal/day, and our average *use* in 1960 was 270 billion gal/day, or 22 percent of the total supply. Except in the Southwest, we have plenty of water to meet our needs. Elsewhere, our problems arise because we use our streams and rivers as sewers to carry wastes out to sea. To improve the quality of our water, we must make sure that the 22 percent of our total supply that we circulate through our homes, farms, and factories is returned to our streams and rivers with a minimum amount of pollution. If we achieve this goal, we shall have plenty of pure water to meet our needs (except for shortages in the Southwest) for many years to come.

13.2 What Do We Mean by "Pure Water"?

Because liquid water is such an excellent solvent, it is never found in nature as a pure substance. It always contains some solutes. Natural waters also contain microorganisms. If these contents of water do not impair our bodily functions when we drink it, we often say that the water is "pure," meaning that it is "fit to drink." We are so accustomed to drinking water containing some solutes that really pure (distilled) water is unpleasantly "flat," and often we do not like the taste of it.

Water that is of acceptable quality for one use may not be for another. Water that is unsuitable for man to drink may be quite all right for animals. Water that will sustain fish and other aquatic animals may be unfit for human drinking but quite acceptable for swimming, waterskiing, boating, etc. Water that is too contaminated to support aquatic life may be acceptable for cooling, as in a power plant. Water that is too salty (contains too many dissolved salts) to drink may be quite useful for irrigating food crops. Ships may navigate in water that is too polluted for any other purpose. Clearly, our definition of polluted water depends on what we want to use it for. We might, then, generalize our definition of a pollutant as *any substance that prevents the use of water for some specified purpose.*

13.3 Kinds of Pollutants

The signs of pollution are obvious. Poor drinking water tastes and smells bad. Scums of algae cover the surface of ponds and slow-moving streams. Matted masses of weeds clog waterways. Lakes, rivers, and ocean shore waters emit foul odors. Beaches are oil-stained and covered with debris. Game fish are replaced by less desirable species, and commercial fisheries fail. The flavor of fish is contaminated, and the meat of some becomes toxic. Dead fish and rotting vegetation clog surface waters. Many different substances can be pollutants. For convenience in studying their effects, we may classify them as follows.

protozoa
virus

1 agents that cause diseases: bacteria, *protozoa* (one-celled animals), *viruses*
2 wastes that use up dissolved oxygen in water
3 solutes that cause excessive growth of aquatic plants

organic compound
inorganic compound

4 dissolved *organic compounds* (carbon compounds)
5 dissolved *inorganic compounds* (compounds of elements other than carbon)
6 oily scums
7 suspended solids (sediments) and colloids
8 radioactive materials

thermal pollution

The heating of streams and lakes (called *thermal pollution*) by power plants and factories gives rise to undesirable changes in aquatic life. Usually, the effluent from a given source of pollution contains several different types of pollutants. Untreated sewage often contains the first three listed above, along with sediments. An industrial waste may well contain organic and inorganic solutes, oily insoluble material, and substances that use up dissolved oxygen.

13.4 Waterborne Diseases

pathogen

coliform

For thousands of years man has been plagued by diseases spread by drinking impure water: typhoid and paratyphoid fevers, dysentery, cholera, poliomyelitis, infectious hepatitis. The feces discharged by a person suffering from one of these waterborne diseases contain large numbers of the organisms that cause the illness. If they get into drinking water used by others who have not been immunized by vaccines or "shots," an epidemic of the disease will break out. These disease-producing organisms (called *pathogens*) do not live for long in water, and so they may not be discovered even in samples taken at regular intervals. But all feces contain large numbers of *coliform* bacteria that live in the large intestine without any harm to the host. These bacteria are easily detected, and their presence in water indicates that it is polluted with feces and *might* contain pathogenic organisms. Their absence indicates that fecal material is not present in the water, and pathogens are *probably* absent.

Cholera epidemics are still common in some parts of the world, but waterborne diseases have been sharply reduced in countries where public water supplies are well developed and are given modern treatment. As public water supplies and public sewer systems were constructed in the late nineteenth and early twentieth centuries in the United States, deaths from typhoid fever fell from 45 per 100,000 persons a year in 1880 to practically zero in 1970. This dramatic drop is due very largely to the sterilization of public water supplies by chlorine, which kills all microorganisms when added to water in carefully controlled amounts. Unpleasant tastes and smells of properly chlorinated drinking water come from the presence of various organic pollutants that combine with chlorine during the treatment. These undesirable qualities do not come from the killing of microorganisms. To further decrease the spread of waterborne disease organisms, the chlorination of municipal sewage is now being encouraged. Today we have the technology for eliminating waterborne diseases from the face of the earth. Time, money, and political ineptitude are the only remaining obstacles to this desirable achievement.

13.5 Wastes That Use Up Dissolved Oxygen in Water

invertebrate

Since the oxygen dissolved in water is used for respiration by aquatic plants and animals, any added material that uses up this oxygen will interfere with the normal and natural growth of the aquatic organisms. Fish require the largest amounts of dissolved oxygen, *invertebrate* animals (protozoa, worms, clams, insects, shrimps, etc.) can live with less, and bacteria can function with least of all. For a diversified population, including game fish, the amount of dissolved oxygen in water should be at least 5 ppm (parts per million), that is, 5 g oxygen for each 1,000,000 g water. If the dissolved oxygen present in a sample of water is lower than this amount, fish suffer most and tend to die out. The populations of invertebrates and bacteria then rise to abnormal levels. This imbalance in living species, marked by the disappearance of fish, is a sign of pollution.

Almost all organic (carbon- and hydrogen-containing) compounds in water serve as food for bacteria, whose metabolic processes convert the carbon to CO_2 and the hydrogen to H_2O:

$$C \quad + \quad O_2 \quad \longrightarrow \quad CO_2 \qquad\qquad (13.1)$$

(present in   (dissolved
organic compounds)   in water)

$$4H \quad + \quad O_2 \quad \longrightarrow 2H_2O \qquad\qquad (13.2)$$

(present in   (dissolved)
organic compounds)

When any organic pollutant is added to water, the concentration of dissolved oxygen is quickly lowered by this bacterial action to levels at which more complicated animals like fish and the higher invertebrates cannot live.

Equation (13.1) shows that every 12 g carbon uses up 32 g dissolved oxygen. Thus each gram of carbon in a pollutant will use up roughly 3 g dissolved oxygen to form CO_2; or we can say that 3 ppm of carbon will use up 9 ppm of dissolved oxygen. The amount of carbon in *one drop of oil* will use up about the amount of oxygen dissolved in *1 gal of water*. Hence organic pollutants take a heavy toll of dissolved oxygen. Such pollutants are present in human and animal wastes and in effluents from food canneries, meat packing plants, slaughterhouses, paper mills, chemical factories, and tanneries. When the dissolved oxygen falls to low levels, algae and other water plants die and add to the burden of organic matter requiring oxygen for its removal.

biochemical oxygen demand (BOD)

The amount of dissolved oxygen used up during the oxidation by bacterial action of the organic matter present in a sample of water is called the *biochemical oxygen demand*, or *BOD*. The BOD is determined as follows. The amount of dissolved oxygen present in one portion of a given sample of water is determined by a chemical analysis. Another portion of the same water is incubated for 5 days at 20°C in the presence of a measured excess of gaseous oxygen. At the end of the incubation, the remaining oxygen (both gaseous and dissolved) is measured. The difference in dissolved oxygen (DO) present before and after incubation and the amount of gaseous oxygen consumed during the incubation are used to calculate the BOD.

Water is considered pure if its BOD is 1 ppm or less, fairly pure with a BOD of 3 ppm, and of doubtful purity when the BOD is as much as 5 ppm. Discharge of waste water with a BOD of 20 into a stream is considered undesir-

TABLE 13.1 *Biochemical Oxygen Demand of Various Effluents*

Effluent	BOD
Untreated municipal sewage	100–400 ppm
Runoff from barnyards and feedlots	100–10,000 ppm
Food-processing wastes	100–10,000 ppm

able by public health authorities. The burdens that some common pollutants place on streams and rivers are summarized in Table 13.1. Obviously, such effluents must be diluted with huge quantities of water by the streams into which they flow if the dissolved oxygen is not to fall to dangerously low levels.

We can more easily understand the problem of pollutants from various industrial processes by comparing their amounts with pollution from human wastes (Table 13.2). From these data we see that domestic animals—most of which are raised for meat—impose a tremendous burden on the environment, a burden comparable to the wastes from almost 2 billion people (10 times the present population of the United States!). If animals being raised for meat are fed in large numbers in a small space (in feedlots) instead of farm pastures, they produce considerably more pounds of meat per pound of feed. This process cuts the cost of meat, but the disposal of manure from feedlots is much more difficult than the use of it to fertilize farmlands. Gain in efficiency of meat production is offset by loss of manure as fertilizer and by almost intolerable pollution of rivers and streams. Furthermore, when manure is not used as fertilizer, more artificial fertilizer must be manufactured, using

TABLE 13.2 *Wastes from U.S. Industries and Number of People Producing Equivalent Amounts of BOD*

Industry	Equivalent number of people
Canneries	5 million
Cotton processing	8 million
Dairy processing	12 million
Meat slaughtering	14 million
Paper processing	216 million
Domestic animal production	1900 million

up irreplaceable mineral resources. Some nonpolluting processes for the disposal of animal wastes will be discussed later in this chapter.

13.6 Pollutants That Cause Excessive Growth of Aquatic Plants

The growth of green plants requires large amounts of carbon in the form of CO_2, large amounts of hydrogen and oxygen in the form of H_2O, and smaller amounts of many other elements that may be present in various kinds of compounds. Fifteen to twenty elements are necessary for the growth of most green plants, including algae. Usually plenty of CO_2 and H_2O is available, so that the rate of growth is controlled by the limited amounts of nitrogen and phosphorus in the water. Other elements required for growth are usually present in natural waters in more than adequate concentrations. When larger than normal concentrations of nitrogen and phosphorus are present in a lake, several species of algae grow very rapidly. These produce a *bloom* when the count of individual algal cells in the water is more than 500 per cm³. This bloom is an unsightly green scum accompanied by unpleasant odors and tastes in the water. When the bloom dies and decays, it makes heavy demands on dissolved oxygen. The consequent lowering of the oxygen content of the water upsets the desirable balance of aquatic organisms (Sec. 13.5).

bloom

The relative amounts of the most important of the 20 elements required for plant growth are reflected in the percentage composition of algae shown in Table 13.3.

Under natural conditions the plant nutrients in a lake increase in concentration as streams bring in dissolved compounds of nitrogen, phosphorus, potassium, and sulfur from decaying plant and animal refuse on land. Compounds of calcium and other elements needed as nutrients are leached (dis-

TABLE 13.3 *Approximate Composition of the Algae That Form Blooms*

Substance	Percent	Substance	Percent
Water	45–50	Iron	
Carbon	35–50	Magnesium	0.3–0.04
Nitrogen	2–10	Sodium	
Potassium	1–6	Manganese	
Phosphorus	0.5–1	Zinc	
Sulfur	0.3–2	Copper	0.1–0.0004
Calcium	0.3–2	Boron	

Excessive growth of water plants such as the American lotus is encouraged by the dumping of pollutants into our waterways.

solved by percolation of water through the soil) from the soil and enter the lake water. This enrichment of the water by nutrients is called *eutrophication* (from the Greek words meaning "well nourished"). Such enrichment leads to the natural aging of lakes. As water plants grow more luxuriantly, animal populations increase; as plants and animals die, the deposits of organic matter on the bottom of the lake build up. Addition of sediments from muddy streams in spring floods helps build up shallows where the water becomes warmer and plants and animals grow faster. Plants take root in these shallows and gradually occupy a larger and larger part of the lake, which becomes edged with marshland. Finally the marsh is overrun by land plants, and the lake is completely filled in. This process of aging normally takes thousands of years, but it is visibly accelerated when large amounts of plant nutrients are dumped into a lake by man's activities. The immediately visible results of the pollution of natural waters are the development of algal blooms, the large-scale death of fish, and the replacement of game fish by species able to live in water with lower concentrations of dissolved oxygen.

The effect of adding a nutrient to a lake depends partly on the nutrients already present from natural sources. If the water of the lake is low in nitrogen but well supplied with dissolved CO_2, phosphorus, and the other elements listed in Table 13.3, then the addition of nitrogen compounds in pollutants will stimulate the growth of algae, but compounds of phosphorus and other elements will have no effect. On the other hand, if the water has a deficiency of phosphorus but plenty of the other nutrients, the addition of phosphorus compounds as pollutants will stimulate the growth of algae, but the addition of compounds of nitrogen and other elements will have no effect. A shortage of carbon appears to limit plant growth in only a few lakes that happen to contain large amounts of all the other nutrients. There is little evidence that any elements other than nitrogen, phosphorus, and carbon are likely to be present in quantities small enough to limit the growth of algae.

Because the balance of plant and animal species in a lake depends on so many interrelated variables, such as temperature, input of sediments, and relative amounts of nutrients, it is hard to determine just what pollutant causes eutrophication and upsets the biological balance. Environmental scientists generally agree that nitrogen is a limiting nutrient in many estuaries and other coastal waters. In the past few years as much as 80 percent of the nitrogen and 75 percent of the phosphorus present in surface waters in the United States originated from manmade sources. Municipal sewage is the main source of phosphorus, of which 70 percent comes from the use of household detergents. Much of the controversy over the use of phosphate detergents arises because we lack knowledge of the effects of other pollutants added simultaneously. Nitrogen in various compounds comes about equally from sewage and from the runoff of the fertilizer used on intensively cultivated land. Although we cannot say that the addition of these pollutants is universally detrimental to our waterways, we have experienced enough problems with the dying of fish and the clogging of streams with excessive growth of water plants to be sure that we should cut down on the quantities of nitrogen and phosphorus compounds that we are dumping into our waters.

13.7 Pollution by Dissolved Organic Compounds

Millions of compounds containing carbon, hydrogen, oxygen, and some other elements exist in natural materials, and millions more have been synthesized by man. These compounds are referred to collectively as organic compounds. It is useful to define organic compounds as those composed predominantly of carbon. Natural organic materials have been present in the environment for millions of years, and living organisms have adapted to using them in their life processes or to tolerating them if they are not useful. Many of the chemicals synthesized by man are quite harmless to living organisms, but many others interfere with normal biochemical processes or even stop them altogether. Many of these synthetic organic chemicals are deliberately introduced into the environment to kill insects, fungi, weeds, and other unwanted living species. They are intended to make man's life more livable by protecting him against other organisms that cause him trouble. To understand some of the hazards involved in the use of these pesticides (killers of pests), we need more background in chemistry. These and other chemical agents used in agriculture will be considered in more detail in Chap. 21, which deals with the impact of modern agriculture on the environment.

Here we are concerned with chemicals that are not deliberately added to the environment but that get into our waterways by way of the wastes from chemical industries. Between 1943 and 1970 the production of synthetic organic chemicals increased 1400 percent. These substances are used as fuels, plastics, fibers for cloth, solvents, detergents, paints, inks, dyes, insecticides, *herbicides*, *fungicides*, food and beverage additives, drugs, and pharmaceuticals. In 1970, 138 billion lb of synthetic organics was manufactured, and thousands of pounds escaped into industrial wastes either by accident or by design.

herbicide
fungicide

Many of these synthetics cannot be broken down by bacteria or other biological organisms in soil or water (i.e., they are not biodegradable), and so they accumulate in the environment. Some of them, like DDT, are concentrated by passage through food chains. For example, bacteria and protozoa ingest the compound and are then eaten by larger organisms which prey upon them. These larger organisms are eaten by still larger ones, and finally fish in the seas and animals on land carry concentrations of DDT considerably greater than the level in the environment as a whole.

The processes by which synthetic organic chemicals that are not biodegradable can be removed from liquid wastes will be considered in Sec. 13.14. Because they are very difficult to remove, our main effort should be to keep them from contaminating the environment in the first place.

In 1970 about 3 billion barrels of petroleum were produced in the United States, and additional billions were imported. It is not surprising that the production, transportation, and refining of petroleum and its products are marked by spills of considerable magnitude. Keeping these out of our waterways is a great problem, and much research is being done to find solid materials that will absorb spills and yield a product that can be used for some constructive purpose (Sec. 4.7).

13.8 Pollution by Dissolved Inorganic Compounds

Classifying chemicals as "organic" and "inorganic" is somewhat arbitrary, and some compounds do not fall neatly into either category. Yet the categories are generally useful, and it is convenient to define an inorganic compound as one that is not an organic compound.

The most common inorganic pollutants come from the runoff from highly fertilized crop lands and the drainage of water from mines or mine dumps. For example, the common mineral iron pyrite, FeS_2, is acted upon by some bacteria in the presence of air and water. By a process not fully understood, this action produces a solution of sulfuric acid and iron sulfate. The production of these pollutants takes place in both underground and surface mines. In deep mines, artificial fissures are formed, and the water percolating down from the surface through these cracks into deposits of FeS_2 contains enough dissolved oxygen for the bacteria to produce the polluting sulfuric acid and iron sulfate. This acid water then makes its way into groundwaters and eventually emerges in springs and streams. Water flowing from the mouths of underground mines is usually highly acid from these pollutants. Surface waters running from mine dumps are likely to be similarly polluted. Because iron pyrite is present in many coal seams, the streams in coal mining areas become highly polluted with acid mine drainage. It is estimated that the total acid produced each year in mine drainage in the United States is the equivalent of more than 4 million tons of sulfuric acid, and about 60 percent of this pollutant comes from *abandoned* mines!

Most natural freshwaters are very nearly neutral; some are very slightly acid, and some are very slightly basic. The chemical processes involved in the life of water plants and animals proceed smoothly within these limits. But when acid pollutants are dumped into natural waters, the delicate balances of these life processes are upset. Most natural waters contain significant amounts of carbonate ions, CO_3^{2-}, and bicarbonate ions, HCO_3^-. The hydrogen ions of acid pollutants in these waters react thus:

$$CO_3^{2-} + H^+ \longrightarrow HCO_3^- \qquad\qquad (13.3)$$
$$HCO_3^- + H^+ \longrightarrow CO_2(\text{dissolved}) + H_2O \qquad (13.4)$$

Aquatic animals produce CO_2 as one product of their metabolism. This CO_2 is carried by the blood to the respiratory organs (gills in fish), where it diffuses into the water bathing these organs. If there is a high concentration of dissolved CO_2 in the water, the CO_2 in the animal's blood cannot readily diffuse into the water. The resultant accumulation of CO_2 in the blood interferes with the transport of oxygen to the animal's tissues and it dies. Increased acidity interferes not only with the respiration of aquatic organisms but also with other metabolic processes. Exposure to acid water kills plants as well as animals.

Pollution by acids has the following effects:

1 the destruction of most higher plants (only some bacteria and algae survive)
2 the killing of aquatic animals (all *vertebrates*, most invertebrates, and many microorganisms)
3 the rapid corrosion of structures in water (metal pipelines for water and sewage, concrete reinforced with steel, iron piers for bridges and docks, iron gates in canals and locks, metal ships)
4 the damaging of irrigated crops

Much research is being done to find ways of preventing pollution by acid mine water. The addition of solid hydrated lime [calcium hydroxide,

vertebrate

Ca(OH)$_2$] produces a muddy slurry (mixture of water and suspended solids). The sulfuric acid is neutralized thus:

$$H_2SO_4 + Ca(OH)_2 \longrightarrow CaSO_4 + 2H_2O \qquad (13.5)$$

Air is bubbled through the slurry, and the iron in solution is precipitated as iron hydroxide by the hydroxide ions from the calcium hydroxide:

$$Fe^{3+} + 3OH^- \longrightarrow Fe(OH)_3(s) \qquad (13.6)$$

The slurry is pumped into a lagoon where the undissolved iron hydroxide and calcium sulfate settle out. The clear, purified water is run into streams. While this treatment reduces water pollution, it does not solve the problem of what to do with the sludge from the lagoons. A pressing problem of water pollution is replaced by a lesser problem of solid waste disposal.

Effluents from industries carry inorganic compounds in great variety. These effluents must be treated by complex processes determined by the chemical nature of the pollutants. As examples we shall consider mercury and lead.

13.9   Mercury Pollution

Mercury compounds are found in nature at low concentrations in different kinds of rocks and soils, in still lower concentrations in river water, and in minute amounts in air. Apparently, living organisms are not adversely affected by these concentrations in nature. But man's uses of mercury have so increased the concentration of its compounds in some parts of the environment that mercury poisoning has become a hazard.

Between 1953 and 1960, 111 people died or were seriously injured by eating fish from Minamata Bay (in Japan), which is highly polluted by mercury compounds. In 1966, 20 persons died and 45 more were injured in Guatemala from mercury poisoning.

The contamination of foods by mercury seems to come by two routes.

1 Mercury compounds dissolved in water are passed up through food chains from microorganisms to fish, where they are concentrated.
2 Mercury compounds used as fungicides on seed grains are absorbed by the plants that sprout from the seed and are passed on to animals that eat the plants.

The U.S. Food and Drug Administration has established 0.5 ppm mercury as the upper limit acceptable for food and 0.005 ppm for water. Fish which have been exposed to only natural concentrations of mercury in water carry about 0.005 to 0.075 ppm mercury in their tissues. Trout exposed to nonlethal concentrations of mercury compounds for some time acquired from 23 ppm mercury in their blood to 4 ppm in their muscles. Freshwater fish caught from the St. Clair River running from Lake Huron through the heavily industrialized Detroit area into Lake Erie (and some caught from Lake Erie itself) had from 0.08 to 3.57 ppm mercury in their edible tissues. Some samples of tissue from saltwater fish (tuna and swordfish) have shown 0.13 to 0.25 ppm mercury. Chickens fed grain produced from seeds treated with mercury compounds showed 0.030 ppm mercury in their tissues, compared to 0.014 ppm in chickens fed on grain raised from untreated seeds. Eggs from the chickens in the first group contained about 0.025 ppm as compared to about 0.01 ppm for the second group. Meat from pigs and beef cattle fed grain grown from seed treated with mercury compounds contained from 3 to 10 times as much mercury as that found in such animals fed grain from untreated seeds. In none of the cases reported here (except fish caught in the St. Clair River) has the level for food been exceeded. Nevertheless, we must do all we can to keep mercury out of our waters.

Although the damaging action of mercury in the body is not well understood it is clear that:

1 All mercury compounds are poisonous to some degree.
2 Some compounds are more poisonous than others.
3 Mercury compounds in the environment are changed to different compounds of mercury in the body by biological processes.
4 Mercury has a strong tendency to unite with sulfur. In body tissues it ties up sulfur-containing compounds and prevents them from carrying on their proper biological functions.
5 The damage done to living organisms is permanent; it cannot be undone.

Inorganic compounds of mercury tend to concentrate in the tissues of the liver and kidneys. Although the mercury is rapidly removed from these organisms by urinary excretion, the damage to them remains. If metallic mercury vapor is breathed into the lungs, it is quickly picked up by the blood and concentrated in the brain, where it does serious damage. When proper doses of inorganic mercury compounds are used as medicines, the body quickly eliminates them, and there is little accumulation in the tissues.

The most dangerous mercury compounds are those in which the metal is combined into organic compounds. Of these, methyl mercury ions, CH_3Hg^+, and dimethyl mercury, $(CH_3)_2Hg$, are the most deadly. When mercury metal is added to a stream or lake, it settles into the mud at the bottom of the water. Here it is transformed by anaerobic microorganisms into dimethyl mercury. This compound is quite volatile. It soon dissolves from the mud into the water. Here it is transformed into soluble CH_3Hg^+, which is readily absorbed by aquatic animals, concentrating in their fatty tissues. This compound is then concentrated further through the food chain, and it can kill people who eat fish containing large amounts of this toxic substance. The deaths at Minamata Bay in Japan were caused in this way. The U.S. Environmental Protection Agency has issued a strong recommendation that industries which discharge mercury into our water supplies should reduce the concentration of this metal to the concentration found in natural waters. But it will be many years before the mercury-poisoning threat is eliminated, because millions of pounds of the metal have been dumped into streams and lakes and now lie in the bottom muds, slowly being transformed into the highly poisonous methyl mercury compounds. The decontamination of these muds is under study, but it will be a slow and costly process. Clearly, we must prevent the addition of more mercury to our water supplies.

Mercury is produced by roasting the ore cinnabar (mercury sulfide, HgS) in air:

$$HgS(s) + O_2(g) \longrightarrow Hg(g) + SO_2(g) \tag{13.7}$$

The mercury vaporized from the roasting area is condensed to the liquid form that it has at ordinary temperatures. One of the large sources of mercury contamination is the manufacture of two widely used chemicals—sodium hydroxide and chlorine. About 1.5 million lb of mercury is consumed annually in this process. If a solution of NaCl is placed in an electric conductance cell (see Fig. 12.1), which has a negative electrode of liquid mercury and a positive electrode of solid carbon, then metallic sodium appears in the mercury and gaseous chlorine bubbles off from the carbon when electricity flows through the cell. This process is called the *electrolysis* of sodium chloride. In general, electrolysis is the term given to the processes that take place when an electric current passes through a conductance cell, usually called an *electrolytic cell* when used to produce chemicals.

Apparently, sodium ions dissolved in the water solution of NaCl pick up electrons from the negative electrode to become sodium atoms dissolved in the mercury, and chloride ions dissolved in the NaCl solution give up electrons to the positive electrode to become molecules of chlorine that are released as bubbles of Cl_2 gas. The chemical equations are

At the negative electrode: $Na^+ + 1e^- \longrightarrow Na$ (13.8)
At the positive electrode: $2Cl^- \longrightarrow 2e^- + Cl_2(g)$ (13.9)

The mercury carrying the dissolved sodium is drawn off from the cell and passed into water. The dissolved sodium reacts with water to produce sodium hydroxide (called "alkali" by the chemical industry) and hydrogen, which is a valuable by-product:

$$2Na + 2H_2O \longrightarrow 2NaOH(aq) + H_2(g) \tag{13.10}$$

After the sodium has been removed by this process, the mercury is returned to the electrolytic cell. Theoretically, there should be no loss of mercury from the electrolytic process, but actually about 800,000 lb is lost each year, mostly in the waste waters from the process. During 1970, the chlor-alkali plants were able to reduce these discharges by 86 percent, mainly by running waste waters into lagoons where the mercury settles out and from which the purified water can be recirculated through the plant.

The manufacture of electrical apparatus consumes 1,400,000 lb mercury annually for use in mercury vapor lights, mercury switches for electric circuits, and mercury batteries, which last longer and are less damaged by high temperature and high humidity than ordinary batteries. There is mercury in fluorescent light fixtures. They should be handled carefully when discarded; breaking them releases mercury into the environment.

Additional millions of pounds of mercury have been used in the manufacture of compounds to prevent fungi and mildew from growing on seeds used for raising food grains, on paints used in damp climates, and on paints especially formulated for use on ships and other structures exposed to water.

Coal contains various amounts of mercury, averaging around 1 ppb (part per billion). Although this proportion is small, the amount of mercury released into the atmosphere from burning fossil fuels is estimated to be more than 11 million lb/yr.

Sludge from municipal sewage-treatment plants contains about 1 ppb mercury even in cities where no mercury is used in industry. This contamination is thought to come from pharmaceuticals, disinfectants, and paints used in homes. Each year about 1000 lb mercury makes its way into the waste waters from a population of about 1 million people.

Because mercury gets into our environment from so many different sources, it seems unlikely that a single line of attack will solve the problem. But we must be continually on our guard to make sure that mercury and its compounds are not used for any purpose for which an alternative, less poisonous substance is available. We cannot eliminate mercury from our environment, but we can greatly reduce the amounts we deliberately introduce into it.

13.10 Lead Pollution

Undesirable and unnatural concentrations of lead are found in our air, our water, and our soil in some areas, particularly near lead smelters and heavily traveled automobile freeways. Long vertical columns of ice have been drilled from deep deposits in Greenland and sections of them analyzed for lead. Ice deposited by snow that fell before 1750 contained less than 0.01 ppb lead. The concentration of lead in ice from snow falling in the two centuries from 1750 to 1940 increased fairly regularly to about 0.08 ppb in 1940, when tetraethyl lead was first introduced into gasoline on a fairly wide scale. Snow that fell in 1950—only 10 years later—contained $2\frac{1}{2}$ *times* this concentration of lead! Obviously, the increased number of cars and trucks manufactured and driven during that decade and the use of leaded gasoline had a tremendous effect on the quality of the air. In large cities we now seem to have reached plateaus of lead contamination as the winds and rains continually spread lead through the countryside. The concentration of lead in our atmosphere as a whole continues to increase rapidly.

The burning of leaded gasoline contributes 98 percent of the lead pollution in our air. Most of it comes out of the tailpipes of our cars as lead bromide–chloride, $PbBrCl$, and a more complex compound $PbBrCl \cdot 2PbO$. The compounds $Pb(CH_3)_4$ (tetramethyl lead) and $Pb(C_2H_5)_4$ (tetraethyl lead) are added to gasoline to make it burn more smoothly under high compression and thus reduce the "knock" that comes from too rapid explosion of the gasoline-air mixture ignited by the spark plug inside the cylinder of an engine. Some ethylene dichloride, $C_2H_4Cl_2$, and ethylene dibromide, $C_2H_4Br_2$, are also added to the gasoline, so that the lead released by the burning of the tetramethyl and tetraethyl compounds will react to form volatile compounds $PbBrCl$ and $PbBrCl \cdot 2PbO$ and thus escape from inside the engine. If the lead remained inside, it would clog the valves in the combustion chambers.

Children living in slum housing have been poisoned by eating lead-containing paints that peel from interior walls. Lead solder in tin cans containing foods has caused some poisoning. The consumption of "moonshine" whiskey has caused lead poisoning in some parts of the country. The pipes and automobile radiators used in homemade stills contain considerable amounts of lead solder. Lead gets into the whiskey in poisonous amounts—sometimes as much as 20 times the maximum allowed in food and beverages by the U.S. Public Health Service.

The lead released into the atmosphere by motor vehicles, by lead smelters, by brass manufacturing and casting, and by other processes involving the use of lead is washed from the atmosphere onto the ground and into our lakes and streams. To date, the only soils and waters that are sufficiently contaminated with lead to be dangerous to humans are found near lead smelters.

Beautiful and deadly colors are produced on ceramics by glazes containing lead compounds. Unfortunately, these glazes are soluble in the acids of many common foods and beverages (like fruit juices.) Poisonous concentrations of lead are found in foods allowed to stand in lead-glazed containers for as little as 1 hr. In 1970 a two-year-old child in Montreal died from drinking apple juice kept in a lead-glazed jug. This tragedy stimulated the study of ceramic dishes as sources of lead poisoning, and half of those examined were found to be dangerous. These included pieces of imported pottery, dishes commercially manufactured in the United States, and handcrafted specimens. The U.S. Food and Drug Administration has established 7 ppm as the maximum allowable amount of lead that may be released from glazes on dishes used for food and drink.

About 30 percent of the lead in the air we inhale is absorbed, but five-sixths of this is soon excreted. About 5 to 10 percent of the lead consumed in food or drink is absorbed in the gastrointestinal tract; the rest is excreted. The presence of lead in the form of lead ions, Pb^{2+}, in the body interferes with the action of a chemical compound necessary for the production of hemoglobin. This interference arises from the strong tendency of Pb^{2+} to combine with sulfur in the compound and so to inhibit the activity of this vital substance.

13.11 Oily Scums, Suspended Solids, and Colloids

Millions of barrels of crude petroleum are pumped from the earth every year through wells drilled on land and in shallow seas. This crude oil is transported to refineries through pipelines and in tanks aboard ships, railroad cars, and trucks. The gasoline, diesel fuel, furnace oils, lubricating oils, and greases produced in the refineries are transported by rail and highway to consumers. With all this handling of crude oil and products refined from it, there is bound to be spillage. Large slicks (floating islands) of oil have been created on the oceans by the wrecking of tankers and by leakage from oil wells drilled in shallow seas. About 4 to 5 million tons of oil are added to the ocean as a pollutant each year; some of this is deliberately done by bilge pumping and tanker washing at sea. Small slicks of oil are produced by the discharge of oily wastes from the engines of ships in fresh and salt waters.

The insoluble oil forms a very thin layer on the water which inhibits the exchange of O_2 and CO_2 between the air and water. By upsetting the respiration of organisms in the water, these oil spills kill many waterfowl and fish. They can be cleaned up by absorbing them in solid material (like straw, sawdust, peat moss, shredded automobile tire rubber) and burning or burying the oily mass. But burning it pollutes the atmosphere, and burying it pollutes the earth. About the only use that can be made of it is to incorporate it into bituminous (asphaltic) paving materials. Small spills of oily products can be tolerated by the organisms living in soil, but heavy soaking kills them. The only way to dispose of such sterile soil is to use it for paving or to bury it where its pollutant will not be leached into groundwater. Obviously, we must make every effort to minimize oil spills and to enforce their cleaning up by the agencies that make them.

Washing soil into waterways is bad not only because of the loss of the land that produces the sediment but also because heavy sediments smother the growth of valuable organisms in the bottom mud. The death of these organisms then upsets the biological balance in the water, and many desirable fish are lost. The presence of solids that do not settle out but that form colloidal suspensions cuts down on the amount of light that penetrates the water and is used in photosynthesis. This, too, upsets the biological balance, because reduced photosynthesis leads to reduced dissolved oxygen in the water. A colloidal suspension like whey from cheese factories is particularly troublesome, because it not only reduces production of dissolved oxygen by photosynthesis but also requires much dissolved oxygen to transform it into harmless solutes. Reduced visibility in clouded water also prevents fish from finding their food, so that the biological balance is further disturbed. Cloudy water is undesirable for domestic uses and has to be clarified in city water-treating plants; this adds to the cost of potable water.

13.12   Radioactive Pollutants

The dangers of radioactive pollutants in the atmosphere were discussed in Sec. 9.8. Radioactive compounds also get into water supplies by dissolving from the dumps at uranium mines and ore-processing mills. Only about 0.1 to 0.3 percent uranium is present in ores of this element. To produce the uranium we need requires handling huge amounts of ore, which has to be mined, crushed, ground to a fine powder, and leached with acid or alkali. The recovery of the uranium from these solutions involves precipitation of insoluble uranium compounds and the *extraction* of the precipitates with various solvents. After the useful uranium compounds have been removed from the rock, the refuse that is left is piled up in dumps of *tailings*. These tailings contain significant amounts of radioactive thorium, $^{230}_{90}$Th, and radium, $^{226}_{88}$Ra. Radium is in the same family in the periodic table as calcium; when it is present in food or drink, it tends to replace some of the calcium in calcium phosphate, $Ca_3(PO_4)_2$, in the bones. Thorium also is somewhat similar chemically to calcium, and it forms compounds that deposit in the bones.

extraction

tailings

Rain and melting snow soaking into dumps of tailings leach out some radioactive compounds and run into water supplies. There are about 12 million tons of tailings piled up in the Colorado River basin. Some waters in the basin have contained concentrations of radium twice as high as the maximum allowable for human consumption. The best way to prevent the leaching of radioactive wastes from tailings dumps is to grade them so that the rain and melting snow run off quickly and do not penetrate far into the mass. Planting vegetation on the dumps also reduces the dispersion of radioactive pollutants by wind or water.

13.13   Thermal Pollution

Most of our electric power is generated by dynamos (electric generators) driven by steam turbines. When the steam leaves the turbine, it is liquefied by cooling in a condenser, and the condensate is returned to the boiler. The condenser is cooled by water pumped from a lake or river near the power plant. This cooling water may be warmed as much as 20°F by its passage through the condenser. When it is returned to its source, it often causes serious biological disturbances in the lake or river. The addition of excess heat giving rise to these disturbances is called *thermal pollution.*

Thermal pollution has three bad effects:

1 The amount of oxygen dissolved in the water is reduced because the gas is less soluble at high temperatures. This situation is particularly bad in deep lakes. Because warm water is less dense than cold water, it tends to float at the surface, where it impedes the dissolving of oxygen into the deeper layers, thus stifling the respiration of many organisms at all levels. Fish may suffocate and die in these circumstances.
2 Biochemical processes are speeded up by increases in temperature, and the metabolic cycles of many organisms are upset by such increases. In warm water, fish respire more rapidly to meet the oxygen demands of their tissues. Their shortage of oxygen is aggravated by the decreased concentration of dissolved oxygen at high temperatures.
3 Temperatures may be so high that the biochemical reactions of affected organisms are drastically upset, and they die immediately.

The spawning, fertilization, and hatching of fish eggs are closely geared to water temperatures. When these processes are speeded up in warmer water, the food required by the newly hatched young may not be present; this food shortage will have disastrous effects on the population of fish. Some organisms flourish in warm waters and others in cold; changing the temperature of a stream or a lake by thermal pollution changes the nature of the dominant populations of both plants and animals.

Water to be used as a coolant for condensers is often chlorinated to prevent the growth of undesirable slimes (algae) that build up in the pipes and impede the flow of heat from the steam to the cool water. When this chlorinated water is discharged into a river or a lake, it kills many desirable organisms.

Water from steam condensers can be cooled by allowing it to trickle down over porous material in a tower with a stream of air blown upward through the packing. This process transfers the thermal pollution from the water to the atmosphere where it is less troublesome, although in very damp climates, dense fogs may be produced by the additional water vapor in the air. In England, where the climate is generally damp, there have been serious problems with fogs from cooling towers. In dry climates, evaporative cooling towers are quite successful in reducing thermal pollution of water. Under these conditions, the heated water from the condenser may be circulated through pipes over which atmospheric air is blown, but the cooling so achieved is much more costly.

Another alternative is to run the warm water from the condensers into lagoons. These may attract waterfowl in cold winters and may also be planted with fish that flourish in warm water. If a power plant is situated in rich farmland, the cooling water may be used for irrigation. The warm water speeds the growth of many crops, but is expensive to pipe for more than a very short distance. In the future, specially designed greenhouses may be built next to power plants to use the hot water from steam condensers profitably.

13.14 The Treatment of Municipal Wastes

In the United States our public water supplies provide about 16 billion gal/day of water. After it is used, most of this water passes through sewers and is ultimately returned to waterways, from which it is later drawn again and reused by other communities. The preservation of high quality in our waters depends heavily on how well municipalities remove pollutants from sewage before discharging it into a river or lake. Industries use an additional 177 billion gal/day. This too must be treated to remove pollutants before it is returned to waterways. The treatment required by municipal sewage does not vary greatly from city to city. The treatment of industrial wastes must be highly individualized, because quite different pollutants are present in different effluents. Only the treatment of municipal sewage will be considered here.

Sewage from 1300 communities and 10 million people is dumped without treatment into our waterways. Sewage from about 50 million people receives only primary treatment; that is, it is run through coarse screens to remove large solids, then into a grit chamber where sand, cinders, and small stones settle out. The partially clarified fluid is then run into large tanks, where the finer suspended solids settle out. The mud (called raw sludge) that accumulates on the bottom of the sedimentation tank is periodically removed and

FIGURE 13.1 Primary sewage treatment

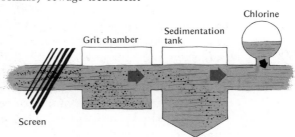

buried or burned. The effluent from the sedimentation tank is treated with chlorine to kill all disease-producing bacteria before it is discharged into a natural waterway (Fig. 13.1).

Sewage from about 80 million people receives secondary treatment also. In this process the effluent from the grit chamber of the primary process flows into an aeration tank where air or oxygen is bubbled up through the suspension. The intimate contact of the oxygen with suspended organic solids stimulates the rapid growth of bacteria that eat the solids and produce soluble compounds. The effluent from the aeration tank runs into a sedimentation tank from which clarified liquid flows to a chlorinator and then to a waterway. The sludge which settles out in the clarifier is returned to the aeration tank. Because it carries a large population of the bacteria that eat the organic

activated sludge solids, it is called *activated sludge*. Sewage, oxygen, and activated sludge are in intimate contact for several hours in the aeration tank, and 90 percent of the suspended solids and BOD is removed. Figure 13.2 shows the activated sludge process diagrammatically.

Unfortunately, some organic compounds and inorganic nitrogen and phosphorus compounds (chiefly nitrates and phosphates) can be eliminated only by tertiary treatment. The organic compounds can be absorbed from the water by passing the effluent from the sedimentation tank of the secondary treatment through a bed of granular carbon. When the carbon can absorb no more of these compounds, it can be removed and reactivated by heating it to a high temperature in the absence of oxygen. The heat drives off the accumulated organic compounds, and the carbon can be returned to the system for reuse. Inorganic phosphorus compounds can be removed from the secondary treatment effluent by adding quicklime (calcium oxide, CaO) to the liquid. A suspension of calcium hydroxide, $Ca(OH)_2$, is formed; it reacts with the phosphorus present to form insoluble calcium phosphate, $Ca_3(PO_4)_2$. Some of the undissolved calcium hydroxide absorbs organic phosphorus compounds,

FIGURE 13.2 The activated sludge process

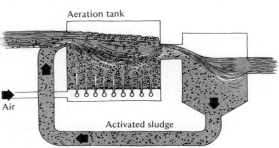

and filtration of the milky suspension removes most of the phosphorus. Because all nitrates are soluble, nitrogen can best be removed from secondary treatment effluents by growing organisms in the water. These use the nitrate for food, and they can be filtered out. This process is still in the experimental stage and is not used commercially. Other tertiary treatment processes for removing dissolved inorganic salts are being studied but are not yet well enough developed to be considered here.

Because all the tertiary treatments developed to date are quite costly, they will be widely adopted only when their efficiency is improved and their cost brought down. Meantime, we must make every effort to minimize the amounts of pollutants we put into our water so that we can minimize the cost of removing them.

Summary

Our water problem is twofold: to minimize pollution in regions with good water supplies, and to allocate water resources among homes, industries, and farms in regions with limited supplies. A pollutant is any substance that prevents the use of water for some specified purpose: for drinking, bathing, recreation, irrigation, or industry. Pollutants may be dangerous to the health or detrimental to the ecological balance in bodies of water. Pollutants dangerous to health include disease-producing organisms, pesticides that make their way into drinking water, mercury and lead compounds, and radioactive materials. Pollutants ecologically detrimental include oily scums, suspended solids, wastes that use up dissolved oxygen in water, and those that cause excessive growth of aquatic plants.

The chlorination of drinking water supplies and of sewage effluents can wipe out waterborne diseases. Making water suitable for recreational uses will require drastic reduction of pollutants in industrial and municipal effluents. Tertiary treatment of these effluents must be developed as rapidly as possible.

Pollution by sulfuric acid in coal-mine water must be controlled by blocking the entrances to abandoned mines and keeping water out of contact with coal beds in active mines. Acid water from active mines should be treated with lime. Mercury pollution can be largely controlled by recycling water through plants using mercury. Chemical treatment to recover the mercury and its compounds from the water will prevent pollution and save this costly material.

Lead pollution, which comes mostly from burning leaded gasoline, can be largely eliminated by keeping the lead out of motor fuels. Paints and pottery glazes containing lead must be replaced by other materials. The recovery of petroleum products from spills must be made the legal and financial responsibility of the agent producing the spill. Research on the technology of recovering and using spilled petroleum products should be given high priority.

Radioactive pollutants in water supplies are confined to regions where the ores of uranium and thorium are mined and processed. This pollution can be minimized by grading tailings dumps so that water does not penetrate them but runs off the surfaces. Thermal pollution from power plants and other industries can be minimized by cooling towers and lagoons. Heated water can probably be used profitably for growing crops with high cash value.

All municipal sewage should be given primary, secondary, and tertiary treatment. We are moving so slowly in this direction that pollution is growing faster than the facilities for treatment. We must step up our rate of expanding the facilities, or our water supply will get progressively more polluted.

Industries that discharge wastes into public waters must be required to remove all but traces of pollutants in their effluents. The recycling of water through industrial plants should be encouraged.

The problems of water pollution in our highly urban and industrialized culture are becoming more acute. Each year, more than 1000 communities outgrow their sewage treatment systems. Waste production from municipal systems is expected to increase fourfold in the next 50 years. The U.S. National Report on the Human Environment prepared for the June 1972 United Nations Conference on Human Environment in Stockholm noted that at least $12 billion needs to be invested in water treatment in the immediate future. Will we pay the price?

New Terms and Concepts

ACTIVATED SLUDGE: Solids from sewage inoculated by microorganisms that metabolize the organic compounds present.

BIOCHEMICAL OXYGEN DEMAND (BOD): The amount of dissolved oxygen used up by bacterial action to oxidize the organic matter present in a sample of water.

BLOOM: A heavy growth of algae on the surface of a body of water.

COLIFORM: Relating to the colon bacillus, a type of bacterium normally found in vertebrate intestines.

ELECTROLYSIS: The passage of electric current through a solution or melted compound to transform those substances present into new substances.

ELECTROLYTIC CELL: A device in which the process of electrolysis is carried on.

EUTROPHICATION: The process of becoming so rich in dissolved nutrients that the respiration of dense aquatic populations produces a continuing oxygen deficiency.

EXTRACTION: The process of separating constituents of a substance by treatment with a solvent.

FUNGICIDE: A substance that kills fungi.

HERBICIDE: An agent used to destroy unwanted plants.

INORGANIC COMPOUND: A chemical compound that does not include carbon and hydrogen bound together.

INVERTEBRATE: Without a backbone; an animal without a backbone.

ORGANIC COMPOUND: A chemical compound containing atoms of carbon and hydrogen bound together.

PATHOGEN: An agent that causes a disease.

PROTOZOA: Single-celled animals.

TAILINGS: Refuse material resulting from the treatment of ores.

THERMAL POLLUTION: The warming of natural water supplies above natural temperature.

VERTEBRATE: Having a backbone; an animal with a backbone.

VIRUS: A submicroscopic disease-causing agent.

Testing Yourself

13.1 Is the presence of coliform bacteria in water a sign that it is infected with pathogenic organisms? Why is the coliform test so widely used?

13.2 Discuss biodegradation of pollutants in water in terms of DO and BOD.

13.3 Why are pollutants that are converted to harmless impurities by biodegradation so undesirable in water?

13.4 What factors affect the rate of aquatic plant growth? Why is it difficult to single out the effect of one nutrient present in a pollutant? How can such an effect be singled out?

13.5 What is an algal bloom? Why is it undesirable?

13.6 What is eutrophication and how is it related to the aging of lakes?

13.7 Why is acid pollution from coal mines so detrimental? How can we deal with this problem?

13.8 How does mercury get into the aquatic environment? Why is it particularly bad for food production? How can we prevent it from building up in the environment?

13.9 How does lead get into our water supplies? How can contamination from this source be reduced?

13.10 Why is the heat from industrial processes considered to be thermal pollution? How does thermal pollution degrade the environment?

13.11 How might thermal damage be converted to thermal benefit?

13.12 What are the principal problems involved in the treatment of domestic and municipal sewage?

13.13 As sewage treatment plants now operate, they produce considerable quantities of sludge (a thick mud of water and solids) which is very rich in plant nutrients. What do you think ought to be done with this?

The photomicrograph shows sodium thiosulfate, a salt crystal. In saturated solutions
of salts, ionic equilibrium is achieved. The schematic diagram shows this situation.

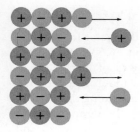

14. Processes in Dynamic Equilibrium

Up to now, when we have considered chemical reactions, we have assumed that if the reactants are in certain definite proportions, they will be used up to form the products. Thus when zinc reacts with sulfur, we find

$$Zn(s) + S(s) \longrightarrow ZnS(s)$$
$$65.4 \text{ g} + 32.1 \text{ g} \longrightarrow 97.5 \text{ g}$$

No matter how much Zn and S we have, if the ratio of weight of zinc to weight of sulfur is 65.4 : 32.1, or 2.04, the only substance present after reaction is ZnS, because the Zn and S have all been used up. Under these circumstances we say that the reaction has "gone to completion."

When some substances are mixed together, they react immediately, but when all evidence of reaction is over, some of the reactants remain. No matter what relative weights of them we mix together, the reaction does not go to completion. Furthermore, if we mix some of the products of the reaction, they form the original substances by a chemical reaction which is the reverse of the original incomplete reaction.

When a reaction does not go to completion, and we can start with the products and generate the reactants, we say that the reaction is reversible and that the system is approaching equilibrium.

Many processes in the environment are reversible and show the characteristics of systems approaching equilibrium. In this chapter we shall study some rather simple processes in equilibrium to see how such systems work. In Chap. 15 we shall discuss some much more complex equilibrium systems in the environment.

14.1 Equilibrium between the Liquid and Gaseous States of Water

When an uncovered dish of water is exposed to the air, the water level gradually lowers, and all the liquid eventually evaporates. In Sec. 6.6 we accounted for this behavior in terms of the kinetic-molecular theory of matter by assuming that some of the molecules in liquid water are moving faster than others. They tend to fly off from the liquid (Fig. 14.1a), reducing the average speed of the molecules remaining and slightly decreasing the temperature of the water. However, heat from the air in the room flows into the water and restores the average speed of the molecules, so that more of them fly off from the liquid. Molecules keep flying off from the liquid and heat keeps flowing into it until all the water has evaporated.

If we place a jar over the dish (Fig. 14.1b), very little water evaporates. Can we explain this observation in terms of the kinetic-molecular theory? It seems likely that the water still evaporates at the same rate in the enclosed dish, but that the molecules rapidly moving about in the gaseous state cannot escape into the room. Because of their erratic motion in all directions, some of them will plunge back into the liquid, i.e., will condense from the gaseous state into the liquid state. As the population of molecules in the gaseous state increases with continuing evaporation, the rate at which molecules of gas will condense (plunge into the liquid) will increase. Eventually, the rate of condensation will equal the rate of evaporation (Fig. 14.2).

We see no reason why the molecules in the liquid should stop jostling about and evaporating when the water level in the dish in Fig. 14.1b stops falling, and so we assume that a dynamic equilibrium is established—that the water

FIGURE 14.1 Evaporation of water from the liquid to the gaseous state: (a) Evaporation of molecules from the surface of water in an open dish; (b) evaporation of molecules from the surface of water and return of molecules to this surface in a dish covered with a jar.

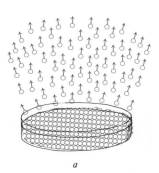

a   b

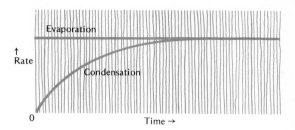

FIGURE 14.2 Rates of evaporation and condensation of water in a closed space

molecules are evaporating from the liquid and entering the gas at exactly the same rate as they are condensing from the gas into the liquid. We symbolize this situation by writing an equation with a double arrow, indicating that the reaction is proceeding in both directions at the same rate.

$$H_2O(l) + heat \rightleftharpoons H_2O(g) \tag{14.1}$$

Whenever we write an equation with double arrows as in Eq. (14.1), we mean that the system is in equilibrium. The forward and reverse processes are going on simultaneously and at the same rate, so that we can see no change in the system. The system in Fig. 14.1a consists of water in an open dish exposed to the air; this is called an open system. The system in Fig. 14.1b consists of water in an open dish exposed to air enclosed by the large cover; no matter can enter or leave this closed system.

14.2 Equilibrium between the Liquid and Solid States of Water

If we add a few ice cubes to some water at room temperature, we note that the ice melts and the temperature of the water falls. As we continue to add ice, the temperature continues to fall until it reaches 0°C; then the addition of more ice has no further effect on the temperature. If we heat the mixture, some of the ice melts, but the temperature remains constant at 0°C as long as some ice is present. If we cool the mixture, more ice forms, but the temperature stays at 0°C as long as there is some liquid water present. If we allow a mixture of ice cubes and water to stand quietly for awhile in an insulated container, the temperature stays at 0°C, but the cubes agglomerate into irregular chunks. This indicates that melting and freezing must be going on simultaneously in the mixture; the cubes melt in some spots, but the water between them freezes in others. The heat absorbed by the melting process is balanced by the heat liberated by the freezing process, so that the overall temperature remains constant. We may symbolize this dynamic equilibrium by the equation

$$H_2O(s) + heat \rightleftharpoons H_2O(l) \tag{14.2}$$

14.3   Equilibrium in Saturated Solutions of Salts

In Sec. 12.2 we considered the preparation of a saturated solution of NaCl, which may be summarized in Eq. (14.3):

$$NaCl(s) \longrightarrow Na^+ + Cl^- \tag{14.3}$$

If we pour the contents of a test tube containing such a solution into a filter, the NaCl(s) is trapped on the filter, and the clear saturated solution passes through. When a concentrated solution of HCl (which contains H^+ and Cl^-) is now added to the clear filtrate, tiny white crystals of NaCl(s) are precipitated. An equation for the process of precipitation may be written thus:

$$Na^+ + Cl^- \longrightarrow NaCl(s) \tag{14.4}$$

Equation (14.4) is the reverse of Eq. (14.3). We can symbolize the reversibility of the dissolving process by combining the two equations thus:

$$NaCl(s) \rightleftharpoons Na^+ + Cl^- \tag{14.5}$$

This equation may be expressed in words: In a saturated solution of sodium chloride, solid sodium chloride is in equilibrium with dissolved sodium ions and chloride ions. If we wish to emphasize that we are concerned with a saturated solution of NaCl in water, we may write the equation:

$$NaCl(s) \rightleftharpoons Na^+(aq) + Cl^-(aq) \tag{14.6}$$

where $Na^+(aq)$ means a sodium ion in water solution.

Many salts are only very slightly soluble. A saturated solution of silver chloride, AgCl, contains only 0.00009 g of solute per liter of solution. Nevertheless, with carefully designed apparatus we can measure electrical conductivity in even such very dilute solutions, which indicates that ions must be present. The equation for the dissolving of AgCl is

$$AgCl(s) \longrightarrow Ag^+ + Cl^- \tag{14.7}$$

If a saturated solution of AgCl is filtered to remove the undissolved salt and a drop of sodium chloride solution (Na^+ and Cl^-) is added to the clear filtrate, a cloudy white precipitate forms. Since AgCl(s) is white, we deduce that Eq. (14.7) must be reversible. To express the reversibility and the consequent state of chemical equilibrium, we write

$$AgCl(s) \rightleftharpoons Ag^+(aq) + Cl^-(aq) \tag{14.8}$$

In this equation we have written a small arrow pointing to the right and a large one to the left. This indicates that when AgCl is shaken up with water, most of the salt remains in the solid state, and only a tiny bit dissolves.

We can explain the precipitation of AgCl(*s*) thus. When solid silver chloride is added to water,

$$AgCl(s) \rightleftharpoons Ag^+ + Cl^- \qquad (14.7)$$

When sodium chloride is added to water,

$$NaCl(s) \longrightarrow Na^+ + Cl^- \qquad (14.3)$$

Since the NaCl is all dissolved, we do not have an equilibrium in Eq. (14.3). When we add the NaCl solution to the clear, saturated solution of AgCl, the increase in the concentration of Cl^- ions increases the opportunities for Ag^+ ions to encounter and unite with Cl^- ions to form AgCl(*s*). Since the solution is already saturated with AgCl, the formation of more AgCl is signaled by the appearance of the cloudy white precipitate of this substance. We can indicate this effect by making the arrow to the left in Eq. (14.8) still larger:

$$AgCl(s) \xleftarrow{\rightharpoonup} Ag^+ + Cl^-$$

in the presence of dissolved sodium chloride.

14.4 The Common Ion Effect

common ion effect

If we have a solution saturated with any slightly soluble electrolyte, it is easy to precipitate this substance (drive it out of solution) by increasing the concentration (adding some more) of one of the ions involved in the equilibrium. The displacement of an equilibrium by adding more of an ion that is already present in the solution is called the *common ion effect*.

We can use the common ion effect to remove poisonous lead compounds from polluted water. Lead sulfide, PbS, is very insoluble; the equilibrium in its saturated solution is

$$PbS(s) \xleftarrow{\rightharpoonup} Pb^{2+}(aq) + S^{2-}(aq) \qquad (14.9)$$

If we add some sodium sulfide, Na_2S, to any solution containing lead ions, the equilibrium in Eq. (14.9) is displaced to the left, and the lead is driven out of solution as black solid PbS, which can be filtered off. The presence of lead in a solution is easily detected by adding a solution of sodium chromate, Na_2CrO_4, which forms a bright-yellow precipitate:

$$Pb^{2+} + CrO_4{}^{2-} \rightleftharpoons PbCrO_4(s) \qquad \text{bright yellow} \qquad (14.10)$$

Adding excess Na_2CrO_4 to the solution furnishes many extra $CrO_4{}^{2-}$ ions, and the common ion effect drives the Pb^{2+} out of solution as the yellow precipitate.

In nature, solutes in bodies of fresh and salt water often establish systems involving chemical equilibrium with the solids of rocks and soils. Understanding equilibrium is necessary for interpreting many processes in the environment, such as the distribution of carbon dioxide between the air, the sea, and carbonate rocks, the distribution of water between the ice caps and the oceans, the formation of hard water in the ground, the carving of limestone caves, the removal of pollutants from mine waters, the loss of fertility from the soil, and so on.

14.5 Evidence That a Solid Is in Dynamic Equilibrium with Its Saturated Solution

When lead iodide, PbI_2, is added to water, only a very little is required to form a saturated solution; the salt is only very slightly soluble. We would write the equation for this system of saturated solution in contact with excess solute thus:

$$PbI_2(s) \rightleftharpoons Pb^{2+} + 2I^- \tag{14.11}$$

When iodine, I_2, is dissolved in a mixture of alcohol and water, it forms some iodide ion, I^-, in solution. If we dissolve a tiny bit of radioactive $^{131}_{53}I$ in some of the alcohol-water solvent, the solution becomes radioactive. If this solution containing radioactive I^- ions is added to the saturated solution of lead iodide in contact with excess solid PbI_2, the *solid* lead iodide soon becomes radioactive. This indicates that at equilibrium Eq. (14.11) is still proceeding from *right to left*. That the reaction in Eq. (14.11) is still proceeding from *left to right* when the system is at equilibrium can be shown by preparing some radioactive $PbI_2(s)$ and adding it to the saturated solution. Soon the *solution* becomes radioactive. In a saturated solution the equilibrium is dynamic; the opposing processes have not stopped.

14.6 Equilibrium in Solutions of Weak Electrolytes

Carbon dioxide gas is very slightly soluble in water. If we bubble CO_2 through pure water, a very weakly conductive solution is formed. We call this soda water. The solution tastes very slightly sour, just barely turns blue litmus pink, and effervesces only slightly when a strip of magnesium is placed in it. Evidently a very weak acid has been produced; we call this carbonic acid. The equations for its production and ionization are

$$CO_2(g) + H_2O \longrightarrow H_2CO_3(aq) \tag{14.12}$$
$$H_2CO_3(aq) \longrightarrow H^+(aq) + HCO_3^-(aq) \tag{14.13}$$
$$HCO_3^-(aq) \longrightarrow H^+(aq) + CO_3^{2-}(aq) \tag{14.14}$$

The stepwise ionization is indicated by the existence of hydrogen carbonates (like sodium hydrogen carbonate, $NaHCO_3$) and carbonates (like sodium carbonate, Na_2CO_3). The degree of dissociation of H_2CO_3 must be very slight, since the acidity of the solution of CO_2 is barely detectable. When we add any acid to any carbonate or bicarbonate, effervescence is produced by the liberation of $CO_2(g)$. Evidently Eqs. (14.12), (14.13), and (14.14) are reversible, and so we may describe carbon dioxide–carbonate systems in terms of three equilibria:

$$CO_2(g) + H_2O \Longleftrightarrow H_2CO_3 \tag{14.15}$$
$$H_2CO_3 \Longleftrightarrow H^+ + HCO_3^- \tag{14.16}$$
$$HCO_3^- \Longleftrightarrow H^+ + CO_3^{2-} \tag{14.17}$$

Ammonia, $NH_3(g)$, has a powerful, penetrating odor and is extremely soluble in water. Even concentrated solutions of the gas are weak conductors of electricity; evidently not many ions are present. Ammonia solutions taste bitter, feel slippery to the touch, and turn red litmus blue. These properties indicate the presence of OH^- ions. An equation for their production is

$$NH_3(g) + H_2O \longrightarrow NH_4^+(aq) + OH^-(aq) \tag{14.18}$$

The existence of ammonium ions, NH_4^+, in the solution is apparent because the addition of HCl to an ammonia solution produces the salt ammonium chloride, NH_4Cl. If we dissolve some ammonium chloride in water, we get a highly conductive solution, showing the presence of many ions. This solution has no odor. The equation for the dissolving is

$$NH_4Cl \longrightarrow NH_4^+(aq) + Cl^-(aq)$$

When we add a strong base like NaOH (which has no odor) to the odorless solution of ammonium chloride in an open dish, we immediately smell ammonia. We conclude that ammonia is produced and write the equation thus:

$$NH_4^+(aq) + OH^-(aq) \longrightarrow NH_3(g) + H_2O \tag{14.19}$$

Equation (14.19) is the reverse of Eq. (14.18), and so we write a statement for chemical equilibrium in the system involving ammonia and ammonium ion:

$$NH_3(g) + H_2O \Longleftrightarrow NH_4^+(aq) + OH^-(aq) \tag{14.20}$$

14.7 Effects of Changes in Concentration on Systems in Chemical Equilibrium

In Sec. 14.3 we noted that the addition of sodium chloride to a clear saturated solution of silver chloride caused the precipitation of $AgCl(s)$ by inducing the right-to-left reaction of Eq. (14.8). We observe the same effect when any other soluble chloride is added in place of the sodium chloride. Evidently the substance that reverses Eq. (14.8) is the added chloride ion, Cl^-. The other ions do not participate in the reaction; they are merely spectators. If we add any soluble silver salt (which will furnish silver ions, Ag^+) to the clear saturated solution of AgCl, the reaction represented by Eq. (14.8) proceeds from right to left. We might summarize these observations by stating: When a system is in chemical equilibrium, the addition of a substance whose formula appears to the right of the double arrow in the equilibrium equation causes an increase in the amounts of the substances whose formulas appear to the left of the double arrow. An alternative statement might be: *Increasing* the concentration of a *product* in a system at equilibrium causes an *increase* in the amount of *reactant* present.

A similar effect is noted when we increase the concentration of a reactant that is involved in a system at equilibrium. If we bubble more $NH_3(g)$ through a dilute solution of this gas involved in the system described by Eq. (14.20), the concentration of ammonia increases, and immediately the basicity (concentration of OH^-) of the solution increases. To summarize these phenomena we might say: *Increasing* the concentration of a *reactant* in a system at equilibrium causes an *increase* in the amount of *product* present.

We may combine the last sentence of each of the preceding paragraphs into a single statement: Increasing the concentration of a substance in a system at equilibrium causes a shift in the relative amounts of all substances present. The shift is in the direction that tends to *use up* the increase in concentration of the substance added.

14.8   Effects of Heating and Cooling a System in Equilibrium

To determine the relation between the vapor pressure of a liquid and its temperature, we set up the apparatus shown in Fig. 5.9 and collected the data in Table 6.1. We noted that when we heated the liquid, the vapor pressure increased. The equilibrium between the liquid and gaseous forms of water was expressed in Eq. (14.1).

$$H_2O(l) + \text{heat} \rightleftharpoons H_2O(g) \qquad (14.1)$$

If the liquid and gaseous water in Fig. 14.1b have come to equilibrium and we add heat to the liquid (warm it up), more gaseous water forms. If we remove heat from the water (cool it down), more liquid forms. These effects are quite analogous to those produced by adding substances to a system in chemical equilibrium. We can extend the conclusion of Sec. 14.7 and say: Adding heat to a system in equilibrium causes a shift in the relative amounts of all substances present, in the direction which tends to use up the added heat. Also: Removing heat from a system in equilibrium causes a shift in the relative amounts of all substances present, in the direction which tends to replace the heat removed.

Many solids are more soluble in hot than in cold water. The equilibrium for the system potassium nitrate–water may be symbolized thus:

$$KNO_3(s) + \text{heat} \rightleftharpoons K^+(aq) + NO_3^-(aq) \qquad (14.21)$$

If heat is added to this system, more solid dissolves; if heat is removed, more solid crystallizes. These observations accord with the last two sentences in the preceding paragraph. Many experiments substantiate the wide applicability of these statements.

14.9 Effects of Changes in Pressure on Systems in Chemical Equilibrium

A sample of any gas dissolved in water occupies much less volume in the dissolved than in the gaseous form. Pressure applied to a gas confined above water increases the solubility of the gas. We can describe this situation for carbon dioxide:

$$CO_2(g) \xrightarrow{\text{pressure increased}} CO_2 \text{ (dissolved)}$$

If a solution is saturated with $CO_2(g)$ at a given pressure and the pressure is then increased, more $CO_2(g)$ goes into solution. If the pressure on a saturated solution of $CO_2(g)$ is diminished, the gas comes out of the solution, producing effervescence. When you remove the cap from a bottle of soda water and release the pressure on the liquid, it bubbles and fizzes. The equation is

$$CO_2 \text{ (dissolved)} \xrightarrow{\text{pressure decreased}} CO_2(g)$$

When $CO_2(g)$ is dissolved in water under pressure, an equilibrium is established:

$$CO_2(g) \underset{\text{pressure decreased}}{\overset{\text{pressure increased}}{\rightleftarrows}} CO_2 \text{ (dissolved)} \tag{14.22}$$

Increased pressure increases the solubility of many other gases, and decreased pressure produces effervescence.

To summarize: In a system at equilibrium involving a gas, an increase in pressure causes a shift in the relative amounts of all substances present, in the direction that tends to relieve the increased pressure. Also in such a system, a decrease in pressure causes a shift in the relative amounts of all substances present, in the direction that tends to counteract the decrease.

The increased solubility of gases with increased pressure has important physiological effects. At high altitudes, the partial pressure of oxygen in air is so reduced that the transfer of oxygen from air in the lungs into the blood is impaired.

$$O_2(g) \rightleftarrows O_2 \text{ (dissolved)} \tag{14.23}$$

When one first reaches a high altitude, he has difficulty in getting enough oxygen into his blood and suffers the dizziness and nausea of "mountain sickness." This can be temporarily eased by breathing a mixture of air and oxygen that has a higher partial pressure of oxygen than the air alone. The increased pressure of O_2 shifts the equilibrium represented by Eq. (14.23) toward the right, and the shortage of oxygen in the blood is relieved. If you live at a high altitude for several months, your body responds to the oxygen shortage by producing a higher concentration of red corpuscles in your blood, so that a larger fraction of the decreased amounts of dissolved oxygen can be used.

When a person dives into deep water and breathes ordinary air under pressure, the amounts of both oxygen and nitrogen dissolved in his blood increase:

$$N_2(g) \rightleftharpoons N_2 \text{ (dissolved)}$$

When he comes to the surface, the decrease in pressure on his blood stream causes the dissolved nitrogen to come out of solution in the blood vessels. The bubbles of nitrogen so formed are damaging to the circulatory system, so that the diver suffers from severe pain—known as "the bends" because he doubles up in his agony. Severe cases of the bends lead to death. If the diver breathes a mixture of oxygen and helium (which is quite insoluble in blood) when he is under pressure, he can be spared the problems of decompression, since the helium will not come out of solution at the lower pressure.

14.10 LeChatelier's Principle

LeChatelier's
principle

The French chemist Henri Louis LeChatelier summarized all the effects we have observed in *LeChatelier's principle*: If a stress is applied to any system in chemical equilibrium, the system reacts in a way to relieve this stress.

If we add more of a substance to a system, it uses up some of the excess; if we add heat, it uses up some of the added heat; if we apply pressure, it produces substances that occupy less volume and so relieves the pressure. Conversely, if we remove some of a substance from a system, it produces more of what we removed; if we remove heat, it produces more heat; if we reduce pressure, it produces substances of greater volume so that the drop in pressure is partly offset.

LeChatelier's principle is very useful in selecting the best concentrations, temperatures, and pressures to promote a reaction in the desired direction. For example, nitrogen can be combined with hydrogen to form ammonia, a valuable fertilizer. The equilibrium involved is

$$N_2(g) + 3H_2(g) \rightleftharpoons 2NH_3(g) + \text{heat} \qquad (14.24a)$$
$$1 \text{ mole} + 3 \text{ moles} \rightleftharpoons 2 \text{ moles} \qquad (14.24b)$$
$$\begin{matrix} 1 \text{ molar} \\ \text{volume} \end{matrix} + \begin{matrix} 3 \text{ molar} \\ \text{volumes} \end{matrix} \rightleftharpoons \begin{matrix} 2 \text{ molar} \\ \text{volumes} \end{matrix} \qquad (14.24c)$$

Since the molar volume for all gases is the same at a given temperature and pressure, we can generalize Eq. (14.24c) to

$$1 \text{ volume } + 3 \text{ volumes} \rightleftharpoons 2 \text{ volumes} \qquad (14.24d)$$

Total volume of reactants $= 4$ Total volume of products $= 2$

If the reaction goes from left to right, there is a decrease in volume; if it goes from right to left, there is an increase.

The forward reaction—the formation of ammonia—takes place so slowly at room temperature that the reaction vessel must be heated to a high temperature to produce any appreciable amount of ammonia in a reasonable time. But heat favors the reverse reaction! However, if the gaseous mixture is put under high pressure (to favor the reaction to the right) and high temperature (to speed the attainment of equilibrium), the formation of ammonia will be promoted. So the mixture of nitrogen and hydrogen is passed into one end of the reaction vessel, comes to equilibrium with ammonia at the high temperature and pressure, and then passes out the other end of the vessel, where it is quickly cooled to "freeze" the system when much of the nitrogen is combined with hydrogen as ammonia. Fortunately, the reverse reaction takes place only very slowly at room temperature, so that the ammonia can be separated from the residual nitrogen and hydrogen before it is lost. The hydrogen and nitrogen are then recycled through the reaction vessel. Table 14.1 shows how the qualitative, theoretical predictions of LeChatelier's principle are substantiated by quantitative, experimental results.

By applying LeChatelier's principle to the reaction mixture as shown in Eq. (14.24a), we predict that adding heat will shift the equilibrium toward the left (which absorbs heat) and reduce the amount of $NH_3(g)$ present. Our prediction is corroborated by the data in the vertical columns of Table 14.1, showing the six different pressures. In each column, the percentage of $NH_3(g)$ present at equilibrium decreases markedly as heat is added (the temperature is increased).

TABLE 14.1 *Percentage of $NH_3(g)$ Present at Equilibrium in a Mixture Consisting of 1 Mole $N_2(g)$ per 3 Moles $H_2(g)$ before Reaction*

Temperature, °C	Pressure, atm					
	1	50	100	200	600	1000
200	15.3%	74.4%	81.5%	85.8%	95.4%	98.3%
400	0.48	15.3	25.1	36.3	65.2	79.8
500	0.13	5.6	10.6	17.6	42.2	57.5
600	0.05	2.25	4.5	8.2	23.1	31.4
800	0.022	0.57	1.19	2.2	—	—
1000	0.004	0.21	0.45	0.9	—	—

Again, by applying LeChatelier's principle to the reaction described in Eq. (14.24d), we predict that increased pressure at a given temperature will shift the equilibrium toward the right (which reduces volume and thus the pressure exerted by the mixture), resulting in an increase in the percentage of $NH_3(g)$. Again, our prediction is corroborated by the data: the percentage of $NH_3(g)$ at equilibrium increases markedly as the pressure increases.

Applications of LeChatelier's principle to industrial processes have had profound effects. Before the outbreak of World War I in 1914, practically all the nitrogen compounds used in fertilizers and explosives in Europe came from the enormous deposits of potassium and sodium nitrates in Chile. Early in the war, the British navy blockaded all German seaports and prevented the importation of nitrates, so that German stockpiles were soon exhausted. But Fritz Haber, a German chemist, perfected a process for synthesizing ammonia from hydrogen and the nitrogen in the air, and the ammonia was used to make the needed fertilizers and explosives. Had it not been for the *Haber process*, Germany might have been defeated 2 or 3 years earlier than 1918.

Haber process

14.11   A General Statement of Dynamic Chemical Equilibrium

Many chemical changes do not go to completion but come to a dynamic chemical equilibrium between reactants and products. When substances A and B react to form C and D, the equation is

$$A + B \longrightarrow C + D \tag{14.25}$$

As soon as A and B are mixed, they begin to react with one another to form C and D. As A and B are used up, the particles of these substances have less opportunity for coming together, and so the rate of their reaction slows down with the passage of time (Fig. 14.3)

When substances C and D are formed, they tend to react to form A and B; thus

$$C + D \longrightarrow A + B \tag{14.26}$$

At time zero (when A and B are mixed), no C and D are present, and so the

FIGURE 14.3 The change in rate of reaction between A and B as time passes

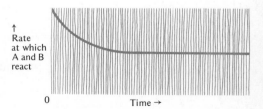

↑
Rate
at which
A and B
react

0   Time →

rate of their reaction is zero. As time passes, the amounts of C and D build up, and they begin to react faster and faster to form A and B (Fig. 14.4).

As the reaction between A and B slows down and that between C and D speeds up, a time will come when the rate at which A and B form C and D will just equal the rate at which C and D form A and B (Fig. 14.5).

When the rate of reaction of Eq. (14.25) equals the rate of reaction of Eq. (14.26), we may write

$$A + B \rightleftharpoons C + D \tag{14.27}$$

This is a general equation for a reacting system in chemical equilibrium. Reaction has not stopped when equilibrium is reached, but the rate of the forward (left-to-right) reaction just equals the rate of the reverse (right-to-left) reaction, so that we see no change in the system.

Summary

Many chemical reactions do not go to completion but come to a dynamic chemical equilibrium between reactants and products. Such a situation may be symbolized thus:

$$A + B \rightleftharpoons C + D$$

This is chemical shorthand for the statement:

When substance A reacts with substance B to produce substances C and D, the reaction does not go to completion. When all evidence of reaction ceases,

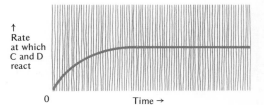

FIGURE 14.4 The change in rate of reaction between C and D as time passes

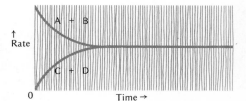

FIGURE 14.5 The rate of reaction between A and B becomes equal to the rate of reaction between C and D.

1 both reactants and products are present
2 both the forward (left-to-right) and reverse (right-to-left) reactions are continuing
3 these two reactions are proceeding at the same rate so that there is no net change in the amounts of substances present

The behavior of systems in chemical equilibrium is summarized in LeChatelier's principle: If a stress is applied to any system in chemical equilibrium, the system reacts in a way to relieve this stress. If more of a substance is added to the system, it reacts in the direction that will use up some of the added material. If the system is heated, it reacts in the direction that will absorb heat. If pressure is applied to the system, it reacts in the direction that will produce a decrease in volume and thus relieve the pressure.

Some kinds of dynamic equilibria that are important in environmental processes are

1 equilibria between gaseous, liquid, and solid water
2 equilibria between solid substances and their saturated solutions
3 equilibria between gases and their saturated solutions
4 equilibria between molecules and ions in solutions of weak acids and bases

New Terms and Concepts

COMMON ION EFFECT: The reversal of a chemical reaction by the addition of one of its products.

HABER PROCESS: The industrial process in which nitrogen and hydrogen are made to combine to form ammonia.

LECHATELIER'S PRINCIPLE: If a stress is applied to any system in chemical equilibrium, the system reacts in a way to relieve this stress.

Testing Yourself

14.1 By applying LeChatelier's principle, predict what would happen to each of the following systems at equilibrium when heat is supplied to the system.

$$2SO_2(g) + O_2(g) \rightleftarrows 2SO_3(g) + \text{heat}$$
$$PCl_5(g) + \text{heat} \rightleftarrows PCl_3(g) + Cl_2(g)$$
$$N_2O_4(g) + \text{heat} \rightleftarrows 2NO_2(g)$$

14.2 By applying LeChatelier's principle, predict what would happen to each of the systems in problem 14.1 if the pressure were increased.

14.3 Acetic acid, $HC_2H_3O_2$, is a weak acid. What effect would you expect on the acidity of an acetic acid solution if you added some ammonium acetate, $NH_4C_2H_3O_2$, to it? Why would you expect this?

14.4 Aqua ammonia, a water solution of NH_3, is a weak base. What effect on the basicity of aqua ammonia would you expect if you added some ammonium acetate to it? Why would you expect this?

14.5 Barium sulfate, $BaSO_4$, is a slightly soluble white salt. A saturated solution of it is prepared and the excess solid filtered off. If some sodium sulfate, Na_2SO_4, were added to this clear saturated solution, what would you expect to see? Why?

14.6 Methyl alcohol, CH_3OH (wood alcohol), can be prepared by the reaction

$$CO(g) + 2H_2(g) \rightleftharpoons CH_3OH(g) + \text{heat}$$

with the system in chemical equilibrium. The forward reaction (forming CH_3OH) proceeds so slowly at room temperature that it is useless for preparing CH_3OH commercially, but it goes rapidly at 400°C. Thousands of tons of CH_3OH are prepared each year by this reaction. How do you suppose this industrial production is achieved?

14.7 The volume of a gram of ice is greater than the volume of a gram of liquid water. If you put very high pressure on ice, how would you expect this to affect the equilibrium $H_2O(s) \rightleftharpoons H_2O(l)$? How would high pressure affect the melting point of ice? Can you suggest what the lubricant is that makes the blade of an ice skate glide so smoothly over ice?

14.8 The nitrogen and oxygen of the air can combine to form nitric oxide, as indicated in the following equation:

$$N_2(g) + O_2(g) + \text{heat} \rightleftharpoons 2NO(g)$$

where the system at equilibrium consists mostly of the elements and very little NO. Would you expect to find much NO in the air under ordinary conditions? During a severe lightning storm, traces of NO are readily detected in the atmosphere. Can you account for this phenomenon?

14.9 The gastric juices in the human stomach normally have an H^+ ion concentration of about 0.01 mole/liter. How do sodium bicarbonate tablets relieve "sour stomach"?

14.10 The concentration of H^+ ions in beer is about 0.00005 mole/liter, while that in some soft drinks is 0.001. Which beverage would you expect to have the more sour taste?

14.11 Cabbage juice contains about 0.00001 mole/liter of H^+ ions and sauerkraut juice about 0.001 mole/liter of H^+ ions. What kind of substance must be produced when cabbage is fermented to produce sauerkraut?

With minor exceptions, all life on earth depends on photosynthesis by green plants. The chlorophyll in these plants brings about the combination of carbon dioxide (CO_2) from the air to form plant tissues and liberate oxygen.

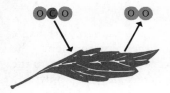

15. Dynamic Equilibria in Environmental Processes

The crux of the environmental crisis building in our time lies in the imbalance between man's activities and nature's processes. Suddenly we have come to the realization that the earth's riches are not inexhaustible, that there is not room for a never-ending growth in population, that we cannot fight nature and win in the long run, that we must learn to cooperate with her and live in balance with the other creatures that share our earthly home.

To learn to live in cooperation with nature, we must learn more about nature. We must understand the fundamental principles involved in the transformation of matter if we are to understand environmental processes.

So far in this book we have studied some of the most basic of these principles and have organized them into mental models to facilitate our learning about them. Our concern with atomism, the kinetic-molecular theory, subatomic particles of matter, the interplay of matter with heat and electricity, the periodic table of the elements, the properties of solutions, and the nature of dynamic chemical equilibrium have given us the background for a better understanding of some of the very complex changes that underlie the stability of our environment. In this chapter we shall use this background to interpret some fundamental cycles of substances throughout the environment.

15.1 The Water Cycle

biosphere

Water is the most abundant of all compounds in the *biosphere*, the name given to the complex systems of living organisms and their environment—the "world of life." The thickness of the biosphere is very slight compared to the diameter of the earth and its surrounding atmosphere. Life exists only as a very thin skin along the surface of the earth. It is most luxuriant in the equatorial regions, which abound in warmth, sunshine, and water. It is found in the cold of Arctic regions, in burning deserts, even in dark caves and ocean depths where no light shines. But it never exists without water. The earth's soils, glaciers, lakes and rivers, oceans, and atmosphere contain about 1.5 billion km^3 (cubic kilometers) of water, which represents about 100 billion gal for every person in the world. It is present as a solid, a liquid, or vapor, and it is found in loose chemical combination with many substances in the soil. Ninety-seven percent of the earth's water is in the oceans, about 2 percent in glaciers, a little less than 1 percent in fresh waters, and a small fraction of 1 percent in the atmosphere (Fig. 15.1).

Water has several unusual properties. It has the largest specific heat, the largest heat of vaporization, and the largest heat of melting (Sec. 5.4) of any common substance at ordinary atmospheric temperatures. At ordinary temperatures it conducts heat better than any other common liquid except mercury. These special properties of water arise from the structure of water molecules as discussed in Chap. 19. Water is the only natural substance that expands when it freezes. Hence the density of ice is less than that of water; and it floats. Because of this peculiarity, when the temperature of the air falls far enough, a layer of ice forms on the surface of a body of water and floats there. This serves as a barrier that reduces the rate of heat flow from the liquid water into the atmosphere. As freezing continues, the increasing thickness

FIGURE 15.1 Distribution of the world's water supply

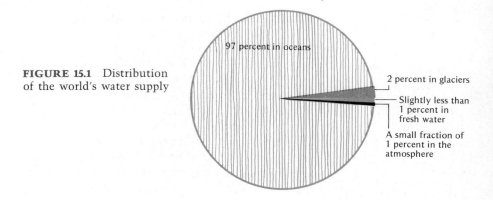

97 percent in oceans

2 percent in glaciers

Slightly less than 1 percent in fresh water

A small fraction of 1 percent in the atmosphere

of the ice between the atmosphere and the water slows down freezing greatly. A layer of ice only a few feet thick protects the rest of the water from freezing, so that aquatic plants and animals can carry on their life processes in the water below the ice.

If water were to contract when it froze, ice would be more dense than liquid water and would sink to the bottom. The whole body of water would soon become solid ice, killing off all but the simplest forms of life. Many of the aquatic creatures we know would never have evolved if water did not have this peculiar property of expanding when it freezes. Furthermore, the high heat of melting means that large amounts of heat must be extracted from water during the freezing process. Water freezes more slowly than other liquids. The dynamic equilibrium between water and ice is shifted toward ice as heat is removed (as one would predict from LeChatelier's principle). This shift furnishes more heat to the atmosphere and tempers the cold. These unique thermal properties of water have played a vital role in the evolution of living organisms and the continuance of their life processes.

The specific heat of water is much greater than that of rocks and soils. When the sun comes up in the morning, bodies of water warm up more slowly than the land; when it goes down at night, they cool off more slowly. Water also is an effective insulator of electric charges—it reduces the attraction between unlike charges and the repulsion between like charges. Indeed, water is a better insulator than any other common liquid. This characteristic makes it excellent for forming solutions containing ions. The structure of water discussed in Chap. 19 accounts for this property also. Water as found in nature is never pure—it always contains some solutes. Water in soils may vary from moderately acidic to moderately basic because of the solutes present. Terrestrial plants grow best when groundwater is very slightly acidic. Marine organisms grow best in a solution which is slightly basic. In a neutral or slightly acid solution many marine animals die.

Water enters the atmosphere by the respiration of plants and animals and by evaporation from vegetation, moist soils, oceans, rivers, and lakes. It leaves the atmosphere by condensation to rain or snow. It may fall as precipitation near the place where it evaporated, or thousands of miles away. Its residence time in the atmosphere may vary from a few hours to a few weeks; a general average is 9 or 10 days. For the entire earth, evaporation and precipitation must be equal. The average precipitation for the earth as a whole is equivalent to about 100 cm/yr of rainfall. Precipitation at sea averages about 110 cm, whereas the average annual evaporation from the oceans is the equivalent of about 120 cm rain. The balance is restored by the flow of rivers into the oceans, the equivalent of about 10 cm rain.

The average precipitation over land is equivalent to 71 cm rain; the average evaporation is equivalent to 47 cm rain. The balance is restored by the flow of rivers, which amounts to the equivalent of 24 cm rain over the whole land mass. Since the area of land on the earth's surface is considerably less than that of the oceans, this 24 cm over land is equivalent to the 10 cm over seas. The total annual precipitation over the United States is about 6000 km³ (about a million billion gallons!), but the water vapor that passes over the United States in a year due to the circulation of the atmosphere is 10 times this amount. The overall circulation of water from the oceans to the air to the land and back to the oceans is indicated by the arrows in Fig. 15.2.

If all the water in the world (Table 15.1) were condensed into a layer of uniform thickness on its surface, the layer would be about 2765 m (meters) thick. Of this, 2700 m would represent the oceans. Other depth equivalents are shown in the table and in Fig. 15.2.

Changes in the dynamic equilibrium between water in the oceans, the atmosphere, glaciers, the polar ice caps, and the rivers of the world can be predicted by applying LeChatelier's principle. Recalling that

$$H_2O(s) + \text{heat} \rightleftharpoons H_2O(l)$$

and

$$H_2O(l) + \text{heat} \rightleftharpoons H_2O(g)$$

we would predict that if the climate of the world warmed up, the increased available heat would displace both equilibria to the right, thus decreasing the amount of water in glaciers and the polar ice caps and increasing the amount in the rivers and oceans. The vapor pressure of water would also increase, so that the partial pressure of water vapor in the atmosphere would rise, resulting in a humid climate. If the ice caps of Greenland and Antarctica were to melt completely, the level of the sea would rise some 300 ft.

The amount of heat received by the earth from the sun varies slightly over

FIGURE 15.2 Distribution of the world's water supply, expressed in terms of meters of depth of water were all of it condensed and spread evenly over the surface of the earth, forming a layer 2765 m thick

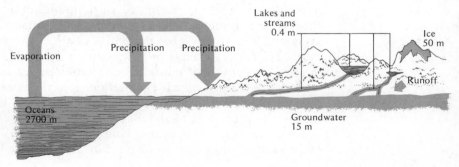

TABLE 15.1 *Depth of the World's Waters If Spread Evenly over the Earth*

Present location	Depth, m
Oceans	2700
Ice caps	50
Groundwater table	15
Lakes and streams	0.4
Atmosphere	0.03
Total (approximate)	2765

long periods of time, and these variations produce changes in the average temperature of the earth's climate. In the past 300,000 years reductions in the heat received from the sun have produced cooling in the climate, resulting in four great "ice ages" when glaciers covered much of North America. Following each ice age was a period in which increasing amounts of heat were received from the sun. The glaciers then partly melted and the level of the sea rose about 100 ft. The floods that were produced when the ice melted and the sea level rose could possibly have been the basis of the story of Noah and his Ark as well as of similar tales in the myths, legends, and religious lore of many other people besides the Hebrews.

The crucial factor in the water cycle is the amount of water available for growing food. With minor exceptions, all life on earth depends on photosynthesis by green plants. The *chlorophyll* in these plants brings about the combination of water with CO_2 from the air to form plant tissues and liberate oxygen. However, only a small portion of the water which passes through the plant is combined in plant tissue. The rest serves as the carrier of nutrients from the soil into the roots, stems, and leaves. Some is used for photosynthesis, but most of it evaporates into the atmosphere. To produce 20 tons of a typical green crop in a growing season, some 2000 tons of water will be drawn from the soil. This 20-ton crop will have a dry weight of only 5 tons; 15 tons of water will be "in transit" in the plants. The dry weight of the plants will include 3 tons of water chemically bound into the plant tissues. These relationships (Fig. 15.3) show the overwhelming importance of water for the production of food; we must husband the world's water supplies if we are to avoid famine.

chlorophyll

15.2 The Carbon Cycle

The element carbon is involved in many dynamic equilibria as it moves through the biosphere in the form of living plants and animals, of dissolved

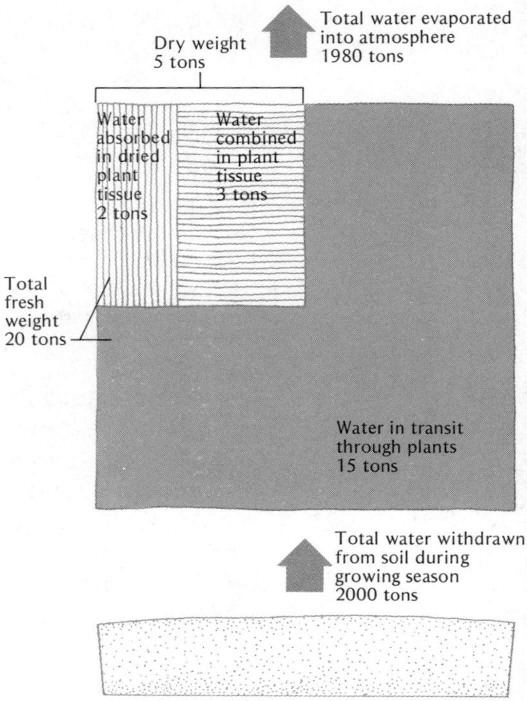

FIGURE 15.3 Utilization of water to grow a crop of green plants

In the figure:
- Dry weight 5 tons
- Total water evaporated into atmosphere 1980 tons
- Water absorbed in dried plant tissue 2 tons
- Water combined in plant tissue 3 tons
- Total fresh weight 20 tons
- Water in transit through plants 15 tons
- Total water withdrawn from soil during growing season 2000 tons

carbonates and bicarbonates in lakes, rivers, and oceans, and of solid carbonates in limestones and the shells of animals. In addition, huge amounts of carbon are buried in the earth in the form of coal, petroleum, and natural gas. The amount of carbon locked in these forms is more than 50 times the total amount contained in the total bulk of all living organisms, the *biomass*. Man's burning of coal, oil, and gas from the depths of the earth has introduced large amounts of carbon oxides into the biosphere, and the increasing concentration of CO_2 in the air will affect many of the natural equilibria involving carbon. This effect is of great ecological importance.

There are two biological processes that have a great impact on the amounts of carbon in various forms in the biosphere. They are the processes of photosynthesis and respiration.

Photosynthesis produces *carbohydrates* (compounds of carbon, hydrogen, and oxygen) in green plants by combining dissolved atmospheric CO_2 and H_2O and simultaneously releasing O_2.

$$xCO_2(aq) + yH_2O(l) + \text{light energy} \longrightarrow C_x(H_2O)_y + xO_2(g)$$

When x and y are 6, the carbohydrate formed is $C_6H_{12}O_6$, the sugar glucose.

biomass

carbohydrate

$$6CO_2(aq) + 6H_2O(l) + \text{light energy} \longrightarrow C_6H_{12}O_6 + 6O_2(g)$$

Carbohydrates found in the tissues of green plants include cellulose and many kinds of *sugars* and *starches*. These compounds are the principal foods for the production of energy by the respiration of plants and animals.

Respiration is essentially the reverse of photosynthesis, producing CO_2 and H_2O by the combustion of compounds like carbohydrates, using up O_2, and producing energy in the form of heat.

$$C_6H_{12}O_6 + 6O_2(g) \longrightarrow 6CO_2(g) + 6H_2O(g) + \text{heat energy}$$

Respiration may take place in the leaves of plants, in their roots, or in the bodies of animals. Dead plants and animals are metabolized by various microorganisms that produce CO_2.

In green plants, photosynthesis is predominant by day and respiration by night. This leads to daily changes in the concentration of CO_2 in the air around luxuriantly growing vegetation. The level is higher at night than in the day. The CO_2 concentration in the air near the ground also varies seasonally. In the Northern Hemisphere it is highest in April (at the onset of spring) and lowest in September (at the end of the growing season). The main cycles of CO_2 over a land mass are shown in Fig. 15.4.

Very small marine organisms called phytoplankton carry on photosynthesis in the upper layers of the water in the sea. They use CO_2 dissolved in seawater, and the O_2 they release also dissolves in water. Like green plants, phyto-

FIGURE 15.4 The cycling of CO_2 through the terrestrial biosphere

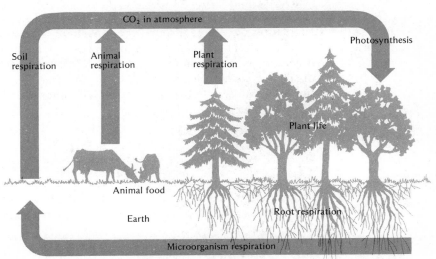

plankton use some carbohydrates in respiration and tissue building. They also serve as food for small marine animals. These, in turn, become the food of larger animals such as crustaceans and fish. The respiration of all these organisms restores CO_2 to the seawater.

The major part of the oceanic biomass consists of short-lived microorganisms. When they die, their bodies disintegrate and their tissues decompose to carbon compounds soluble in seawater. Many marine organisms grow shells composed of calcium carbonate, $CaCO_3$, which is very slightly soluble. When these organisms die, their shells settle to the ocean bottom, there "fixing" large amounts of carbon in the sediments and sedimentary rocks they form. Coal, petroleum, and natural gas probably are produced when large deposits of dead organic material at the bottom of bodies of water are covered by layers of rock and subjected to high temperatures and pressures.

hydrosphere Only a few tenths of 1 percent of the mass of carbon at or near the surface of the earth is in rapid circulation in the biosphere—which includes the atmosphere, the *hydrosphere* (all bodies of water), the upper portions of the earth's crust, and the biomass itself. Most of the carbon near the surface exists in deposits of carbonate rocks and of fossil fuels—oil shale, coal, petroleum, and natural gas. These have required hundreds of millions of years to reach their present magnitude.

The circulation of carbon through the atmosphere is coupled with that through the oceans by the transfer of CO_2 across the surface between air and water. The rate at which radioactive $^{14}_6C$ in the CO_2 produced by nuclear weapons tests in the atmosphere has passed into the oceans indicates that the residence time of CO_2 in the atmosphere is between 5 and 10 yr. Every year about 100 billion tons of CO_2 pass from the atmosphere into the oceans and are replaced by an equal amount of CO_2 liberated from the oceans. The overall circulation of carbon through the biosphere is summarized in Fig. 15.5.

When carbonate rock is lifted out of the water by a rise in the land mass, it is exposed to erosion by running water. Water percolating through the soil dissolves considerable amounts of CO_2 that has been liberated by the respiration of organisms living in the soil. The chemical equilibria between CO_2 and carbonates are

$$CO_2(g) + H_2O \rightleftharpoons H_2CO_3 \tag{14.15}$$
$$H_2CO_3 \rightleftharpoons H^+ + HCO_3^- \tag{14.16}$$
$$HCO_3^- \rightleftharpoons H^+ + CO_3^{2-} \tag{14.17}$$

The carbonic acid formed by the dissolving of CO_2 in ground water reacts with insoluble limestone, $CaCO_3(s)$, forming the more soluble calcium hydrogen carbonate, $Ca(HCO_3)_2$. This reaction is reversible, and so we write it as an equilibrium:

$$CaCO_3(s) + H_2CO_3(aq) \rightleftharpoons Ca^{2+}(aq) + 2HCO_3^-(aq) \tag{15.1}$$

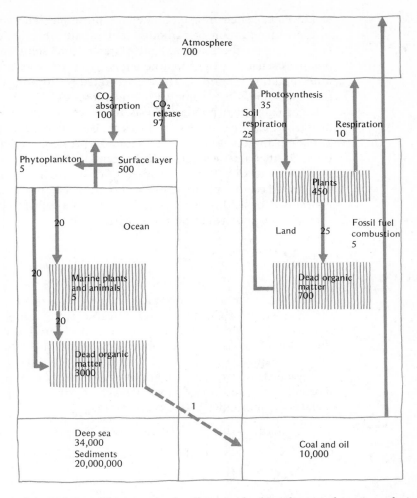

FIGURE 15.5 The storage of carbon in the biosphere and its annual circulation through the biosphere. All figures are in billions of metric tons; 1 metric ton is 1000 kilograms. The figures in boxes show tonnages in storage; those next to arrows are tonnages circulating each year.

The partial pressure of $CO_2(g)$ in the gases trapped in the ground is greater than its partial pressure in air, so that the concentration of dissolved CO_2 builds up in groundwater. When groundwater, saturated with CO_2 and the slightly soluble $Ca(HCO_3)_2$, emerges into the air under the roof of a cave, some of the dissolved CO_2 comes out of solution as a gas, reversing Eq. (14.15), as would be expected from LeChatelier's principle. The consequent reduction in the concentration of carbonic acid in the water leads to the reversal of Eq. (15.1), and the $CaCO_3(s)$ precipitates from solution. This precipitation produces stalactites hanging like icicles from the roof of the cave and stalagmites building upward where the saturated solution of $Ca(HCO_3)_2$ drops on the cave's floor, loses some CO_2, and forms $CaCO_3(s)$.

Recognizing that photosynthesis in seawater depends on the presence of dissolved CO_2, we can see why the slight basicity of seawater favors the growth of phytoplankton. Applying LeChatelier's principle to Eqs. (14.15) to (14.17) indicates that a basic solution (in which H^+ ions have been neutralized) will dissolve more CO_2 than an acid solution (with its higher concentration of H^+), thus favoring photosynthesis. If ocean water in the mouths of rivers is neutralized by acid pollutants, the equilibria are displaced to the left, CO_2 is evolved, and much marine life is extinguished.

15.3 The Oxygen Cycle

The movement of oxygen through the biosphere is intimately connected with the circulation of CO_2 and H_2O. Organic compounds predominate in the biomass. About three million such compounds are known, and thousands of them contain oxygen. In fact, about one-fourth of the total number of atoms in living matter are oxygen.

Oxygen also exists free (uncombined) in nature as oxygen atoms, O, oxygen molecules, O_2, and ozone molecules, O_3. Ultraviolet light from the sun dissociates O_2 molecules into oxygen atoms:

$$O_2(g) + \text{ultraviolet light} \longrightarrow 2O(g)$$

Oxygen atoms then combine with oxygen molecules:

$$O(g) + O_2(g) \longrightarrow O_3(g)$$

The ozone collects in the atmosphere in a layer about 10 mi thick with its lower surface about 15 mi above the earth. Ozone absorbs most of the ultraviolet light in sunlight but transmits visible light. Were the ozone layer not present, life as we know it would be impossible on the earth because the ultraviolet light filtered out is lethal to living organisms. One reason for being extremely cautious about using a supersonic transport (SST) is the possibility that its engines might produce exhaust gases that would contaminate the atmosphere with pollutants that could react with and thus destroy the protective layer of ozone, with disastrous effects on the earth's inhabitants.

The average water molecule, H_2O, endures for about 2 million years before it is split apart by photosynthesis and then reunited (most likely with hydrogen and oxygen from other water molecules) to form new water molecules.

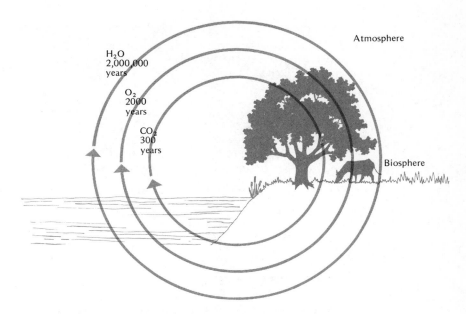

FIGURE 15.6 The circulation of oxygen through the biosphere

Within a period of about 2000 years, all the oxygen generated by photosynthesis passes into the atmosphere and is recycled. Carbon dioxide respired by animals and plants enters the atmosphere and is fixed again by photosynthesis after an average atmospheric residence time of about 300 yr. The circulation of oxygen through the biosphere as H_2O, free O_2, and CO_2 is shown in Fig. 15.6.

Oxygen is so reactive that it is present in more inorganic compounds than any other element. Most of the inorganic compounds in nature are found in the *lithosphere* (the rocky outer portion of the earth, about 1800 mi thick, which surrounds a core composed largely of iron and nickel). The ocean contains huge amounts of inorganic compounds in solution. The oxygen cycles are shown in Fig. 15.7.

lithosphere

15.4 The Nitrogen Cycle

protein

The element nitrogen is a constituent of all *proteins*, the chemical compounds of which all living organisms are made. Though the biosphere is bathed by an atmosphere containing 79 percent free gaseous nitrogen, N_2, this uncombined form cannot be metabolized by most organisms. In order to grow (synthesize protein), most organisms require nitrogen in some suitable compound. Such nitrogen is often called *fixed nitrogen*. Nitrogen is fixed by a few organisms that are able to combine atmospheric nitrogen into compounds in their

fixed nitrogen

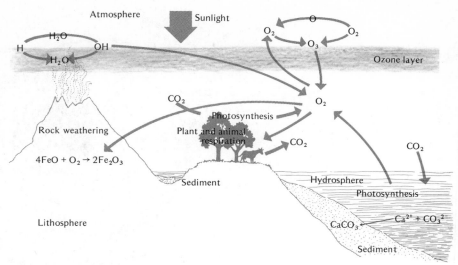

FIGURE 15.7 Oxygen cycles

tissues. Small amounts of nitrogen are fixed by lightning, cosmic rays, and meteor trails, which ionize the gases of the atmosphere and provide the energy needed to form nitrogen oxide NO:

$$N_2 + O_2 + energy \longrightarrow 2NO$$

NO reacts with O_2 thus:

$$2NO + O_2 \longrightarrow 2NO_2$$

This reacts with water vapor in the atmosphere and falls to the earth in rain:

$$3NO_2 + H_2O \longrightarrow 2HNO_3 + NO$$

Man's greatest impact on the cycles of nature has been on that of nitrogen. One of nature's important agents for the fixation of nitrogen is the group of plants called the legumes, which include peas, beans, clover, alfalfa, and soybeans. The roots of legumes serve as the hosts for microorganisms in the soil which can fix nitrogen in their tissues. This fixed nitrogen then finds its way into the soil and is available for the growth of other plants. Today, huge crops of legumes are planted to build up the available nitrogen in the soil for other crops. Nitrogen can also be fixed by the Haber process [Eq. (14.23)], in which hydrogen from petroleum is combined with atmospheric nitrogen to form ammonia. These two processes carried on by man produce each year about 10 percent as much fixed nitrogen as the entire world's supply of fixed nitrogen from natural fixation in all the time before modern agriculture was developed. This increase in the nitrogen available to the biosphere is having complicated, far-reaching, and dramatic effects on the balance of organisms in our environment.

denitrification

Before man intervened in the nitrogen cycle, it is likely that the rate of fixation of nitrogen from the atmosphere by natural processes was just about balanced by the rate of its return to the atmosphere by *denitrification*, a process carried out by certain microorganisms. Today's rate of fixation, which is higher than the rate of denitrification, is already causing trouble in water supplies, as noted in Chap. 13. The excessive runoff of nitrogen compounds from highly fertilized soils stimulates excessive growth of aquatic plants and animals. The extensive blooms of algae and the consequent upsetting of the biological balance in lakes and rivers destroy fish and other oxygen-dependent organisms. The accelerated aging of Lake Erie is only the most dramatic of many examples of the results of eutrophication. Even more alarming is the seepage of the soluble nitrates in water from highly fertilized farmlands into wells supplying human drinking water, causing nitrate poisoning, to which young children are especially susceptible. We must learn how to stimulate food production by using nitrogen-bearing fertilizers that can be more efficiently absorbed by plants with less loss to the soil, or nitrate poisoning will become even more widespread.

The full story of the biological fixation of nitrogen is not yet known. The first step involves splitting the N_2 molecule and combining the atoms with hydrogen to form ammonia, NH_3, thus:

$$N_2 + 3H_2 \longrightarrow 2NH_3$$

This process takes place in many complicated steps and requires large amounts of energy, which is supplied by the respiration of the nitrogen-fixing organisms or their hosts if they live in close association with the roots of legumes. In a slightly acid soil the ammonia is transformed into ammonium ions:

$$NH_3 + H^+ \longrightarrow NH_4^+$$

amino acids

Either ammonia or ammonium ions can be absorbed by the roots of plants, and the nitrogen can be incorporated into *amino acids* (Chap. 20) and then into proteins. When the plant is consumed by an animal, its proteins are used to build animal tissues. When a plant or animal dies, bacteria decompose the protein to amino acids, and other organisms metabolize the amino acids to carbon dioxide, water, and ammonia. Still other organisms convert the ammonia to nitrite, NO_2^-, and yet others change the nitrite to nitrate, NO_3^-. The denitrifying bacteria then use the nitrate in their metabolic processes, and release nitrogen as the gaseous element, thus completing the cycle. The principal pathways in the nitrogen cycle are shown in Fig. 15.8. With the various fixing and denitrifying processes in ecological balance, the distribution of nitrogen in nature is about as shown in Fig. 15.9.

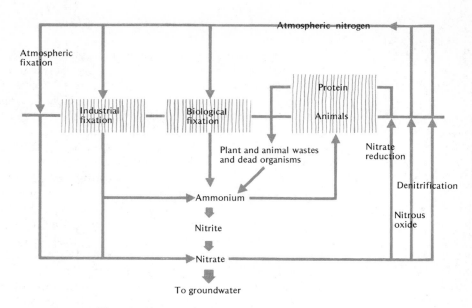

FIGURE 15.8 The circulation of nitrogen through the biosphere

15.5   The Cycling of Sulfur and Phosphorus

Although the principal constituents of the biomass are oxygen, carbon, hydrogen, and nitrogen, 11 other elements are present in more than trace amounts. They are calcium, potassium, silicon, magnesium, sulfur, aluminum, phosphorus, chlorine, iron, manganese, and sodium. Of these, sulfur and phosphorus are crucial in tissue building. The elements oxygen, carbon, hydrogen, nitrogen, sulphur, and phosphorus are often called "the essential 6" because of their overwhelming preponderance in the chemical reactions fundamental to vital processes. Because sulfur readily forms the gaseous compounds hydrogen sulfide, H_2S, and sulfur dioxide, SO_2, it can be cycled through the biosphere by volatilization from the land or sea and transportation through the atmosphere. Considerable amounts of hydrogen sulfide are generated by microorganisms living in the mud at the bottoms of swamps, marshes, and eutrophic lakes. When this gas is released into the atmosphere, it is oxidized to sulfur dioxide, which is brought to earth dissolved in rain:

$$2H_2S(g) + 3O_2(g) \longrightarrow 2H_2O(g) + 2SO_2(g)$$
$$SO_2(g) + H_2O(l) \longrightarrow \quad H_2SO_3(aq)$$

Phosphorus, however, forms no gaseous compounds in natural processes; once it enters the hydrosphere as a compound, it is no longer available for use

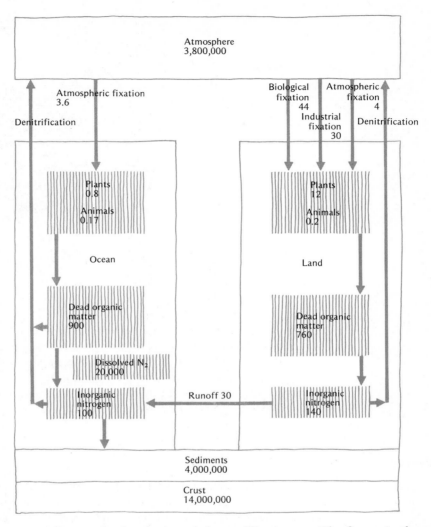

FIGURE 15.9 The distribution of the world's nitrogen. The figures in the various reservoirs are expressed in billions of metric tons; those showing the transfer of nitrogen (arrows) are expressed in millions of metric tons per year.

in the terrestrial portions of the biosphere. For this reason, there is considerable concern among scientists that we may be depleting our terrestrial sources of phosphorus so rapidly that there will be a shortage of the element for agricultural fertilizer in the future. The current arguments about the use of phos-

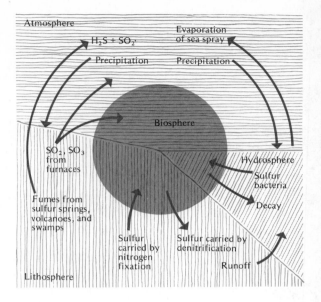

FIGURE 15.10 The transfer of sulfur through the biosphere

phates in detergents are arising because, on the one hand, we are polluting our rivers and lakes with so much phosphorus that eutrophication is proceeding rapidly, and, on the other hand, we are depleting the world's supply of phosphorus needed to grow increasing amounts of food for mankind. The movement of sulfur through the biosphere is shown in Fig. 15.10 and that of phosphorus in Fig. 15.11.

Summary

The biological balance in the environment is maintained through many chemical processes that recycle the chemical elements involved in the reactions essential to life through the atmosphere, the hydrosphere, and the lithosphere. These cyclic processes involve chemical equilibria, and if the equilibria are displaced, the cycles are disturbed.

The most important cycles involve the circulation of the elements hydrogen, oxygen, carbon, nitrogen, sulfur, and phosphorus. The principal compounds involved are water, carbon dioxide, ammonia, nitrates, nitrites, oxides of nitrogen, hydrogen sulfide, sulfur oxides, phosphate salts, carbohydrates, amino acids, and proteins.

These cyclic processes are interrelated, and disturbances in one of them have repercussions on others. As we continue to develop technologies for the

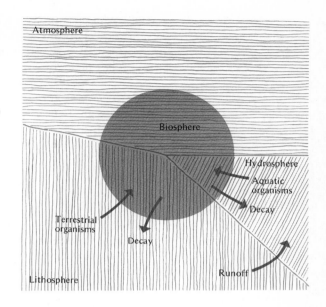

FIGURE 15.11 The cycling of phosphorus. Note that there is no natural pathway for phosphorus to move from the hydrosphere back to the lithosphere except by mountain building from the sea.

benefit of man, we must continually study their impact on the environment so that we do not upset the cycles in nature that are essential to our life on earth.

New Terms and Concepts

AMINO ACIDS: The building blocks of which proteins are made.

BIOMASS: The total mass of all living organisms in a given system.

BIOSPHERE: All living organisms together with their environment.

CARBOHYDRATE: A compound of carbon, hydrogen, and oxygen having the general formula $C_x(H_2O)_y$.

CHLOROPHYLL: The green coloring matter of plants that functions in photosynthesis.

DENITRIFICATION: Reduction of nitrates or nitrites by living organisms and their environment.

FIXED NITROGEN: Nitrogen in compounds (as contrasted to elemental nitrogen).

HYDROSPHERE: The aqueous envelope of the earth including fresh and salt bodies of water.

LITHOSPHERE: The solid part of the earth.

PROTEIN: Any of numerous naturally occurring extremely complex combinations of amino acids that (1) contain the elements carbon, hydrogen, nitrogen, oxygen, usually sulfur, occasionally phosphorus, iron, and other elements, (2) are essential constituents of all living cells, and (3) are synthesized from raw materials by plants but assimilated as separate amino acids by animals.

STARCH: One of a class of complex carbohydrates with the general formula $(C_6H_{10}O_5)_x$.

SUGAR: One of a class of sweet carbohydrates.

Testing Yourself

15.1 Water has some unusual properties when compared to other liquids. What are these properties and why are they important to living creatures?

15.2 Summarize the processes involved in the carbon cycle in nature. What are the relative amounts of carbon stored in various "stations" in the cycle? What kinds of compounds are present in these stations? Where are the great reserves of carbon? Is the flow of carbon compounds into and out of the atmosphere greater over land or over sea? Can you account for this? How is fossil-fuel burning affecting the carbon cycle?

15.3 Why does the air near the ground in a cornfield in summer contain less CO_2 by day than by night? Why does the concentration of CO_2 in the atmosphere near the earth increase during the winter and decrease during the summer?

15.4 Is the rate of fixation of carbon by photosynthesis greater on land or at sea? Is there much difference? Bearing in mind the relative areas of the land and the sea, are terrestrial or marine plants the more effective photosynthesizers?

15.5 Carbon is present in compounds of one kind or another in the atmosphere, the hydrosphere, and the lithosphere. In which region would you expect its residence time to be greatest? Least? Why?

15.6 Would you expect the rate of photosynthesis to be greater in surface layers of seawater (which is slightly basic) or in freshwater (which is neutral or slightly acidic) flowing from a river into the sea? Assume that all factors other than the solubility of CO_2 in the two different kinds of water are constant.

15.7 Summarize the processes involved in the nitrogen cycle in nature. What are the relative amounts of nitrogen stored in various "stations" in the cycle? What kinds of compounds are present in these stations? Where are the great reserves of nitrogen? What reserves are most accessible to us? In what ways is modern man upsetting the distribution of nitrogen?

Are we likely to run out of nitrogen? What, then, is the problem with respect to nitrogen in a modern technological society?

15.8 How is sulfur cycled through the biosphere? Why does it accompany the cycling of nitrogen when microorganisms are involved?

15.9 Why does the recycling of phosphorus through the biosphere present special problems? How is man upsetting the balance of phosphorus compounds in the lithosphere and hydrosphere? Is this detrimental to the environment? To man?

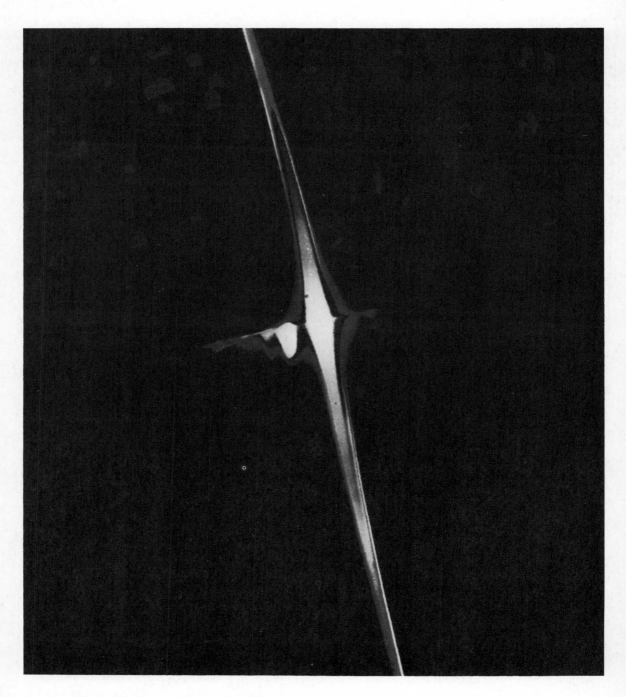

*The photograph taken in ultraviolet light shows the spectrum of
the upper atmosphere of the earth and geocorona.*

Ultraviolet   3645.6 Å
(convergence
limit)

16. Light and Other Forms of Radiant Energy

thermodynamics

electrochemistry

photochemistry

Dalton's atomic theory, Avogadro's hypothesis, the kinetic-molecular theory of matter, Mendeleev's periodic table, Arrhenius' theory of ionization, and the concept of chemical equilibrium—these were milestones in the interpretation of chemical reaction in the nineteenth century. Toward the end of that period chemists discovered many new elements and prepared tens of thousands of new compounds. They developed many useful theories to explain the interactions of these elements and compounds as gases and liquids and in solutions. From studies of the effect of heat on chemical systems grew the science of chemical *thermodynamics*. From investigations of the interactions of electric charges with chemical systems evolved the science and art of *electrochemistry*. Research on the stimulation of chemical reactions by light and the production of light in chemical systems developed into a body of knowledge called *photochemistry*. None of these studies gave immediate insight into the ways atoms interact to form molecules, but all contributed later to our understanding of how elements are held together in compounds—an understanding that developed in the first half of the twentieth century.

Of the many problems still unsolved at the beginning of the twentieth century, three major ones attracted the attention of scientists. One of the problems was the nature of radiant energy, another was the structure of atoms, and a third was how to explain the brilliant light of different colors emitted by different elements present in electric discharges through gases in partially evacuated tubes (Sec. 8.2), or by flames into which the elements or their compounds are sprayed. The story of these problems and their intertwined solutions is a fascinating example of how science grows. The first problem will be studied in this chapter, the other two in Chap. 17.

16.1   Introduction

To understand how atoms are linked to form molecules, we need to know the structures of the atoms themselves. From the experiments described in Chap. 8 we learned that an atom is composed of a very tiny positively charged nucleus surrounded by a much larger space occupied by negative electrons. What keeps these electrons from falling into the positive nucleus? In what sort of pattern are they arranged?

Since individual atoms have a diameter of about one ten-millionth of a centimeter, they are too small to be seen directly by microscope. We can investigate their structure only by indirect means. The first contribution to this investigation was made by scientists who were studying the colored light emitted by electric discharges through gases at low pressures (Sec. 8.2). The discharge through air is rosy violet, that through sodium vapor yellow, and through hydrogen bright blue. You are familiar with the yellow light of sodium lamps on highways and the wide range of colors in "neon" signs, which are electric discharge tubes containing different gases. How can we account for these brilliant colors?

Other scientists who were studying the chemistry of various minerals noticed that handling some of them in powdered form near a candle produced bright colors in the flame—red and orange, green and blue. You may have seen the varicolored flames when driftwood from the sea is burned. Old timbers soaked in waste water from a copper mine burn with green and blue-green flames. You can buy tablets of wax impregnated with various chemicals that will burn with brilliant colors in your fireplace. What causes this variety of colors?

Scientists also noted that the white light generated when an electric arc jumps between two carbon rods can be colored by introducing various metals and their compounds into the carbon.

The energizing of atoms in these various ways causes them to emit light. Can this fact be explained? What significance does it have?

When a metal anode is bombarded with highly energetic electrons in a cathode ray tube (Fig. 16.6), x-rays which will penetrate all but the densest matter are emitted from the anode. In order to explain the production of light and x-rays by energized atoms, we need to know something about the basic nature of light.

16.2 Light—Wave or Particle?

For 2500 years men have speculated about the nature of light and how it interacts with matter. The ancient Greeks thought that the sun and all bodies that emit light must shoot off minute particles or "corpuscles" which strike the eye or skin to produce the sensation of seeing light or feeling heat. This corpuscular concept was generally accepted until the end of the seventeenth century, when the Dutch scientist Christian Huygens showed that the reflection and *refraction* of light (Sec. 16.4) could be explained rather simply by conceiving a ray of light as a train of waves. However, Isaac Newton espoused the corpuscular theory so strongly that Huygens' ideas were ignored by the scientific community throughout the eighteenth century.

refraction

Early in the nineteenth century Thomas Young was studying the rainbow of colors produced by thin films of oil on water and in soap bubbles. These colors could not be explained by the corpuscular theory of light but were readily understandable if light were assumed to be a train of waves. By the middle of the nineteenth century additional observations of the behavior of light had strengthened support for the *wave theory*. And in 1865 James Clerk Maxwell combined all the laws describing the electric and magnetic behavior of matter then known into a set of mathematical equations. Observing that these equations resembled those describing wave motion, Maxwell used them to calculate a theoretical value for the velocity of light (through a vacuum),

wave theory of light

James Maxwell, 1831–1879

assuming it to be a composite of electric and magnetic (electromagnetic) waves. His calculated value turned out to be the observed velocity of light! Thus was born the electromagnetic theory of the nature of light.

Maxwell was able to show that the ultimate source of electromagnetic waves is an accelerated electric charge. In 1888 Heinrich Hertz verified Maxwell's theory by accelerating electric charges in a circuit which generated waves that were propagated through space and could be detected by a similar circuit some distance away. These electromagnetic waves we now call *radio waves*. This confirmation of Maxwell's theory by a totally new phenomenon resulted in the triumph of the wave theory of light in the nineteenth century. To understand this wave theory of light, we need to know a few simple properties of all waves.

radio waves

16.3   The Nature of Waves

If one end of a rope is fastened to a solid wall and the other end is jerked quickly up and down once, a wave travels from the loose end to the fixed end, as shown in Fig. 16.1*a*. If the loose end is jerked up and down several times in rapid succession, waves travel one after another along the rope, as shown in Fig. 16.1*b*. The high point of a wave is called a *crest*, the low point a *trough*. The maximum displacement of the rope from its central position when straight is called the *amplitude* of the wave. The difference in height between a crest and a trough is twice the amplitude. In a series of identical waves, the distance between crests or between troughs is called the *wavelength* and given the symbol λ. The number of waves that pass a given point on the rope each second is called the *frequency* of the waves, symbolized by *f*. The product of the wavelength and the frequency gives the *velocity v* of the waves:

wave amplitude

wavelength

wave frequency
wave velocity

$$v = (\lambda \text{ cm})\left(\frac{f}{\text{sec}}\right) = \lambda f \, \frac{\text{cm}}{\text{sec}} \tag{16.1}$$

FIGURE 16.1 Patterns of waves generated in ropes: (*a*) A single wave moves from left to right in a rope fixed to a wall; (*b*) a succession of waves moves from left to right in the rope.

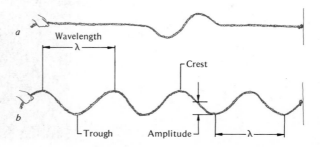

16.4 Some Evidence for the Wave Nature of Light

Some further characteristics of waves can be illustrated by observing the behavior of water waves in a small glass tank (Fig. 16.2). If we hold a flat strip of wood parallel to one end of the tank so that it dips into the surface of the water, and if we then move the wood back and forth, we set up waves like those shown in Fig. 16.3a and b. In part a we are looking horizontally through

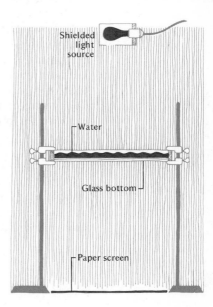

FIGURE 16.2 Apparatus for showing the behavior of waves

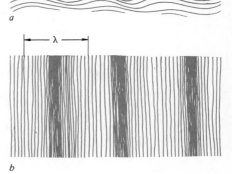

FIGURE 16.3 Waves of water viewed (a) from the side, and (b) from above

the side of the tank. If we shine a light down through the water and look at the image of the waves on the paper screen, we see the pattern shown in part *b*, where the troughs are the dark regions.

If we put a flat piece of glass (which is just a little thinner than the depth of the water in the tank) on the bottom of the tank in Fig. 16.2, the direction in which the waves move (shown by the arrows) is bent to the right when they pass over the plate (Fig. 16.4). We call this change in direction *refraction* of the waves.

wave refraction

Light waves have characteristics similar to those of waves in water. When a beam of light traveling from left to right and grazing the surface of a table passes into the side of a rectangle of glass on the table (Fig. 16.5), part of the light is refracted to the right of the incident direction as the beam enters the glass, and part of the light is reflected. The similarity of the *refraction* of the light beam to the refraction of the waves in water indicates that light has some of the properties of waves.

16.5 Light and Color

When a triangular glass prism is stood on end and a ray of sunlight is directed obliquely against one side of it, the ray is refracted once as it passes from the air into the glass and again as it passes from the glass into the air. The light emerging from the prism is spread into a rainbow (Fig. 7.12). When Isaac

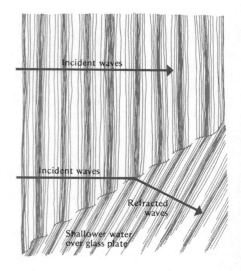
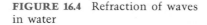

FIGURE 16.4 Refraction of waves in water

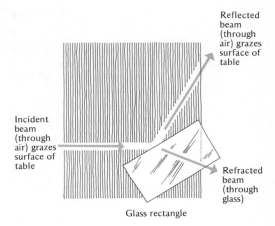

FIGURE 16.5 Refraction of a beam of light passing from air into glass

Reflected beam (through air) grazes surface of table

Incident beam (through air) grazes surface of table

Refracted beam (through glass)

Glass rectangle

spectrum

continuous spectrum

Newton performed this experiment, he called the rainbow band of color a *spectrum*, literally, a "visible range." The colors in the spectrum of sunlight merge gradually from one into the next; there are no sharp lines between colors. We call this a *continuous spectrum*.

Another kind of spectrum is produced when common table salt (or any other compound containing sodium) is sprinkled into a gas flame and the bright-yellow light from the flame passes through the slit and the prism as in Fig. 7.12. Only one color is observed on the screen, a bright-yellow line which is the image of the slit. If a compound containing lithium is sprinkled into the flame, a bright-red line and a yellow line appear on the screen at some distance to the left of the yellow line of sodium. If a thallium compound is sprinkled into the flame, a bright-green line is seen on the screen at some distance to the right of the yellow line. If a mercury arc lamp is used as the source of light, the spectrum contains several distinguishable lines—red, orange, yellow, green, blue, and violet. The spectra of sodium, lithium, thallium, and mercury are called *line spectra* because they consist of discrete lines that do not merge into neighboring colors.

line spectrum

If we think of a ray of white light as a train of waves, what causes this train to be split up into the rays of colored light seen in the rainbow spectrum? Why is violet light refracted more than red light (Fig. 7.12)? What is the difference between red light and violet light? We are familiar with three differences in the waves we observed in a rope and in water: velocity, amplitude, and wavelength. Light waves are described in terms of the same variables. Which variable is associated with differences in color?

Precise measurements of velocity indicate that light of all colors has the same velocity, and so the difference between violet light and red light cannot be a difference in velocity. Incidentally, the velocity is tremendous: 3×10^8 m/sec or 186,000 mi/sec through a vacuum. That's 7.5 times around the world in 1 sec!

When the wind blows hard over a lake, it produces some pretty big waves. If we paddle a canoe through these, we get thrown about pretty roughly. If we paddle the canoe through some little waves, even if they are of the same wavelength, they don't disturb us so much. The intensity of waves seems to be related to their amplitude. If we shine a very bright (intense) ray of light against a prism, we get a very bright rainbow spectrum. If we reduce the brightness of the light source, the spectrum doesn't change; it just gets fainter. Therefore the difference between violet light and red light cannot be a difference in amplitude.

Since the color of light is not dependent on either the velocity or the amplitude of the waves, then it must depend on their third characteristic—wavelength. By measuring the wavelengths of the colors of light emitted by sodium, lithium, thallium, and mercury (Table 16.1), the relation between wavelength and color can readily be seen.

angstrom unit

The wavelengths of all visible light are so small that it is inconvenient to express them in centimeters or millimeters. So a length called 1 *angstrom unit* (abbreviated 1 Å) was chosen for measuring wavelengths. One angstrom unit is 10^{-8} cm; it was named in honor of A. J. Ångström, a Swedish physicist who in 1868 published a monumental catalog of wavelengths of about 1000 different spectral lines from different elements.

Ångström's catalog of spectral lines helped greatly in making qualitative analyses to determine the metallic elements present in a substance. If a solid or a liquid is injected into a flame or heated to a very high temperature in a furnace, the light it gives off will contain the spectral lines of all the metals present. If an electric discharge is passed through an unknown gas at low pressure, the lines observed in the spectrum will identify the elements present in the gas. Spectral analysis has been used to identify rocks and minerals which

TABLE 16.1 *Wavelengths in Angstroms of Some Prominent Lines in Spectra of Sodium, Lithium, Thallium, and Mercury*

	Violet	Blue	Green	Yellow	Orange	Red
Sodium				5893		
Lithium						6708
Thallium			5351			
Mercury	4358	4960	5461	5791	6234	6907

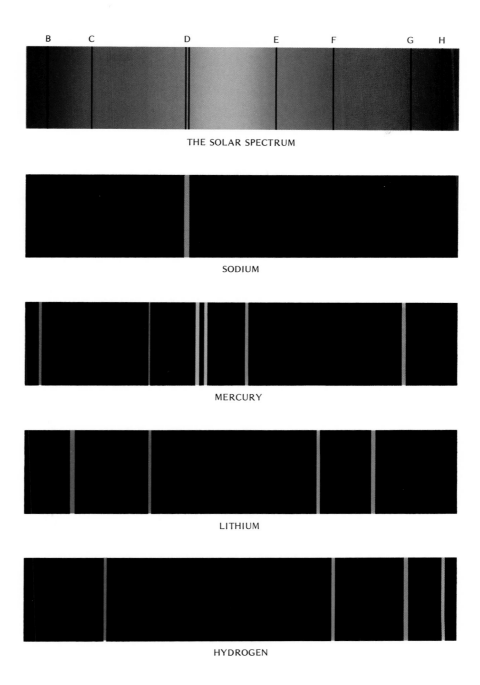

FIGURE 16.7 The continuous spectrum of the sun and some typical line emission spectra. The black lines in the solar spectrum are explained in Sec. 16.7.

PERIODIC TABLE OF THE ELEMENTS

Transition elements are shown in color

Noble gases

Period	IA	IIA	IIIB	IVB	VB	VIB	VIIB	VIII	VIII	VIII	IB	IIB	IIIA	IVA	VA	VIA	VIIA	VIIIA
1	1 H 1.00797 s^1 Hydrogen																	2 He 4.0026 s^2 Helium
2	3 Li 6.939 [He]$2s^1$ Lithium	4 Be 9.0122 [He]$2s^2$ Beryllium											5 B 10.811 $s^2 2p^1$ Boron	6 C 12.0111 $s^2 2p^2$ Carbon	7 N 14.0067 $s^2 2p^3$ Nitrogen	8 O 15.9994 $s^2 2p^4$ Oxygen	9 F 18.9984 $s^2 2p^5$ Fluorine	10 Ne 20.183 $s^2 2p^6$ Neon
3	11 Na 22.9898 [Ne]$3s^1$ Sodium	12 Mg 24.312 [Ne]$3s^2$ Magnesium											13 Al 26.9815 $s^2 3p^1$ Aluminum	14 Si 28.086 [Ne]$3s^2 3p^2$ Silicon	15 P 30.9738 [Ne]$3s^2 3p^3$ Phosphorus	16 S 32.064 [Ne]$3s^2 3p^4$ Sulfur	17 Cl 35.453 [Ne]$3s^2 3p^5$ Chlorine	18 Ar 39.948 [Ne]$3s^2 3p^6$ Argon
4	19 K 39.102 [Ar]$4s^1$ Potassium	20 Ca 40.08 [Ar]$4s^2$ Calcium	21 Sc 44.956 [Ar]$3d^1 4s^2$ Scandium	22 Ti 47.90 [Ar]$3d^2 4s^2$ Titanium	23 V 50.942 [Ar]$3d^3 4s^2$ Vanadium	24 Cr 51.996 [Ar]$3d^5 4s^1$ Chromium	25 Mn 54.938 [Ar]$3d^5 4s^2$ Manganese	26 Fe 55.847 [Ar]$3d^6 4s^2$ Iron	27 Co 58.933 [Ar]$3d^7 4s^2$ Cobalt	28 Ni 58.71 [Ar]$3d^8 4s^2$ Nickel	29 Cu 63.54 [Ar]$3d^{10} 4s^1$ Copper	30 Zn 65.37 [Ar]$3d^{10} 4s^2$ Zinc	31 Ga 69.72 [Ar]$3d^{10} 4s^2 4p^1$ Gallium	32 Ge 72.59 [Ar]$3d^{10} 4s^2 4p^2$ Germanium	33 As 74.922 [Ar]$3d^{10} 4s^2 4p^3$ Arsenic	34 Se 78.96 [Ar]$3d^{10} 4s^2 4p^4$ Selenium	35 Br 79.909 [Ar]$3d^{10} 4s^2 4p^5$ Bromine	36 Kr 83.80 [Ar]$3d^{10} 4s^2 4p^6$ Krypton
5	37 Rb 85.47 [Kr]$5s^1$ Rubidium	38 Sr 87.62 [Kr]$5s^2$ Strontium	39 Y 88.905 [Kr]$4d^1 5s^2$ Yttrium	40 Zr 91.22 [Kr]$4d^2 5s^2$ Zirconium	41 Nb 92.906 [Kr]$4d^4 5s^1$ Niobium	42 Mo 95.94 [Kr]$4d^5 5s^1$ Molybdenum	43 Tc (99) [Kr]$4d^5 5s^2$ Technetium	44 Ru 101.07 [Kr]$4d^7 5s^1$ Ruthenium	45 Rh 102.905 [Kr]$4d^8 5s^1$ Rhodium	46 Pd 106.4 [Kr]$4d^{10}$ Palladium	47 Ag 107.870 [Kr]$4d^{10} 5s^1$ Silver	48 Cd 112.40 [Kr]$4d^{10} 5s^2$ Cadmium	49 In 114.82 [Kr]$4d^{10} 5s^2 5p^1$ Indium	50 Sn 118.69 [Kr]$4d^{10} 5s^2 5p^2$ Tin	51 Sb 121.75 [Kr]$4d^{10} 5s^2 5p^3$ Antimony	52 Te 127.60 [Kr]$4d^{10} 5s^2 5p^4$ Tellurium	53 I 126.904 [Kr]$4d^{10} 5s^2 5p^5$ Iodine	54 Xe 131.30 [Kr]$4d^{10} 5s^2 5p^6$ Xenon
6	55 Cs 132.905 [Xe]$6s^1$ Cesium	56 Ba 137.34 [Xe]$6s^2$ Barium	57 La† 138.91 [Xe]$5d^1 6s^2$ Lanthanum	72 Hf 178.49 [Xe]$4f^{14} 5d^2 6s^2$ Hafnium	73 Ta 180.948 [Xe]$4f^{14} 5d^3 6s^2$ Tantalum	74 W 183.85 [Xe]$4f^{14} 5d^4 6s^2$ Tungsten	75 Re 186.2 [Xe]$4f^{14} 5d^5 6s^2$ Rhenium	76 Os 190.2 [Xe]$4f^{14} 5d^6 6s^2$ Osmium	77 Ir 192.2 [Xe]$4f^{14} 5d^7 6s^2$ Iridium	78 Pt 195.09 [Xe]$4f^{14} 5d^9 6s^1$ Platinum	79 Au 196.967 [Xe]$4f^{14} 5d^{10} 6s^1$ Gold	80 Hg 200.59 [Xe]$4f^{14} 5d^{10} 6s^2$ Mercury	81 Tl 204.37 [Xe]$4f^{14} 5d^{10} 6s^2 6p^1$ Thallium	82 Pb 207.19 [Xe]$4f^{14} 5d^{10} 6s^2 6p^2$ Lead	83 Bi 208.980 [Xe]$4f^{14} 5d^{10} 6s^2 6p^3$ Bismuth	84 Po (210) [Xe]$4f^{14} 5d^{10} 6s^2 6p^4$ Polonium	85 At (210) [Xe]$4f^{14} 5d^{10} 6s^2 6p^5$ Astatine	86 Rn (222) [Xe]$4f^{14} 5d^{10} 6s^2 6p^6$ Radon
7	87 Fr (223) [Rn]$7s^1$ Francium	88 Ra (226) [Rn]$7s^2$ Radium	89 Ac†† (227) [Rn]$6d^1 7s^2$ Actinium	104 Ku (260) [Rn]$6d^2 7s^2$ Kurchatovium*	105 Ha Hahnium													

† 6

58 Ce 140.12 [Xe]$4f^1 5d^1 6s^2$ Cerium	59 Pr 140.907 [Xe]$4f^3 6s^2$ Praseodymium	60 Nd 144.24 [Xe]$4f^4 6s^2$ Neodymium	61 Pm (147) [Xe]$4f^5 6s^2$ Promethium	62 Sm 150.35 [Xe]$4f^6 6s^2$ Samarium	63 Eu 151.96 [Xe]$4f^7 6s^2$ Europium	64 Gd 157.25 [Xe]$4f^7 5d^1 6s^2$ Gadolinium	65 Tb 158.924 [Xe]$4f^9 6s^2$ Terbium	66 Dy 162.50 [Xe]$4f^{10} 6s^2$ Dysprosium	67 Ho 164.930 [Xe]$4f^{11} 6s^2$ Holmium	68 Er 167.26 [Xe]$4f^{12} 6s^2$ Erbium	69 Tm 168.934 [Xe]$4f^{13} 6s^2$ Thulium	70 Yb 173.04 [Xe]$4f^{14} 6s^2$ Ytterbium	71 Lu 174.97 [Xe]$4f^{14} 5d^1 6s^2$ Lutetium

†† 7

90 Th 232.038 [Rn]$6d^2 7s^2$ Thorium	91 Pa (231) [Rn]$5f^2 6d^1 7s^2$ Protactinium	92 U 238.04 [Rn]$5f^3 6d^1 7s^2$ Uranium	93 Np (237) [Rn]$5f^4 6d^1 7s^2$ Neptunium	94 Pu (242) [Rn]$5f^6 7s^2$ Plutonium	95 Am (243) [Rn]$5f^7 7s^2$ Americium	96 Cm (247) [Rn]$5f^7 6d^1 7s^2$ Curium	97 Bk (247) [Rn]$5f^8 6d^1 7s^2$ Berkelium	98 Cf (251) [Rn]$5f^{10} 7s^2$ Californium	99 Es (251) [Rn]$5f^{11} 7s^2$ Einsteinium	100 Fm (254) Fermium	101 Md (253) Mendelevium	102 No (254) Nobelium	103 Lw (257) Lawrencium

Key

Atomic number — 1
Atomic weight — 1.00797
Symbol — H
Electronic configuration (showing noble gas structure kernel) — s^1
Name — Hydrogen

* Also called Rutherfordium, Rf (259); see Sec. 10.7.

were good sources of metals. Some metals previously unknown were discovered by the presence of spectral lines which had not been observed before. The line spectrum of an element is like the fingerprint of a person. Just as every person's fingerprints are different from those of every other person, so the line spectrum of one element is different from that of every other element.

16.6 The Prism Spectroscope

prism spectroscope

An instrument widely used to measure wavelengths of light is the *prism spectroscope*. We note from Fig. 7.12 and Table 16.2 that the angle through which a ray of light is deflected, as it passes through a triangular glass prism, depends on the color (wavelength); the greater the wavelength, the less the bending. The prism spectroscope uses this relationship between wavelength and bending to measure the wavelengths of light emitted from various atoms. When sunlight is passed through such a spectroscope, we see the colors in the continuous rainbow spectrum in the ranges of wavelengths noted in Table 16.2.

16.7 Absorption Spectra

spectroscopy

When scientific inquiry moves from qualitative observation to quantitative measurement, great advances are frequently made. This is exemplified by the advances made in *spectroscopy* when quantitative studies were made of the positions of the lines in a spectrum. Joseph Fraunhofer, a German optical-glass maker, examined the continuous spectrum of sunlight in a spectroscope and found that there were hundreds of black lines distributed along the continuum of color. These lines had the same relative position no matter what kind of slit and prism system he used to separate the sunlight into its components. In 1814 Fraunhofer counted more than 700 dark lines and determined the wavelengths at which they appeared in the sun's spectrum. Forty-five years later, G. R. Kirchhoff, a German physicist, proved conclusively that any substance which emits a radiation of a given wavelength will also *absorb* radiation of the same wavelength. He showed that if a beam of intense sunlight passes through sodium vapor, the continuous spectrum will carry a dark line at exactly the same position as that occupied by the yellow line in the

emission spectrum

emission spectrum of sodium. Thus Kirchhoff established the idea that the

TABLE 16.2 *Wavelengths of Colors Visible to the Human Eye*

Color	Violet	Blue	Green	Yellow	Orange	Red
λ in Å	4000–4500	4500–5000	5000–5700	5700–5900	5900–6100	6100–7500

absorption spectrum *absorption spectrum* of a substance, as well as its emission spectrum, may be used to identify it.

As soon as Kirchhoff saw Fraunhofer's spectrum of the sun, with its dark lines distributed along the continuum, he concluded that these lines were caused by the absorption of radiation by the outer layers of the sun's atmosphere, and that these substances must be the elements which produced bright lines in the same positions in their emission spectra. The German chemist R. W. Bunsen worked with Kirchhoff in examining the emission and absorption spectra of pure elements and of various materials to identify the elements present in them. In some materials they found a number of spectral lines unaccounted for by the elements then known. This led to the discovery of cesium, rubidium, gallium, indium, and thallium. The element helium was discovered in the sun from its characteristic spectral lines before it was found on earth. Neon, argon, krypton, and xenon were identified by the characteristics of their spectra.

16.8   The Visible Spectra of the Elements

The usefulness of a spectroscope is greatly enhanced if a photographic film is used to record the image of the spectrum produced. The spectrum then appears on the photographic negative as a series of black lines in positions corresponding to those of the colored lines seen in the eyepiece or on the screen. The wavelengths of the light producing the lines in the photograph of a spectrum can then be determined with ease and great precision. A spectro-

spectrograph   scope producing spectra on film is often called a *spectrograph*.

flame spectrum   When a spectrum is obtained by vaporizing some compound of a given element in a flame, we call it a *flame spectrum*. The spectrum of the light produced when an electric arc is struck between two rods of a given metallic

arc spectrum   element is called an *arc spectrum*. That obtained from the light of electric
spark spectrum   sparks between two electrodes of a metallic element is called a *spark spectrum*. If an element is vaporized in an evacuated tube like that shown in Fig. 8.2, the spectrum of the light emitted when the electric discharge is flowing is called

discharge tube   a *discharge tube spectrum*.
spectrum   Flame spectra are comparatively simple. The sodium spectrum contains one prominent yellow line, the lithium spectrum red, yellow, blue, and violet lines, of which red is the most intense. The spectrum for potassium has red, green, blue, and violet lines of which the violet is the most intense. The barium flame spectrum contains red, yellow, green, and blue lines with the green much more intense than the others.

Arc, spark, and discharge tube spectra are much more complicated. They contain dozens of lines for some elements, hundreds for others, and thousands for some, like iron. Even so, no two elements have the same spectrum. Apparently each spectrum arises from some feature of the structure of the atoms of a given element. As we shall see in Chap. 17, much of our modern theory of atomic structure has developed from a careful study of spectra.

16.9 Ultraviolet, Infrared, and X-ray Spectra

If we put some mercury in an evacuated tube made of clear quartz and containing two electrodes, a brilliant light and considerable heat are produced when an arc is struck between the electrodes. You have seen mercury arc lights along streets and highways. If such a mercury arc is used as the source of light passing through a spectroscope, we see bright lines in the red, orange, yellow, green, blue, and violet regions of the spectrum. When a photographic plate is used to locate the images of the slit, we are not surprised to note that the plate develops a black line at every position where we saw a colored line. What does surprise us is that the photographic plate "sees" lines where we saw only darkness—some lines lying beyond the dark end of the violet part of the visible spectrum, and some beyond the dark end of the red part. The radiation beyond the visible violet is called ultraviolet and that beyond the visible red is infrared.

If we replace the glass parts of a spectrograph with parts made of quartz, we find that we can see even shorter wavelengths of ultraviolet radiation. These are apparently filtered out by glass. If we replace the glass parts with parts made of rock-salt crystal (single crystals of NaCl), and replace the photographic plate with a row of thermometers (which are more responsive to heat rays than the photographic plate), we find even longer wavelengths of infrared radiation. These rays too are absorbed by glass.

When a cathode ray tube is operated with a potential difference of several thousand volts between the electrodes at so low a gas pressure that only cathode ray discharge is evident, and the whole tube is surrounded by a cover of black cardboard, some radiation is produced that passes through the black cardboard with great ease. This radiation was discovered by Wilhelm Röntgen in 1895. It seemed to originate where the cathode rays impinged on the glass of the tube. Röntgen found that the radiation which penetrated the black cardboard could also penetrate wood, thin sheets of metal, and human flesh. He found that these rays were not deflected by magnets, and so he concluded that they are not streams of particles like cathode rays and positive rays but are more like light. He named the radiation x-rays. Figure 16.6 is a diagram of an x-ray tube. The cathode consists of a filament of tungsten wire heated white-hot by an electric current. Such a hot filament emits large numbers of electrons, which are drawn toward the anode by a very large positive charge on it. The surface of the anode carries a tungsten target, and when the rapidly moving electrons strike it, x-rays are emitted as shown. Gamma rays from radioactive elements have the properties of x-rays with extremely short wavelengths.

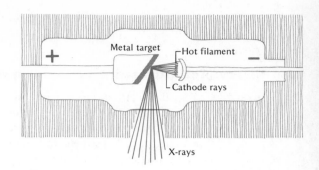

FIGURE 16.6 A cathode ray tube for generating x-rays

16.10   The Electromagnetic Spectrum

So far we have noted that radio signals, infrared light, visible light, ultraviolet light, and x-rays have the characteristics of waves. All these can be reflected like light waves and are refracted when they pass from one medium into another, just as visible light is refracted when it passes from air into glass (Fig. 16.5). The techniques for producing, detecting, and measuring the wavelengths of these different radiations are very different, but we find that all of them travel through free space (a vacuum) at the same velocity, 186,000 mi/sec. All these kinds of radiation consist of electromagnetic waves and differ only in wavelength. The schematic spectrum in Fig. 16.7 (found at the color insert) shows the interrelationships of these waves.

electromagnetic spectrum

Only a very small portion of the *electromagnetic spectrum* is visible, and yet our knowledge of the entire spectrum grew out of studies beginning in this narrow segment. The variations in wavelengths from billionths of a centimeter for gamma rays to millions of centimeters for long radio waves is so enormous as to be almost beyond comprehension. Yet in less than 100 years man has learned how to interpret, generate, direct, and otherwise manipulate these radiations with great skill. This control over one of nature's most useful but elusive forms of energy is a triumph of modern technology.

16.11   Return to the Corpuscular Theory

Maxwell's electromagnetic wave theory of light was one of the great triumphs of nineteenth century theoretical physics. The theory was highly successful in interpreting many different experiments with gamma rays, x-rays, ultraviolet rays, visible light, infrared rays, and radio waves. But near the end of the century all efforts to use Maxwell's theory to interpret the observed intensities of infrared radiation from a heated object failed. This forced scientists to take a new look at the nature of light.

We all know that an object at a high temperature radiates both heat and light, and that the color of the light changes from dark red to bluish-white as the temperature is increased. A hot object that is dark red has a temperature of about 700°C. Bright-red heat indicates about 900°C. Yellowish-white heat occurs at about 1200°C, and blue-white light indicates about 1500°C. As a hot object gradually cools, it seems likely that it continues to radiate energy even at lower temperatures, when we can see no color. Indeed, one can feel "heat rays" being emitted from an object that is only a few degrees above body temperature.

Whan a black object is heated until it is white-hot, it emits radiation of many different wavelengths. The distribution of the amounts of energy emitted at different wavelengths is shown by the solid line in Fig. 16.8. This curve indicates that the maximum intensity of radiation from a body heated to 1500 K is about 2 units at wavelength 20,000 Å. The intensity of radiation falls off rapidly at shorter wavelengths. At 10,000 Å the intensity is only about a quarter of a unit, or one-eighth the intensity (2 units) of the maximum. At wavelengths greater than 20,000 Å the intensity falls off less rapidly. At wavelength 40,000 Å the intensity is about 1 unit, or half the maximum intensity.

Maxwell's theory predicted that the intensity of radiation from an object at 1500 K should become greater and greater at shorter wavelengths, rising steeply between wavelengths 60,000 and 40,000 Å. This prediction is shown by the dotted line in Fig. 16.8. But the experimental data completely disagreed with the prediction. When scientists are faced with facts like these, they have

FIGURE 16.8 The *observed* relation between wavelength and intensity of radiant energy emitted from an object heated to 1500 K (solid curve); the *predicted* relation on the basis of Maxwell's wave theory of light (dotted curve)

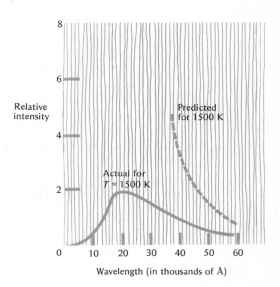

to modify their theories to fit the facts, no matter how successful the theory may have been previously.

corpuscular
theory of light

In 1900 Max Planck proposed a *corpuscular theory of light* which also included some features of the wave theory. He invented the equation

$$E = hf \qquad (16.2)$$

photon
quantum

to describe the radiant energy carried in a particle of light called a *photon* or a *quantum* of light. In Eq. (16.2), E stands for energy, f for frequency of the light in the photon, and h is a proportionality constant. The velocity with which the waves (and hence the photon) are moving is given by the equation:

$$\text{Velocity in } \frac{\text{cm}}{\text{sec}} = \left(\text{frequency in } \frac{\text{waves}}{\text{sec}}\right)(\text{wavelength in cm})$$

or

$$v = f\lambda \qquad (16.1)$$

As noted in Sec. 16.5, the velocity of light is a constant quantity, 3×10^8 m/sec through a vacuum. It is often symbolized by c. Using this in Eq. (16.1), we see that

$$c = f\lambda \qquad \text{or} \qquad f = \frac{c}{\lambda}$$

Substituting this value for f in Eq. (16.2) yields

$$E = h\frac{c}{\lambda} \qquad (16.3)$$

Equation (16.3) tells us that the *larger* the wavelength of light, the *smaller* is the amount of energy carried in one photon. A photon of red light carries less energy than a photon of violet light. Planck's quantum theory correctly predicts the relation between intensity and wavelength of infrared radiation (the solid line in Fig. 16.8).

Now we have come full circle in our query: Is light a wave or a particle (Sec. 16.2)? The answer today is: Light is both a wave and a particle. Under certain conditions it shows wavelike properties; under others, it behaves like a stream of particles. The acceptance of this duality in the nature of light at the beginning of the twentieth century paved the way for spectacular developments in our understanding of the structure of atoms (see Chap. 17), how they combine (see Chaps. 18 and 19), and how they interact with energy.

Summary

When light is refracted or reflected, it behaves like waves. When it is emitted from a hot object, it behaves like particles. A ray of light is both a train of waves and a stream of particles. The wavelike and the corpuscular characteristics of light are brought together in Planck's quantum equation:

$$E = h\frac{c}{\lambda}$$

where E is the energy of a particle (photon or quantum) of light that moves with a velocity of c (the velocity of light, 3×10^8 m/sec in a vacuum) and that has a wavelength of λ. Wavelengths of light are measured in angstroms; $1 \text{ Å} = 10^{-8}$ cm.

The color of light is related to its wavelength. Violet light has a wavelength of 4000 to 4500 Å. Red light has a wavelength of 6100 to 7500 Å. Blue, green, yellow, and orange light have wavelengths between those of violet and red.

Sunlight, or any other white light, is composed of a mixture of light having the colors of the rainbow. These lights of different colors (wavelengths) may be separated into a continuous spectrum (rainbow) by passing white light through a glass prism.

Radiation with a wavelength somewhat shorter than that of violet light is present in sunlight; it is called ultraviolet light. Since it is invisible, a better name is ultraviolet radiation. Sunlight also contains radiant energy with a wavelength somewhat longer than that of red light; it is called infrared light or radiation.

Other kinds of radiant energy exhibit properties very like those of light. Their behavior can be explained by Maxwell's mathematical equations relating the electrical and magnetic behavior of matter. The general term for all such radiant energy is *electromagnetic radiation*. It varies in wavelength from 10^{-11} cm (gamma rays from radioactive substances) to 10^7 cm (long radio waves). All electromagnetic radiation travels with a speed of 3×10^8 m/sec through a vacuum.

When an element or its compounds are heated to a high temperature (by a flame, by an electric arc or spark, or by an electric discharge through a tube containing the element in a gas at low pressure) and the light so produced is passed through a spectrograph, a spectrum is obtained that consists of a series of lines (images of the slit in the spectrograph) ranging from ultraviolet to infrared. The line spectrum for any element is different from that for any other element, so that line spectra can be used to determine the presence or absence of any element in a sample to be analyzed.

New Terms and Concepts

ABSORPTION SPECTRUM: A spectrum consisting of black lines against a continuous background obtained when a cool gas is placed between a source of continuous radiation and a spectrograph; the black lines are found where bright lines would be if the gas were emitting instead of absorbing radiation.

ANGSTROM UNIT: 10^{-8} cm.

ARC SPECTRUM: The spectrum of light produced by an electric discharge in the form of an arc between two electrodes.

CONTINUOUS SPECTRUM: A spectrum with no sharp lines between colors.

CORPUSCULAR THEORY OF LIGHT: The theory that light is composed of a stream of tiny corpuscles called photons or quanta.

DISCHARGE TUBE SPECTRUM: The spectrum of light produced by an electric discharge through a gas contained in a glass tube at low pressure.

ELECTROCHEMISTRY: A study of the transformations of matter that produce electricity or that are brought about by electricity.

ELECTROMAGNETIC SPECTRUM: The entire range of electromagnetic radiation, extending from gamma radiation of very short wavelength to radio waves of great length.

EMISSION SPECTRUM: The series of lines of various wavelengths produced when an arc, spark, or other electric discharge passes between electrodes of a substance or when the substance is heated to incandescence.

FLAME SPECTRUM: The spectrum emitted when a substance is heated in a flame.

LINE SPECTRUM: A spectrum consisting of a series of lines that are the images of the slit of the spectroscope.

PHOTOCHEMISTRY: A study of the transformations of matter that produce light or that are brought about by light.

PHOTON: A packet of electromagnetic energy, hf, defined by the equation $E = hf$.

PRISM SPECTROSCOPE: A device for producing a spectrum by passing light through a slit, then through a glass prism, and finally to an observer's eye.

QUANTUM: A unit quantity of energy.

RADIO WAVES: Electromagnetic waves having a wavelength greater than that of infrared waves.

REFRACTION: Deflection from a straight path.

SPARK SPECTRUM: A spectrum produced by passing an electric spark between two electrodes.

SPECTROGRAPH: An apparatus for dispersing radiation into a spectrum and recording it on photographic paper.

SPECTROSCOPY: The study of spectra.

SPECTRUM: A range or series of images formed when a beam of radiant energy is subjected to dispersion and brought to a focus so that the component waves are arranged in the order of their wavelengths.

THERMODYNAMICS: The study of the changes in energy that accompany any kind of transformation of matter.

WAVE AMPLITUDE: The maximum value of the displacement of a wave.

WAVE FREQUENCY: The number of crests in a moving wave that pass a given point in 1 sec.

WAVE REFRACTION: The bending of the direction of propagation of waves when they encounter some object.

WAVE THEORY OF LIGHT: The theory that light consists of trains of waves traversing any transparent medium.

WAVE VELOCITY: The distance covered by a wave in 1 unit of time.

WAVELENGTH: The distance in the line of advance of a wave from crest to crest, from trough to trough, or between any one point to the next point at which, at the same instant, there is the same phase.

Testing Yourself

16.1 What experimental evidence supports the theory that light is composed of waves?

16.2 What experimental evidence supports the theory that light is composed of particles?

16.3 How are these two concepts brought together into one theory?

16.4 What is the difference in the nature of light having different colors? How are the colors of the rainbow classified in terms of this difference?

16.5 Are the boundaries between colors in the rainbow sharp or diffuse? How can we explain this?

16.6 What is common in the nature of gamma rays, x-rays, ultraviolet rays, visible rays, infrared rays, and radio waves? How do these forms of radiation differ?

16.7 How does the gaseous discharge spectrum of sodium differ from that for lithium, for thallium, and for mercury? What do we call this type of spectrum?

16.8 What is the difference between an emission spectrum and an absorption spectrum for a given element? Of what value are absorption spectra?

16.9 The element helium was discovered in the sun before it was found on earth. How was this possible?

16.10 One way of detecting the presence of very minute quantities of mercury in polluted water is to spray some of the water into a flame and observe the spectrum of light from a mercury lamp passing through the flame. This method is often called "atomic absorption spectroscopy." How would you expect to observe the presence of mercury by such a procedure?

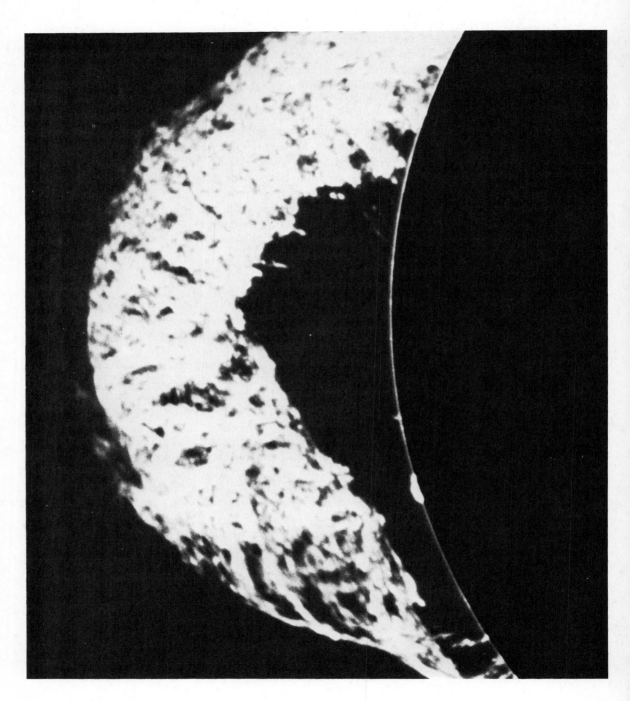

Two views of hydrogen: in the photograph, gaseous eruption of hydrogen from the surface of the sun; in the drawing, its atomic structure.

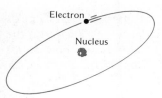

Electron

Nucleus

17. Atomic Spectra and the Bohr Theory of Atomic Structure

At the beginning of Chap. 16 we noted that at the end of the nineteenth century many scientists were working on three problems, the nature of light, how electrons are held around the nucleus of an atom, and what takes place when an atom emits or absorbs light.

By the 1880s spectroscopists had accumulated great encyclopedias of wavelengths for the radiations emitted by atoms under different conditions. Although the lines in electromagnetic spectra clearly show some regularity in spacing, the best spectroscopists of the time were unable to devise a formula that would describe this regularity quantitatively in terms of the wavelengths observed.

In 1885 the long-sought formula was finally discovered by a Swiss mathematics teacher, J. J. Balmer, who studied the spectral data known for hydrogen just as a mathematical puzzle, with no thought of their significance for the nature of light and of matter.

In 1900 Max Planck showed how some of the properties of light which could not be explained by wave theory alone could be explained by assuming that the energy in light is quantized, that it comes in tiny packets of waves which he called photons.

In 1913 the Danish physicist Niels Bohr used Balmer's formula and Planck's quantum equation to develop a revolutionary concept which answered the two remaining questions, how electrons are held in atoms, and how atoms produce electromagnetic radiation when energized. This fascinating development is outlined in this chapter.

17.1 The Spectrum of Atomic Hydrogen

In his search for a formula which would describe the regularity of spectral lines in terms of wavelengths, Balmer studied the spectrum of atomic hydrogen because it contains a convenient number of lines—enough to show a definite pattern, but not so many as to be confusingly complicated. The gaseous discharge tube spectrum for hydrogen shows a bright-red line, a bright-blue line, and two fainter lines in the violet. The wavelengths of these lines are 6562.8, 4861.3, 4340.5, and 4101.7 Å, respectively. If we photograph the discharge tube spectrum, we observe these four lines plus a dozen or so more in the ultraviolet (Fig. 17.1). The most intense line is the red at 6562.8 Å. The others become progressively less intense as their wavelengths decrease. Also, the distance between the lines decreases with the wavelength, until the lines in the ultraviolet are so close together that they seem to merge into a continuum at a wavelength of 3645.6 Å. This point in the spectrum is called the *convergence limit* for this series of lines. Beyond the convergence limit the spectrum is continuous—that is, it shows no structure but is just a gray blur on the photographic plate. Balmer discovered that the wavelengths of all the lines in this spectrum could be *calculated* from the formula

convergence limit

$$\lambda \text{ (angstroms)} = \frac{3645.6 n^2}{n^2 - 4} \tag{17.1}$$

where λ is the wavelength and n is an integer in a series beginning at 3.

EXAMPLE 17.1 What is the calculated wavelength of the line in the H atom spectrum for which the value of n is 3? Does it agree well with the observed value?

FIGURE 17.1 The lines in the Balmer series in the spectrum of atomic hydrogen

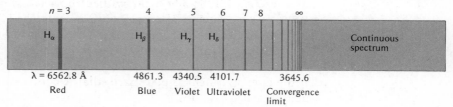

ANSWER

$$\lambda = \frac{3645.6 \times 3^2}{3^2 - 4} = 3645.6 \times \frac{9}{5} = 3645.6 \times 1.8$$
$$= 6562.1 \text{ Å}$$

This quantity differs from the observed value by 0.7 Å, which is the same as $0.7/_{6562.8}$ or 1×10^{-4}. This is 0.01 percent, which is very good agreement. The values of n that give the correct value for the wavelength of the first half-dozen lines in the Balmer series are given above the diagram of the hydrogen spectrum in Fig. 17.1.

Balmer's startling success in deriving a mathematical equation for calculating the wavelengths of the lines in the visible and near ultraviolet regions of the atomic hydrogen spectrum led other investigators to search for similar relationships in other regions of the spectrum. A general equation for calculating the wavelengths of lines found in any region of this spectrum was developed by Walter Ritz and J. R. Rydberg in 1908:

$$\lambda = 911.4 \frac{n^2 m^2}{n^2 - m^2} \tag{17.2}$$

In this equation m is an integer in a series beginning with 1 and n an integer in a series beginning with $m + 1$, as in Table 17.1. The spectral series are named to commemorate the scientists who discovered them.

17.2 The Bohr Model of the Hydrogen Atom

Ritz-Rydberg equation

Theoretical physicists were fascinated by the way laboratory data could be correlated by means of such mathematical expressions as the *Ritz-Rydberg equation*. Five years went by, however, before anyone developed a *theoretical* basis for this excellent *empirical* equation based solely on precise laboratory data and mathematical ingenuity. In 1913 Niels Bohr, a Danish physicist working with Rutherford in England, proposed a model for the hydrogen

TABLE 17.1 *The Spectral Series of Atomic Hydrogen*

Value of m	Values of n	Discoverer of series	Location of series
1	2,3,4, etc.	Lyman	Far ultraviolet
2	3,4,5, etc.	Balmer	Visible
3	4,5,6, etc.	Paschen	Near infrared
4	5,6,7, etc.	Brackett	Infrared
5	6,7,8, etc.	Pfünd	Far infrared

atom that would account for its atomic spectrum. He related both the Ritz-Rydberg equation and Planck's quantum equation to his model to explain why light is not emitted from energized atoms in a continuous spectrum but in a series of lines having certain colors and corresponding wavelengths.

One of the questions confronting Bohr was, "Why don't the negative electrons in an atom fall into the positive nucleus?" It seemed likely to Rutherford and his coworkers that this collapse fails to occur because electrons are moving around the nucleus as the planets move around the sun. There is a strong gravitational attraction between the sun and a planet, but this is just balanced by the tendency of the planet to fly off in a straight line (as a stone whirled in a circle at the end of a string flies off if the string breaks). The balance between the gravitational attraction and this tendency keeps the planet moving in its orbit. Perhaps the balance of the electric attraction between nucleus and electron and the tendency of the electron to fly off would keep it in orbit around the nucleus.

electron orbit

However, Maxwell's equations for the behavior of a moving charged particle predicted that an *electron* revolving in an *orbit* would emit radiant energy and fall into the nucleus. Since hydrogen exists as stable atoms, Maxwell's prediction does not hold. So Bohr made a bold break with Maxwell's theory and built his own theory on the following postulates.

1 When an electron is revolving in a given orbit around the nucleus, it does not emit radiant energy; the atom is considered to be a system in a "stationary state."
2 Each stationary state is characterized by a particular orbit for the electron, and the electron has a specific energy when it occupies that orbit.
3 The dynamic equilibrium of the system in a stationary state is governed by the ordinary laws of physics, but these laws do not hold for the passing of the system from one stationary state to another.
4 When an electron moves from one orbit to another of less energy (when the system passes from a stationary state of given energy to one of less energy), radiant energy of a certain frequency is emitted; the difference in energy between the two states is related to the radiation by Planck's quantum equation: $E_2 - E_1 = hf$.

angular momentum

5 The only orbits that an electron can occupy are those for which the *angular momentum* of the electron is an integral multiple of $h/(2\pi)$, where h is Planck's constant.

The term "angular momentum" in postulate 5 needs some explanation. If you turn a bicycle upside down and work the pedals, the rear wheel can be made to revolve about its axle. When you stop working the pedals, the wheel continues to revolve. We say that the wheel has angular momentum. The faster you get the wheel to revolve, the greater its angular momentum, i.e., the longer it will run before friction stops it. You can vary the angular momentum of the bicycle wheel continuously; that is, you can work the pedals to give any desired amount of angular momentum to the wheel. But the angular momentum of the electron revolving around the atomic nucleus cannot vary continuously. It can take on only the values $1h/(2\pi)$, $2h/(2\pi)$, $3h/(2\pi)$, and so on.

When you walk up a ramp, you can take steps that will lift you as much or as little as you wish. When you walk up a staircase, the amount you go up is quantized. You have to go up 1, 2, or 3 steps; you can't go up a fraction of a step. The electron in the atom can go up from one orbit (*energy level*) to another but cannot remain at any intermediate level.

energy level

17.3   Bohr Orbits and the Locations of Lines in the Hydrogen Spectrum

Figure 17.2 shows the Bohr model of orbits that may be occupied by the electron in a hydrogen atom. Under ordinary conditions, the electron is in the smallest orbit. The angular momentum in this orbit is $1h/(2\pi)$, and so we can say the orbit is described by the quantum number $n = 1$. In the next larger orbit, the angular momentum is $2h/(2\pi)$, and so $n = 2$. As n increases to 3, 4, 5, and so on, the corresponding orbits are larger and larger.

FIGURE 17.2   The simplest orbits for an electron in a hydrogen atom and their relation to spectral series

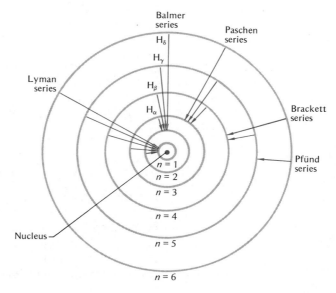

When an electric discharge is passed through hydrogen gas at low pressure, the electron in the smallest orbit of a hydrogen atom is knocked into a larger orbit for which $n = 2$ or 3 or 4 or more. When this electron later falls back to a smaller orbit, one quantum (photon) of light is emitted. The energy in this photon is related to its wavelength thus:

$$E_2 - E_1 = h\frac{c}{\lambda} \tag{17.3}$$

where E_2 is the electron's energy in the larger orbit and E_1 its energy in the smaller one. The wavelength of the line in the hydrogen spectrum (Fig. 17.1) marked H_α can be calculated by Eq. (17.3) from the difference between the energy the electron has when in the orbit for which $n = 3$ and that which it has when $n = 2$. The wavelength of the H_β line can be calculated from the difference between the energy the electron has when in the orbit for which $n = 4$ and that when $n = 2$. The calculated wavelengths agree with those observed.

Recalling that the wavelength of the light emitted when an electron falls into a smaller orbit varies inversely [Eq. (17.3)] with the difference in energy between the two orbits, we see that as the *energies* for the photons in the Balmer series *increase* from H_α through H_γ, the *wavelengths* of the photons *decrease* (go from red toward violet). This shift is in accord with the spectrum recorded in Fig. 17.1. Thus the conclusions reached from Bohr's model fit the observed facts.

When an electron falls from the orbit for which $n = 6$ into that for $n = 1$ (as in the Lyman series in Fig. 17.2), the energy released is larger than that released when the electron falls from the $n = 6$ orbit into the $n = 2$ orbit (as in the Balmer series). All the *energy releases* in the Lyman series will be *larger* than the corresponding releases in the Balmer series. Consequently, the *wavelengths* of the photons in the Lyman series will be *smaller* than those in the Balmer series. Indeed, the Lyman series is found in the ultraviolet region of the hydrogen spectrum. Similar explanations account for the presence of the Paschen, Brackett, and Pfünd series in the infrared region of the hydrogen spectrum.

When hydrogen atoms are energized by the passage of an electric discharge through them, we would expect that large numbers of them would have their electrons boosted into the $n = 2$ orbit, smaller numbers into the $n = 3$ orbit, still smaller numbers into the $n = 4$ orbit, etc. When these electrons fall back into the $n = 2$ orbit, the largest number of photons would be liberated by the fall of the many electrons from the $n = 3$ orbit into the $n = 2$ orbit, a smaller

17.4   The Bohr Model and the Intensities of Lines in the Balmer Series

number of photons would be liberated by the fall of fewer electrons from the $n = 4$ orbit into the $n = 2$ orbit, and still smaller numbers of photons would be liberated by the fall of electrons from larger orbits. In other words, we would expect the intensity of the light emitted by the fall of electrons from the $n = 3$ orbit to the $n = 2$ orbit to be greater than the intensity of the light emitted from the fall of electrons from the $n = 4$ orbit to the $n = 2$ orbit, etc. Thus the intensity of the spectral lines in the Balmer series should be greatest for the H_α (red) line, less for the H_β (blue) line, still less for the H_γ (violet) line, etc. Again, the expectations from the Bohr model are fulfilled in the observed Balmer spectrum in Fig. 17.1, where the H_α line is the blackest and the H_γ is faint.

17.5 Orbits and Energy Levels

If an electron orbiting about a hydrogen nucleus is moved into an orbit of greater radius, some energy will have to be expended to achieve this displacement against the force of attraction between the positive nucleus and the negative electron. Stated another way, the larger the radius of an electron's orbit, the greater the energy of the electron in that orbit. The concepts "orbit" and "energy level" are thus seen to be closely related. As spectra of atoms having greater complexity than H atoms were studied, it became easier to think of the electrons as occupying different energy levels rather than trying to assign them to certain orbits in an atom. Using this symbolism, we talk about the transition of an electron from one energy level to a lower one and how the difference in energy levels is shown by the wavelength of the light emitted.

When hydrogen atoms are given a jolt of energy by the passing of an electric discharge through them, their electrons are kicked into higher energy levels. When they fall back to lower energy levels, they emit photons of light corresponding to their losses of energy. Some electrons are kicked into higher energy levels than others, and some fall back to lower levels than others. The transitions between the various levels give rise to the various wavelengths of radiation in the spectrum. The energy levels involved in transitions producing the series of lines observed in the spectrum of atomic hydrogen are shown in Fig. 17.3.

When the electron of a hydrogen atom is given sufficient energy to separate it completely from the nucleus with which it was associated, we say that the H atom has been ionized:

$$H + \text{energy} \longrightarrow H^+ + e^- \tag{17.4}$$

In Fig. 17.3 we note that the energy levels get closer and closer together as n increases. Consequently, the lines in the hydrogen spectrum get closer and

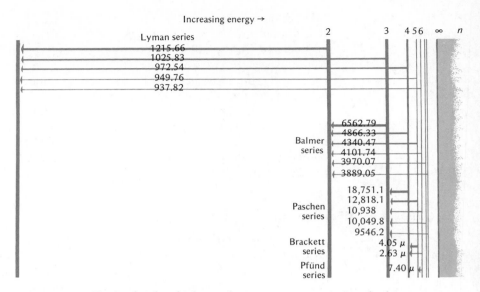

FIGURE 17.3 Energy levels which an electron may occupy in a hydrogen atom

closer together until they finally blur into the continuum noted in Fig. 17.1. The beginning of this continuum is at a wavelength involving the fall of electrons from a level where $n = \infty$. When the electron reaches this level, the atom is ionized. From the convergence limit the *ionization energy* of the H atom (that energy required to remove an electron and produce an H⁺ ion) can be calculated. The calculated value agrees with the value found by direct measurement with electric apparatus, thus strengthening the case for the Bohr model.

ionization energy

17.6 Sublevels within Main Energy Levels

doublet lines
quantum number

Shortly after Bohr postulated his model of the hydrogen atom, more refined spectrographs showed that the lines in the hydrogen spectrum were really *doublets* (two separate lines very close together). This indicated that the energy level that had been described by a given value of the *quantum number*, n, was really two energy levels lying very close to each other. A second quantum number, symbolized by the letter l, was introduced to take into account this difference between the two sublevels that gave rise to the doublets. Although the values of n run from 1 upward in whole-number steps, the values of l for sublevels within a main energy level start with zero and increase in whole number steps up to $n - 1$.

When a gaseous discharge tube containing hydrogen is placed between the poles of a powerful magnet, the spectrum noted in Fig. 17.1 contains many more lines. To account for this complexity, the Bohr model employs a third quantum number m, which is related to the magnetic splitting of the l sublevels of energy into sub-sublevels. For the m sublevels, values of m run in whole-number steps from $-l$ through zero and up to $+l$.

The energy levels, sublevels, and sub-sublevels that account for the fine structure of the hydrogen spectrum are insufficient to account for the fine structure in spectra of atoms containing larger charges on the nucleus. To account for these, it was postulated that an electron in any atom spins about its own axis as it orbits the nucleus—just as the earth rotates once each day around its axis as it orbits the sun once each year. The spin of an electron may be clockwise or counterclockwise. The spin is described by the quantum number s, which may have only one of two values: $-\frac{1}{2}$ or $+\frac{1}{2}$. The complete description of the energy level (now using this as a general term for all levels and sublevels) for a given electron in a hydrogen atom involves the assignment of values for each of the four quantum numbers.

All the lines in the hydrogen spectrum can be accounted for by assigning the quantum numbers with the following limitations on their values:

$$n = 1, 2, 3, \ldots , \infty$$
$$l = 0, 1, 2, \ldots , n - 1$$
$$m = -l, \ldots , 0, \ldots , +l$$
$$s = -\tfrac{1}{2} \text{ or } +\tfrac{1}{2}$$

17.7 Energy Levels in Terms of Quantum Numbers

The lowest energy level is described by the quantum numbers $n = 1$, $l = 0$, $m = 0$, and $s = -\frac{1}{2}$. When $n = 1$, the only value l can have is 0 (because l cannot exceed $n - 1$). Also the only value m can have is 0 (because m ranges from $-l$ through 0 to $+l$). Since two levels cannot be described by the same set of quantum numbers, the next-to-lowest energy level must have the values $n = 1$, $l = 0$, $m = 0$, $s = +\frac{1}{2}$.

Lowest level:	$n = 1$ $l = 0$	$m = 0$	$s = -\frac{1}{2}$
Second level:	$n = 1$ $l = 0$	$m = 0$	$s = +\frac{1}{2}$

Only two levels can be described by different sets of quantum numbers when $n = 1$.

The third and fourth energy levels must be described by the quantum number $n = 2$, because we have exhausted the possibilities for different sets of quantum numbers when $n = 1$. If $n = 2$, then l can have the values 0 or 1. If $l = 0$, the only value m can have is 0, and the energy levels are described as follows:

| Third level: | $n = 2$ | $l = 0$ | $m = 0$ | $s = -\frac{1}{2}$ |
| Fourth level: | $n = 2$ | $l = 0$ | $m = 0$ | $s = +\frac{1}{2}$ |

The fifth and sixth energy levels can be described by $l = 1$ and $m = -1$.

| Fifth level: | $n = 2$ | $l = 1$ | $m = -1$ | $s = -\frac{1}{2}$ |
| Sixth level: | $n = 2$ | $l = 1$ | $m = -1$ | $s = +\frac{1}{2}$ |

The seventh and eighth energy levels can be described by $l = 1$ and $m = 0$.

| Seventh level: | $n = 2$ | $l = 1$ | $m = 0$ | $s = -\frac{1}{2}$ |
| Eighth level: | $n = 2$ | $l = 1$ | $m = 0$ | $s = +\frac{1}{2}$ |

The ninth and tenth energy levels can be described by $l = 1$ and $m = +1$.

| Ninth level: | $n = 2$ | $l = 1$ | $m = +1$ | $s = -\frac{1}{2}$ |
| Tenth level: | $n = 2$ | $l = 1$ | $m = +1$ | $s = +\frac{1}{2}$ |

We have now exhausted the possibilities of different sets of quantum numbers when $n = 2$. The next higher energy levels must have $n = 3$ with $l = 0$, 1, or 2 and $m = -2, -1, 0, +1,$ or $+2$. The sets of quantum numbers for describing all the energy sublevels for $n = 1, 2, 3,$ and 4 are summarized in Table 17.2. The 2 sublevels for which $n = 1$ are said to form the *K shell* of levels, the 8 sublevels for which $n = 2$ are in the *L* shell, the 18 for which $n = 3$ are in the *M* shell, and the 32 for which $n = 4$ are in the *N* shell. These are called *electron shells*.

electron shell

17.8   Energy Levels in Atoms with More Than One Electron

Our study of Bohr theory thus far has been concerned with the mathematical correlation of wavelengths of lines found in the spectrum of atomic hydrogen and the interpretation of these wavelengths in terms of a model involving a central nucleus and a planetary electron. The equations for calculating the energy levels available to a lone electron in a hydrogen atom are equally valid when applied to the lone electron in a singly charged helium ion He^+, a doubly charged lithium ion Li^{2+}, a triply charged beryllium ion Be^{3+}, etc. But when there are two or more electrons in one atom, the interactions among the nucleus and the electrons become so complicated that we cannot develop equations precisely describing the energy levels for the electrons.

However, we need not abandon the Bohr approach simply because we cannot use it to produce precise mathematical equations describing these *polyelectronic* systems. The concept of *levels of energy* that may be occupied by electrons is still useful even though we cannot precisely determine the energy for each level. In considering the structure of more complicated atoms, we assume that the energy levels available to the many electrons in these atoms are essentially the same as those found through the study of the atomic hy-

polyelectronic

TABLE 17.2 *Combinations of Values for Quantum Numbers for Electron Energy Levels in Hydrogen*

n	l	m	s	Total number of electrons	Energy level
1	0	0	$-\frac{1}{2}, +\frac{1}{2}$	2	K
2	0	0	$-\frac{1}{2}, +\frac{1}{2}$	2	L
	1	-1	$-\frac{1}{2}, +\frac{1}{2}$	6	
		0	$-\frac{1}{2}, +\frac{1}{2}$	(8)	
		$+1$	$-\frac{1}{2}, +\frac{1}{2}$		
3	0	0	$-\frac{1}{2}, +\frac{1}{2}$	2	M
	1	-1	$-\frac{1}{2}, +\frac{1}{2}$	6	
		0	$-\frac{1}{2}, +\frac{1}{2}$		
		$+1$	$-\frac{1}{2}, +\frac{1}{2}$		
	2	-2	$-\frac{1}{2}, +\frac{1}{2}$	10	(18)
		-1	$-\frac{1}{2}, +\frac{1}{2}$		
		0	$-\frac{1}{2}, +\frac{1}{2}$		
		$+1$	$-\frac{1}{2}, +\frac{1}{2}$		
		$+2$	$-\frac{1}{2}, +\frac{1}{2}$		
4	0	0	$-\frac{1}{2}, +\frac{1}{2}$	2	N
	1	-1	$-\frac{1}{2}, +\frac{1}{2}$	6	
		0	$-\frac{1}{2}, +\frac{1}{2}$		
		$+1$	$-\frac{1}{2}, +\frac{1}{2}$		
	2	-2	$-\frac{1}{2}, +\frac{1}{2}$	10	
		-1	$-\frac{1}{2}, +\frac{1}{2}$		
		0	$-\frac{1}{2}, +\frac{1}{2}$		(32)
		$+1$	$-\frac{1}{2}, +\frac{1}{2}$		
		$+2$	$-\frac{1}{2}, +\frac{1}{2}$		
	3	-3	$-\frac{1}{2}, +\frac{1}{2}$	14	
		-2	$-\frac{1}{2}, +\frac{1}{2}$		
		-1	$-\frac{1}{2}, +\frac{1}{2}$		
		0	$-\frac{1}{2}, +\frac{1}{2}$		
		$+1$	$-\frac{1}{2}, +\frac{1}{2}$		
		$+2$	$-\frac{1}{2}, +\frac{1}{2}$		
		$+3$	$-\frac{1}{2}, +\frac{1}{2}$		

drogen spectrum. We assume also that in an atom at ordinary temperatures the electrons occupy the lowest energy levels available to them (just as water flows down to the lowest place available to it). When a polyelectronic atom is excited by heating or by an electric discharge, the electrons are elevated to higher energy levels; when electrons fall back into the vacancies thus created in the lower levels, electromagnetic radiation of a wavelength determined by the difference in energy levels is emitted (as in Fig. 17.3). From a careful analysis of the spectra of polyelectronic atoms, Wolfgang Pauli, an Austrian physicist, discovered that transitions of electrons do not take place between all pairs of levels. He concluded that no more than one electron in a given atom could occupy the energy level described by a given set of the four quantum numbers n, l, m, and s. This is analogous to the common saying that two bodies cannot occupy the same space at the same time.

17.9   A Simplified Chart of the Energy Levels in Polyelectronic Atoms

Chemists are interested in the arrangement of electrons in atoms because they want to explain how atoms link to form molecules. The differences in energy among the sub-sublevels that are described by different values of the quantum numbers m and s are very small when compared to the differences between levels with different values of n and l. It is convenient, therefore, to lump the m and s sub-sublevels together when looking at the overall picture of the energy levels occupied by electrons in an atom. What we need is a chart showing the energy levels for electrons with different values of n and l (Fig. 17.4).

To simplify the notation in this chart, we use a number for the value of the quantum number n but a letter to indicate the value of the quantum number l:

Value of l:	0	1	2	3
Symbol:	s	p	d	f

The 1s level is that for which $n = 1$ and $l = 0$. The 2p level is that for which $n = 2$ and $l = 1$. The 3d level is the one for which $n = 3$ and $l = 2$, etc.

The lowest energy level is 1s. The increase in energy from level to level gets smaller and smaller as the total energy gets greater and greater. The levels shown in Fig. 17.4 are approximate, not exact.

17.10   The Building Up of Electron Configurations in Polyelectronic Atoms

If you think of the squares in Fig. 17.4 as water tanks connected by pipes running between levels, you would expect that water poured into the top tanks would run all the way to the bottom tank. A slow trickle would first fill the 1s tank, then the 2s, then 2p tanks, etc. It is not surprising, then, that a study

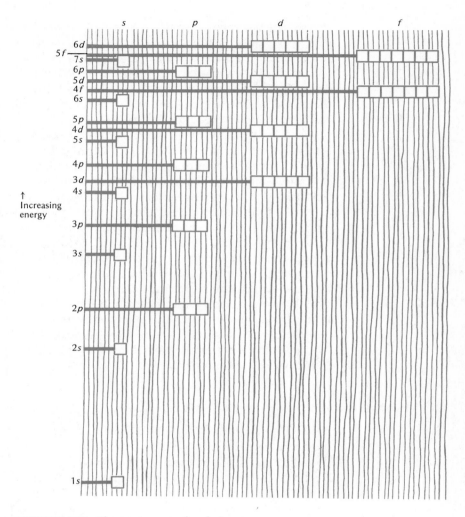

FIGURE 17.4 Electron energy levels in atoms

of spectra of polyelectronic atoms shows that the electrons present tend to fill up the lower energy levels first.

In the hydrogen atom with its nuclear charge of 1, the electron will be found in the $1s$ level unless we energize it by an electric discharge, heat, or radiant energy. We can symbolize this state by putting a small arrow in the $1s$ box. In the helium atom, the nuclear charge is 2 and there are 2 electrons. They must occupy two levels having different quantum numbers. This will be achieved if one level has $s = -\frac{1}{2}$ and the other $s = +\frac{1}{2}$. For simplicity, we indicate the opposite spins by pointing arrows (representing the electrons) in opposite directions.

The lithium atom has 3 electrons. Two of these can occupy the $1s$ level, but the third must go into the higher $2s$ level. The beryllium atom has 4 electrons. The fourth goes into the $2s$ box, but its spin will be opposite to that of the third electron. Using the same reasoning, we put the fifth electron in boron into one of the $2p$ boxes.

But what about the sixth electron in the carbon atom? Does it pair up with the fifth electron already in one of the three $2p$ boxes or does it enter the second $2p$ box? Analysis of the spectrum of carbon atoms shows that the sixth electron is unpaired—that it enters the second $2p$ box. The spectrum of nitrogen shows that it contains three unpaired electrons—that the seventh electron enters the third $2p$ box. The eighth electron in an oxygen atom pairs up with one of the electrons already in a $2p$ box, the ninth electron in a fluorine atom pairs up with a second of the electrons already in a $2p$ box, and the tenth electron in a neon atom completes the pairing in the $2p$ level.

It is convenient to describe the filling of electron levels by the notation that $1s^1$ means there is one electron in the $1s$ energy level, $1s^2$ means there are two electrons in the $1s$ energy level, $2p^6$ means that there are six electrons in the $2p$ energy level, etc. This notation is used in Table 17.3, which shows how the electrons fill up the energy levels of the first 18 elements in the periodic table. The *electron configurations* for the elements with atomic numbers from 1 to 102 are summarized in Table 17.4.

electron
configuration

17.11 Absorption Spectra and the Bohr Model

In Sec. 16.7 we noted that when a sodium compound is sprayed into a colorless flame, a bright-yellow line appears in the spectrum emitted by the flame. We noted also that when white light is passed through sodium in the gaseous form, the observed spectrum has a black line where we observed the yellow line in the emission spectrum for sodium. The Bohr model explains both these observations.

When an element emits light having its characteristic wavelengths, it does so because the atoms present have been given energy and lifted into levels of higher energy. When they fall back to the lower energy levels they normally occupy, they emit photons with energies equivalent to the decreases in the energies of all the electrons.

When an element absorbs light, the electrons in the lower energy levels absorb photons of light with energy just sufficient to boost them into higher energy levels. When photons with these characteristic energies are removed from the white light, there are black spaces in the continuous spectrum of white light at the wavelengths of the photons absorbed.

TABLE 17.3 *Electron Configurations of the First 18 Elements in the Periodic Table*

	Element	Abbreviated notation	Schematic representation				
			1s	2s	2p	3s	3p
First period	$_1$H	$1s^1$	↑				
	$_2$He	$1s^2$	↑↓				
Second period	$_3$Li	$1s^2 2s^1$	↑↓	↑	○ ○ ○		
	$_4$Be	$1s^2 2s^2$	↑↓	↑↓	○ ○ ○		
	$_5$B	$1s^2 2s^2 2p^1$	↑↓	↑↓	↑ ○ ○		
	$_6$C	$1s^2 2s^2 2p^2$	↑↓	↑↓	↑ ↑ ○		
	$_7$N	$1s^2 2s^2 2p^3$	↑↓	↑↓	↑ ↑ ↑		
	$_8$O	$1s^2 2s^2 2p^4$	↑↓	↑↓	↑↓ ↑ ↑		
	$_9$F	$1s^2 2s^2 2p^5$	↑↓	↑↓	↑↓ ↑↓ ↑		
	$_{10}$Ne	$1s^2 2s^2 2p^6$	↑↓	↑↓	↑↓ ↑↓ ↑↓		
Third period	$_{11}$Na	$1s^2 2s^2 2p^6 3s^1$	↑↓	↑↓	↑↓ ↑↓ ↑↓	↑	○ ○ ○
	$_{12}$Mg	$1s^2 2s^2 2p^6 3s^2$	↑↓	↑↓	↑↓ ↑↓ ↑↓	↑↓	○ ○ ○
	$_{13}$Al	$1s^2 2s^2 2p^6 3s^2 3p^1$	↑↓	↑↓	↑↓ ↑↓ ↑↓	↑↓	↑ ○ ○
	$_{14}$Si	$1s^2 2s^2 2p^6 3s^2 3p^2$	↑↓	↑↓	↑↓ ↑↓ ↑↓	↑↓	↑ ↑ ○
	$_{15}$P	$1s^2 2s^2 2p^6 3s^2 3p^3$	↑↓	↑↓	↑↓ ↑↓ ↑↓	↑↓	↑ ↑ ↑
	$_{16}$S	$1s^2 2s^2 2p^6 3s^2 3p^4$	↑↓	↑↓	↑↓ ↑↓ ↑↓	↑↓	↑↓ ↑ ↑
	$_{17}$Cl	$1s^2 2s^2 2p^6 3s^2 3p^5$	↑↓	↑↓	↑↓ ↑↓ ↑↓	↑↓	↑↓ ↑↓ ↑
	$_{18}$Ar	$1s^2 2s^2 2p^6 3s^2 3p^6$	↑↓	↑↓	↑↓ ↑↓ ↑↓	↑↓	↑↓ ↑↓ ↑↓

Both emission spectra and absorption spectra can be used to identify the presence of elements in samples of substances.

17.12 The Bohr Model and the Periodic Table

The configurations of electrons worked out through the Bohr model contribute greatly to the usefulness of the periodic table of the chemical elements. We see that each period in the table marks the complete filling of the energy levels with a given value for *n*. For period 1, $n = 1$, for period 2, $n = 2$, etc. All the elements for group IA have one unpaired electron in an *s* level. All the elements in group IIA have two paired electrons in an *s* level. The elements in group IIIA have two paired electrons in an *s* level and one unpaired electron in a *p* level, those in group IVA have two unpaired electrons in a *p* level, those in group VA have three unpaired electrons in a *p* level. In groups VIA, VIIA, and the noble gas group, pairing proceeds in the *p* levels until it is

TABLE 17.4 *Electron Configurations of the Elements*

Atomic number	Element	1 s	2 s p	3 s p d	4 s p d f	5 s p d f	6 s p d f	7 s
1	H	1						
2	He	2						
3	Li	2	1					
4	Be	2	2					
5	B	2	2 1					
6	C	2	2 2					
7	N	2	2 3					
8	O	2	2 4					
9	F	2	2 5					
10	Ne	2	2 6					
11	Na	2	2 6	1				
12	Mg	2	2 6	2				
13	Al	2	2 6	2 1				
14	Si	2	2 6	2 2				
15	P	2	2 6	2 3				
16	S	2	2 6	2 4				
17	Cl	2	2 6	2 5				
18	Ar	2	2 6	2 6				
19	K	2	2 6	2 6	1			
20	Ca	2	2 6	2 6	2			
21	Sc	2	2 6	2 6 1	2			
22	Ti	2	2 6	2 6 2	2			
23	V	2	2 6	2 6 3	2			
24	Cr	2	2 6	2 6 5	1			
25	Mn	2	2 6	2 6 5	2			
26	Fe	2	2 6	2 6 6	2			
27	Co	2	2 6	2 6 7	2			
28	Ni	2	2 6	2 6 8	2			
29	Cu	2	2 6	2 6 10	1			
30	Zn	2	2 6	2 6 10	2			
31	Ga	2	2 6	2 6 10	2 1			
32	Ge	2	2 6	2 6 10	2 2			
33	As	2	2 6	2 6 10	2 3			
34	Se	2	2 6	2 6 10	2 4			
35	Br	2	2 6	2 6 10	2 5			
36	Kr	2	2 6	2 6 10	2 6			
37	Rb	2	2 6	2 6 10	2 6	1		
38	Sr	2	2 6	2 6 10	2 6	2		
39	Y	2	2 6	2 6 10	2 6 1	2		
40	Zr	2	2 6	2 6 10	2 6 2	2		
41	Nb	2	2 6	2 6 10	2 6 4	1		
42	Mo	2	2 6	2 6 10	2 6 5	1		
43	Tc	2	2 6	2 6 10	2 6 6	1?		
44	Ru	2	2 6	2 6 10	2 6 7	1		
45	Rh	2	2 6	2 6 10	2 6 8	1		
46	Pd	2	2 6	2 6 10	2 6 10			
47	Ag	2	2 6	2 6 10	2 6 10	1		
48	Cd	2	2 6	2 6 10	2 6 10	2		
49	In	2	2 6	2 6 10	2 6 10	2 1		
50	Sn	2	2 6	2 6 10	2 6 10	2 2		
51	Sb	2	2 6	2 6 10	2 6 10	2 3		

Atomic number	Element	1	2		3			4				5				6				7
		s	s	p	s	p	d	s	p	d	f	s	p	d	f	s	p	d	f	s
52	Te	2	2	6	2	6	10	2	6	10		2	4							
53	I	2	2	6	2	6	10	2	6	10		2	5							
54	Xe	2	2	6	2	6	10	2	6	10		2	6							
55	Cs	2	2	6	2	6	10	2	6	10		2	6			1				
56	Ba	2	2	6	2	6	10	2	6	10		2	6			2				
57	La	2	2	6	2	6	10	2	6	10		2	6	1		2				
58	Ce	2	2	6	2	6	10	2	6	10	2	2	6			2?				
59	Pr	2	2	6	2	6	10	2	6	10	3	2	6			2?				
60	Nd	2	2	6	2	6	10	2	6	10	4	2	6			2				
61	Pm	2	2	6	2	6	10	2	6	10	5	2	6			2?				
62	Sm	2	2	6	2	6	10	2	6	10	6	2	6			2				
63	Eu	2	2	6	2	6	10	2	6	10	7	2	6			2				
64	Gd	2	2	6	2	6	10	2	6	10	7	2	6	1		2				
65	Tb	2	2	6	2	6	10	2	6	10	9	2	6			2?				
66	Dy	2	2	6	2	6	10	2	6	10	10	2	6			2?				
67	Ho	2	2	6	2	6	10	2	6	10	11	2	6			2?				
68	Er	2	2	6	2	6	10	2	6	10	12	2	6			2?				
69	Tm	2	2	6	2	6	10	2	6	10	13	2	6			2				
70	Yb	2	2	6	2	6	10	2	6	10	14	2	6			2				
71	Lu	2	2	6	2	6	10	2	6	10	14	2	6	1		2				
72	Hf	2	2	6	2	6	10	2	6	10	14	2	6	2		2				
73	Ta	2	2	6	2	6	10	2	6	10	14	2	6	3		2				
74	W	2	2	6	2	6	10	2	6	10	14	2	6	4		2				
75	Re	2	2	6	2	6	10	2	6	10	14	2	6	5		2				
76	Os	2	2	6	2	6	10	2	6	10	14	2	6	6		2				
77	Ir	2	2	6	2	6	10	2	6	10	14	2	6	7		2				
78	Pt	2	2	6	2	6	10	2	6	10	14	2	6	9		1				
79	Au	2	2	6	2	6	10	2	6	10	14	2	6	10		1				
80	Hg	2	2	6	2	6	10	2	6	10	14	2	6	10		2				
81	Tl	2	2	6	2	6	10	2	6	10	14	2	6	10		2	1			
82	Pb	2	2	6	2	6	10	2	6	10	14	2	6	10		2	2			
83	Bi	2	2	6	2	6	10	2	6	10	14	2	6	10		2	3?			
84	Po	2	2	6	2	6	10	2	6	10	14	2	6	10		2	4?			
85	At	2	2	6	2	6	10	2	6	10	14	2	6	10		2	5?			
86	Rn	2	2	6	2	6	10	2	6	10	14	2	6	10		2	6			
87	Fr	2	2	6	2	6	10	2	6	10	14	2	6	10		2	6			1?
88	Ra	2	2	6	2	6	10	2	6	10	14	2	6	10		2	6			2
89	Ac	2	2	6	2	6	10	2	6	10	14	2	6	10		2	6	1		2?
90	Th	2	2	6	2	6	10	2	6	10	14	2	6	10		2	6	2		2
91	Pa	2	2	6	2	6	10	2	6	10	14	2	6	10	2	2	6	1		2?
92	U	2	2	6	2	6	10	2	6	10	14	2	6	10	3	2	6	1		2
93	Np	2	2	6	2	6	10	2	6	10	14	2	6	10	4	2	6	1		2?
94	Pu	2	2	6	2	6	10	2	6	10	14	2	6	10	5	2	6	1		2
95	Am	2	2	6	2	6	10	2	6	10	14	2	6	10	7	2	6			2?
96	Cm	2	2	6	2	6	10	2	6	10	14	2	6	10	7	2	6	1		2?
97	Bk	2	2	6	2	6	10	2	6	10	14	2	6	10	8	2	6	1		2?
98	Cf	2	2	6	2	6	10	2	6	10	14	2	6	10	9	2	6	1		2?
99	Es	2	2	6	2	6	10	2	6	10	14	2	6	10	10	2	6	1		2?
100	Fm	2	2	6	2	6	10	2	6	10	14	2	6	10	11	2	6	1		2?
101	Md	2	2	6	2	6	10	2	6	10	14	2	6	10	12	2	6	1		2?
102	No	2	2	6	2	6	10	2	6	10	14	2	6	10	14	2	6			2?

complete and the period ends. The similarities in the electron configurations of the elements in a given group help us to understand the similarities in their properties, as we shall see in Chaps. 18 and 19.

17.13   Particles and Waves; Orbits and Orbitals

In Chap. 16 we noted that some of the properties of light indicate that it is composed of a stream of particles, whereas other properties indicate that it is a train of waves. Accepting this duality of behavior has made our theories about light more useful than either alternative by itself.

Today we have the same duality in our theories about electrons. Bohr's model of an atom as composed of electrons (particles) orbiting the nucleus was quite successful in explaining atomic spectra. But electrons also have some of the properties of waves. When a beam of x-rays strikes the surface of a crystal of calcium carbonate at various angles, the beam is strongly reflected at some angles of incidence but very little affected at others. This behavior is nicely explained by wave theory. If a stream of x-ray photons (particles) behaves like waves, might not a stream of electrons (particles) also behave like waves? To answer this question, scientists directed a stream of electrons at a crystal of pure nickel. Sure enough, intense reflection took place at some angles of incidence, but at most angles there was little reflection. The moving electrons had the properties of waves!

Today we find it useful to think about electrons as particles *or* waves. In the Bohr model of atomic structure the electrons are thought of as particles occupying space by circulating in orbits around the nucleus. In more recent theories, the behavior of atoms is more adequately explained by assuming that the electrons are standing waves occupying space in various regions around the nucleus. We call these regions of space *orbitals*. The region occupied by the standing wave of a 1s electron is called the 1s orbital; that occupied by the standing wave of a 2p electron is called the 2p orbital, etc. Again, the dual theory of wave-particle is more useful than either theory alone.

electron orbital

Summary

The wavelengths of lines observed in the spectra of the elements can be explained by the Bohr model of atomic structure.

1 The electrons in the atoms of all elements occupy energy levels described by four quantum numbers: n, l, m, and s.
2 The values of n may be 1, 2, 3, . . . , ∞.
   The values of l may be 0, 1, 2, . . . , $n - 1$.
   The values of m may be $-l$, . . . , 0, . . . , $+l$.
   The values of s may be $-\frac{1}{2}$ or $+\frac{1}{2}$.

3 Every energy level is characterized by a set of four quantum numbers different from the set for every other energy level.

4 Under ordinary conditions the electrons in an atom occupy the lowest possible energy levels, but there can be only one electron in an energy level described by a particular set of quantum numbers.

5 Electrons can be raised to higher energy levels by heating the atoms, passing an electric discharge through them, or exposing them to electromagnetic radiation.

6 When an electron in a given energy level falls to a level of less energy, one quantum of light (one photon) is emitted. The energy of the photon is equal to the difference in the energy of the higher and lower energy levels.

7 When an electron in a given energy level absorbs one photon of light, it is raised to an energy level greater by the amount of energy in the photon.

For purposes of chemical studies, the arrangement of electrons in energy levels for the various chemical elements is simplified by noting that the differences in energy associated with differences in the quantum numbers m and s are negligible, as illustrated in Fig. 17.4. When placing electrons in these energy levels, two guidelines are followed:

1 Electrons are assigned to the lowest unfilled energy levels first.

2 Electrons are assigned to energy levels with unpaired spins until there is one electron in each available level; then the electrons are assigned to available levels by pairing opposite spins.

In the periodic table, all the chemical elements in period 1 have electrons for which $n = 1$, those in period 2 have electrons for which $n = 2$, etc. The elements in chemical families have similar configurations of electrons in their outermost shells. Knowing the electron configurations in the elements helps us understand their properties.

Electrons have the properties of both particles and waves. The wave-particle concept is more useful than either alternative.

New Terms and Concepts

ANGULAR MOMENTUM: The quantity of motion associated with the revolving of an electron in its orbit around the nucleus of an atom or that associated with its spinning about its own axis.

CONVERGENCE LIMIT: The position in the spectrum where individual spectral lines are so close together that they merge into a continuum.

DOUBLET LINES: Two spectral lines very close together.

ELECTRON CONFIGURATION: The locations in space of orbiting and spinning electrons described by specific values of the four quantum numbers.

ELECTRON ORBIT: A path an electron follows around the nucleus of an atom, similar to the path followed by a planet around the sun.

ELECTRON ORBITAL: A region in space occupied by an electron.

ELECTRON SHELL: All electrons having the same value of n. In the K shell, $n = 1$; in the L shell, $n = 2$; etc.

ENERGY LEVEL: The energy associated with an electron that occupies an orbit described by specific values of the four quantum numbers.

IONIZATION ENERGY: The energy required to remove an electron from an atom and produce an ion.

POLYELECTRONIC: Having more than one electron.

QUANTUM NUMBER: A number used to describe the energy level occupied by an electron.

RITZ-RYDBERG EQUATION: $\lambda = 911.4 \dfrac{n^2 m^2}{n^2 - m^2}$

Testing Yourself

17.1 The energies of electrons in the K, L, M, N, O, and P orbits (energy levels) increase in the order given. How are these differences between energy levels demonstrated by the *wavelengths* of the lines in the Balmer series? How are these differences demonstrated by the *intensities* of the Balmer lines?

17.2 How do we account for the fact that the Balmer series in the hydrogen spectrum is in the visible range of radiation, the Lyman series is in the ultraviolet, and the Paschen series is in the infrared?

17.3 What is meant by the ionization energy for the H atom? How is this related to the energy levels in the Bohr model?

17.4 What are the limits on the values of the four quantum numbers required to describe main levels and sublevels of energy for an electron in the Bohr model of the H atom?

17.5 When building up the configurations of electrons in energy levels in polyelectronic atoms, what two rules have been found helpful in determining which energy levels are occupied by the electrons in a given atom?

17.6 What is the significance of the numbers 1, 2, 3, etc. and the letters s, p, d, f in electron energy level notations?

17.7 What atom has the configuration $1s^2 2s^2 2p^6 3s^2 3p^6 3d^6 4s^2$? Translate this symbolic statement of configuration into words.

17.8 How do we account for the fact that the black lines in an absorption spectrum for an element occur at the same wavelengths as the bright lines in the emission spectrum?

17.9 How does the Bohr model of atomic structure help us understand the groupings of the elements in the periodic table?

17.10 Are electrons particles or waves? What evidence supports your statement?

The lattice structure of salt crystals, which are shown
in the photograph, is illustrated in the drawing.

18. Ionic Compounds and Ionic Bonds

In the last two chapters we saw how scientists discovered the nature of radiant energy, the arrangement of electrons in atoms, and the interactions of radiant energy with matter. The deeper insight into matter and radiant energy that came from solving these problems was soon found to be of immense value to chemists. They were trying to comprehend the processes by which atoms join to form molecules and how the arrangements of electrons in atoms determine the shapes and properties of the molecules.

Why does ice melt at 0°C and salt at 800°C? Why does melting sugar decompose to form caramel? Why are crystals of salt harder than crystals of sugar? Why does ice (made from water) melt when you heat it but Dry Ice sublime? Why is pure carbon in the form of diamond hard and sharp and clear white, but pure carbon in the form of graphite soft and slippery and black? Why does glass gradually soften when you heat it, but salt have a sharp melting point? When you smash a crystal of salt with a hammer, it shatters into many tiny replicas of the original piece. When you smash a piece of glass, you get an assortment of pieces with many different shapes that bear no resemblance to the shape of the original. Why is this so? Such questions perplexed chemists at the turn of the century.

The differences in behavior between crystalline electrolytes which yield ions when melted or dissolved in water, and nonelectrolytes in which atoms

bonding

join to form molecules, must be due to differences in the way atoms are held together. *Bonding* is the term chemists use to describe the forces of attraction between atoms in molecules, and theories of chemical bonding represent one of the great achievements of this century.

Many natural and manmade inorganic substances dissociate into ions. Many of the millions of organic compounds around us do not yield ions. Even so, a sharp line cannot be drawn between electrolytes and nonelectrolytes. The first category simply merges into the second. It will be convenient to consider first the bonding in compounds that easily yield ions, and we shall do so in this chapter. In Chap. 19 we shall discuss bonding in nonionic compounds.

18.1 Some Properties of Crystalline Electrolytes

In classifying solutions in Chap. 12, we found two categories of compounds—electrolytes and nonelectrolytes. We defined electrolytes as substances that conduct electricity when melted or dissolved in water. We attributed their conductance to the existence of mobile charged particles (ions) in the molten or dissolved state. What other characteristics do electrolytes have besides the ability to conduct electricity? Most of them are salts and exist at room temperature and pressure in the form of hard crystals. Common table salt is the most plentiful crystalline electrolyte found in nature. NaCl has a high melting point (801°C) and a high boiling point (1413°C). Crystals of NaCl are cubic in shape. When a large cube of NaCl is struck sharply with a hammer, it shatters into small cubes.

18.2 The Structure of Salt Crystals

We know that NaCl is composed of sodium and chlorine, that it yields ions (Na^+ and Cl^-) in solution or when melted, and that it forms crystals. Is there any way we can connect the behavior of a crystal of NaCl, composed of many billions of particles, with the relation between the particles of which it is composed? What can we learn about the basic structure of an NaCl crystal that will explain its behavior on shattering? We cannot see the individual particles in a crystal, but we can guess how they are arranged by shining x-rays on the crystal and observing the pattern produced by their reflection. This is crudely analogous to determining the positions of small mirrors grouped in a box by shining a light on them and observing the reflection pattern, as suggested in Fig. 18.1. In part *a* we cannot see into the box, but the pattern of light reflected from within it can be observed on the screen. This pattern suggests that the mirrors are arranged as shown in part *b*.

When x-rays are directed against an NaCl crystal, they are reflected in a

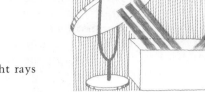

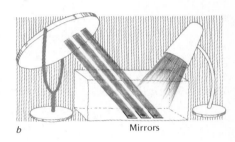

FIGURE 18.1 Reflection of light rays by mirrors

a

b   Mirrors

symmetrical geometric pattern that suggests that the crystal is composed of layers of particles in sheets or planes separated by a constant distance. By directing x-ray beams at different crystal surfaces from different angles, recording the pattern of reflections produced, and then using some fairly complicated mathematical analysis, it appears that solid NaCl exists as a *crystal lattice* with positively charged sodium ions, Na⁺, and negatively charged chloride ions, Cl⁻, arranged alternately in a regular cubical pattern. The particles in the crystal lattice are not atoms but ions. This is illustrated in Fig. 18.2*a*, where small balls represent the ion centers and sticks between them

crystal lattice

FIGURE 18.2 The crystal lattice of sodium ions and chlorine ions in sodium chloride

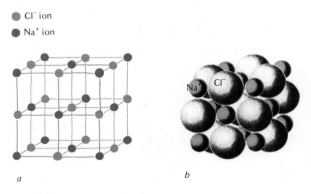

● Cl⁻ ion
● Na⁺ ion

a   *b*

symbolize the bonds necessary to hold them together. In actuality the ions are packed so closely together that they obscure the cubical geometry, as shown in Fig. 18.2b, but visualizing the perspective of the arrangement is easier if we use the first model.

ionic compound

Because the constituents of crystalline salts are ions, such salts are often called *ionic compounds*. It is the electrical attractions between the oppositely charged ions which hold them in a definite pattern in the crystal lattice; these electrostatic attractions are called *ionic bonds*. Some salts consist of lattice arrays of simple *monatomic ions* like Na^+ or Ca^{2+} and Cl^- or O^{2-}; others contain *polyatomic ions* like NH_4^+ and SO_4^{2-} in the crystal lattice. Many salt lattices are much more complicated than the NaCl lattice, but in all cases there is a definite pattern of positive and negative ions.

ionic bond
monatomic ion
polyatomic ion

18.3 The Alkali Metals in Ionic Compounds

How is an ionic bond formed? To shed light on this question we must correlate our discussion of families of elements in Chap. 10 with our discussion of electron configurations in Chap. 17. In Chap. 10 we noted that families of elements exhibit similarities in chemical properties but differences in vigor of chemical reactivity. Why should this be so? Why, for instance, should cesium and sodium be so alike but cesium be so much more vigorously reactive? With the development of theories of atomic structure it is now possible to explain these family similarities and the periodicity of the elements.

In the case of the alkali metal family (group IA in the periodic table) we observed in Chap. 10 some typical reactions, expressed by the following equations, where M′ stands for any alkali metal:

$$4M'(s) \;+\; O_2(g) \;\longrightarrow\; 2M_2'O(s) \tag{18.1}$$
$$2M'(s) \;+\; 2H_2O(l) \;\longrightarrow\; 2M'OH(aq) \;+\; H_2(g) \tag{18.2}$$
$$2M'(s) \;+\; Cl_2(g) \;\longrightarrow\; 2M'Cl(s) \tag{18.3}$$

We find in each case that an electrolyte is produced. X-ray analysis of the solids and the conductivity and freezing points of water solutions of the compounds indicate the M′ is present as M′+ in each case. Therefore, the typical behavior of an alkali metal atom *in a chemical reaction* in which an electrolyte is formed is to give up an electron to another atom or group of atoms, thus:

$$M' \longrightarrow M'^+ + e^- \tag{18.4}$$

Since cesium is generally the most reactive of the naturally occurring alkali metals, we conclude that the cesium atom must have the greatest tendency to give its electron to another element.

TABLE 18.1 *Electron Configurations of Alkali Metal Atoms*

Element	Atomic number	Arrangement of electrons in energy levels
Li	3	$1s^2 2s^1$
Na	11	$1s^2 2s^2 2p^6 3s^1$
K	19	$1s^2 2s^2 2p^6 3s^2 3p^6 4s^1$
Rb	37	$1s^2 2s^2 2p^6 3s^2 3p^6 3d^{10} 4s^2 4p^6 5s^1$
Cs	55	$1s^2 2s^2 2p^6 3s^2 3p^6 3d^{10} 4s^2 4p^6 4d^{10} 5s^2 5p^6 6s^1$
Fr	87	$1s^2 2s^2 2p^6 3s^2 3p^6 3d^{10} 4s^2 4p^6 4d^{10} 4f^{14} 5s^2 5p^6 5d^{10} 6s^2 6p^6 7s^1$

To see why this is so, let's look again at the electron configurations of the alkali metals given in Table 17.4. These may be conveniently symbolized in Table 18.1, where the superscripts indicate the number of electrons in a given energy level in each atom. The electron configurations of the alkali metals are summarized in still briefer form in Table 18.2. From these tables it is apparent that the observed similarities in properties of the alkali metals may be attributed to the similarities in the structures of their atoms. In each case, there is a single electron in an orbital outside a shell that contains an octet of electrons (except for the K shell, which can hold only two electrons). It seems logical to assume that the one electron that each alkali metal atom donates to another element in the typical reaction

$$M' \longrightarrow M'^+ + e^- \tag{18.4}$$

is this single electron. The nucleus and the main shells lying beneath the

TABLE 18.2 *Numbers of Electrons in Main Shells of Atoms of the Alkali Metals*

Element	Atomic number	Main shells and values of n						
		K	L	M	N	O	P	Q
		$n = 1$	2	3	4	5	6	7
Li	3	2	1					
Na	11	2	8	1				
K	19	2	8	8	1			
Rb	37	2	8	18	8	1		
Cs	55	2	8	18	18	8	1	
Fr	87	2	8	18	32	18	8	1

core of an atom
kernel of an atom

valence electrons
valence shell

outermost shell containing the single electron, we call the *core* or *kernel of the atom*; and the electrons in it are called core electrons or kernel electrons. The electron in the outermost shell, which we believe to have been removed in a chemical reaction, we call a *valence electron*; the shell from which it was removed we call the *valence shell*.

A third mode of notation for presenting the electron configurations of atoms is shown in Table 18.3. Here the chemical symbol in parentheses stands for the complete electron configuration of the element symbolized; the additional electrons (described in the notation of Table 18.1) are superimposed on this configuration. The electrons in the atom which are included in the noble gas configuration (He, Ne, etc.) are the core electrons. The outer electrons are the valence electrons that form chemical bonds with other atoms.

We note from Table 18.3 that when an atom of an alkali metal loses its valence electron in a chemical reaction, the M'+ ion so formed has the electron configuration of a member of the family of quite stable gaseous elements in the zero group at the extreme right of the periodic table (Table 10.3). Apparently, any alkali metal atom easily loses its valence electron to achieve the chemically stable configuration of a zero group element.

The 2 electrons present in the helium core of the Li atom partially neutralize the 3 positive charges on the Li nucleus, reducing its effective charge to $1+$. Likewise, the 54 electrons present in the xenon core of the Cs atom partially neutralize the 55 positive charges on the Cs nucleus, reducing its effective charge to $1+$.

The valence electron of the Li atom is a $2s$ electron; its orbit is much closer to the Li nucleus than is that of the $6s$ valence electron in the Cs atom. The force of attraction between positive and negative electric charges decreases rapidly with increasing distance between them. Consequently, the Li nucleus holds onto its $2s$ electron much more strongly than the Cs nucleus holds onto its $6s$ electron. Since it is easier for another atom to pick up an electron from a Cs atom than from a Li atom, the Cs atom is generally more chemically

TABLE 18.3 *Electron Configurations of Alkali Metal Atoms*

Element	Atomic number	Electrons in energy levels	Element	Atomic number	Electrons in energy levels
Li	3	(He) $2s^1$	Rb	37	(Kr) $5s^1$
Na	11	(Ne) $3s^1$	Cs	55	(Xe) $6s^1$
K	19	(Ar) $4s^1$	Fr	87	(Rd) $7s^1$

reactive than the Li atom. The reactivities of the other alkali metals are intermediate between those of Li and Cs. Thus our concept of the configuration of electrons in atoms is of great help in explaining the relative chemical activities of the alkali metals.

18.4 The Alkaline Earth Metals in Ionic Compounds

A parallel analysis of electron configurations enables us to explain the gradations in reactivity within the alkaline earth metal family (group IIA in the periodic table). We let M″ stand for any group IIA element; typical reactions are

$$2M'' + O_2 \longrightarrow 2M''O \tag{18.5}$$
$$M'' + 2H_2O \longrightarrow M''(OH)_2 + H_2 \tag{18.6}$$
$$M'' + Cl_2 \longrightarrow M''Cl_2 \tag{18.7}$$

When hydrogen is generated according to Eq. (18.6), Be and Mg must be exposed to steam in order to make the reaction go. However, hydrogen is produced vigorously when cold water is added to Ca, Sr, or Ba. Furthermore, the rapidity with which hydrogen is generated from cold water increases in the order in which these three metals are listed. (Ra would probably react more vigorously then Ba, but Ra is so dangerously radioactive that we do not handle it in an ordinary laboratory.) The electron configurations of this family of elements are shown in Table 18.4. The products of the three typical reactions noted above are electrolytes. The alkaline earth metal is present in each product as a doubly charged ion, M''^{2+}.

$$M'' \longrightarrow M''^{2+} + 2e^- \tag{18.8}$$

Again it appears that the metal atoms tend to lose electrons to achieve a configuration of electrons like that of a noble gas in the zero group.

Metals other than those in groups IA and IIA also form many salts that are

TABLE 18.4 *Electron Configurations of Alkaline Earth Metal Atoms*

Element	Atomic number	Electrons in energy levels		Element	Atomic number	Electrons in energy levels	
Be	4	(He)	$2s^2$	Sr	38	(Kr)	$5s^2$
Mg	12	(Ne)	$3s^2$	Ba	56	(Xe)	$6s^2$
Ca	20	(Ar)	$4s^2$	Ra	88	(Rd)	$7s^2$

ionic compounds. A metal in the transition series often forms more than one kind of ion. In the case of iron, both Fe^{2+} and Fe^{3+} are common. Copper forms two kinds of ions, Cu^+ and Cu^{2+}. The chemistry of the transition metals is too complicated to be included here. When the positive charge on a metallic ion exceeds 2, it is difficult to separate the oppositely charged ions in an ionic compound, and its ionic nature is obscured. For instance, iron chloride, $FeCl_3$, melts at 282°C, but NaCl, a typical ionic compound, melts at 801°C.

18.5   The Halogens in Ionic Compounds

The halogens (the elements in group VIIA) display interesting gradations in properties, as noted in Sec. 10.13. The similarities and differences in the behavior of the halogens are explained by the similarities and differences in the structures of their atoms. We let X′ stand for any group VIIA element; two typical reactions for these elements are

$$X_2'(g) + 2Na(s) \longrightarrow 2NaX'(s) \tag{18.9}$$
$$X_2'(g) + Mg(s) \longrightarrow MgX_2'(s) \tag{18.10}$$

The product of each reaction is a white, crystalline salt. The electrical conductivity and freezing points of solutions of these salts indicate that the halogen is present as the ion, X'^-. Therefore, the typical behavior of a halogen atom in such a chemical reaction is to take on an electron:

$$X' + e^- \longrightarrow X'^- \tag{18.11}$$

or

$$X_2' + 2e^- \longrightarrow 2X'^- \tag{18.12}$$

Since fluorine is the most reactive of the halogens, it must have the greatest affinity for electrons. The electron configurations of the halogen atoms and zero group atoms shown in Table 18.5 indicate why this is true. From this table we see that each halogen has seven electrons in its outermost (valence) shell. Since it is the electrons in this shell that will interact with those of other atoms, we see why the halogens show similarities in chemical behavior. It is also evident that in the typical reaction in Eq. (18.11) the halogen atom has achieved the electron configuration of the noble gas element adjacent to it in the periodic table; the halide ion has a more stable electron configuration than the halogen atom.

Since the added electron in F^- (fluoride ion) is in the L shell, it is much closer to the nucleus than is the added electron in the O shell of the I^- (iodide ion). Consequently, the added electron is held more firmly by the electrical attraction of the nucleus in F^- than in I^-. Stated another way, once the ions

TABLE 18.5 *Numbers of Electrons in Main Shells of Atoms of the Halogens and of the Noble Gases*

Element	Atomic number	Main shells						
		K	L	M	N	O	P	Q
F	9	2	7					
Ne	10	2	8					
Cl	17	2	8	7				
Ar	18	2	8	8				
Br	35	2	8	18	7			
Kr	36	2	8	18	8			
I	53	2	8	18	18	7		
Xe	54	2	8	18	18	8		
At	85	2	8	18	32	18	7	
Rn	86	2	8	18	32	18	8	

are formed, it is more difficult to remove an electron from F^- than from I^-. The F^- ion is much more stable chemically than the I^- ion, and so the tendency of the F atom to become an ion is greater than the tendency of the I atom to become an ion. The chemical reactivity of the F atom is thus greater than that of the I atom. The reactivities of Cl and Br atoms are intermediate between those of F and I, as one would expect from the electron configurations.

18.6 The Oxygen Family in Ionic Compounds

In the oxygen family we see a situation parallel to that in the halogen family. Adding two to the array of electrons in the oxygen atom yields the neon configuration, two added to the sulfur atom array yields the argon configuration, and so on. It does not surprise us that the equation

$$X'' + 2e^- \longrightarrow X''^{2-} \tag{18.13}$$

typifies the chemical behavior of oxygen and sulfur in combining with the alkali metals and alkaline earth metals to form saltlike crystals of oxides and sulfides. However, the chemical behavior of selenium, tellurium, and polonium is considerably more complicated and not amenable to the simple concepts of ionic bonding.

18.7 The Stable Octet

A graphic way of showing how electrons transfer from one atom to another to form stable ions with an external octet of electrons is by means of electron dot formulas for writing chemical equations. The electrons in the valence shell are indicated by dots.

$$\text{Na} \cdot + \cdot \ddot{\text{Cl}} : \longrightarrow \text{Na}^+ + (: \ddot{\text{Cl}} :)^- \qquad (18.14)$$

$$\text{Mg} : + \ddot{\text{O}} : \longrightarrow \text{Mg}^{2+} + (: \ddot{\text{O}} :)^{2-} \qquad (18.15)$$

$$: \ddot{\text{Br}} \cdot + \text{Ca} + \cdot \ddot{\text{Br}} : \longrightarrow (: \ddot{\text{Br}} :)^- + \text{Ca}^{2+} + (: \ddot{\text{Br}} :)^- \quad (18.16)$$

metal hydride

In the presence of an atom that readily loses an electron, a hydrogen atom will accept an electron, forming an ionic compound known as a *metal hydride*.

$$\text{Na} \cdot + \cdot \text{H} \longrightarrow \text{Na}^+ + (: \text{H})^- \qquad (18.17)$$

The hydride ion, H^-, has the electron configuration of the noble gas element He. The tendency of atoms to achieve noble gas element configurations by gaining or losing electrons is clearly illustrated by the reactions summarized in the following equations and in Table 18.6. With the exceptions of Li and Be (which form the heliumlike, stable, two-electron duo), all the elements in the alkali metal, alkaline earth metal, oxygen, and halogen families have a strong tendency to achieve a *stable octet of electrons* in the outermost shell of the atom when it is combined into a compound:

stable octet
of electrons

$$\text{M}' \longrightarrow \text{M}'^+ + e^- \qquad \text{where M' is Li, Na, K,}$$
$$\text{Rb, Cs, or Fr} \qquad (18.4)$$
$$\text{M}'' \longrightarrow \text{M}''^{2+} + 2e^- \qquad \text{where M'' is Be, Mg, Ca,}$$
$$\text{Sr, Ba, or Ra} \qquad (18.8)$$
$$\text{X}' + e^- \longrightarrow \text{X}'^- \qquad \text{where X' is F, Cl, Br, I, or At} \quad (18.11)$$
$$\text{X}'' + 2e^- \longrightarrow \text{X}''^{2-} \qquad \text{where X'' is O, S, Se, Te, or Po} \quad (18.13)$$

TABLE 18.6 *Electron Configurations of Some Isoelectronic Atoms and Ions; the Stable Octet Is Underlined*

Configuration	Atoms and ions				
$1s^2$		H^-	He	Li^+	Be^{2+}
$1s^2\underline{2s^22p^6}$	O^{2-}	F^-	Ne	Na^+	Mg^{2+}
$1s^22s^22p^6\underline{3s^23p^6}$	S^{2-}	Cl^-	Ar	K^+	Ca^{2+}
$1s^22s^22p^63s^23p^63d^{10}\underline{4s^24p^6}$	Se^{2-}	Br^-	Kr	Rb^+	Sr^{2+}
$1s^22s^22p^63s^23p^63d^{10}4s^24p^64d^{10}\underline{5s^25p^6}$	Te^{2-}	I^-	Xe	Cs^+	Ba^{2+}
$1s^22s^22p^63s^23p^63d^{10}4s^24p^64d^{10}4f^{14}5s^25p^65d^{10}\underline{6s^26p^6}$	Po^{2-}	At^-	Rn	Fr^+	Ra^{2+}

Particles of matter that have the same number and configuration of electrons are called *isoelectronic*. The particles in a horizontal line in Table 18.6 are isoelectronic.

18.8 The Formation of Ionic Crystals

ionic crystal

Now that we have seen how ionic compounds are formed by the donation and acceptance of electrons, we can understand the structure of an *ionic crystal*. The reaction that typifies the behavior of an atom of a metallic element in forming an ionic compound is

$$M \longrightarrow M^{n+} + n\ e^- \tag{18.18}$$

Typical behavior for an atom of a nonmetallic element forming an ionic compound is

$$X + n(e^-) \longrightarrow X^{n-} \tag{18.19}$$

The electrons donated by the atoms of the metallic element are accepted by the atoms of the nonmetallic element. An ionic crystal is produced by the assembly of these oppositely charged ions into a regular geometric pattern, a crystal lattice. When positive ions are formed by the donation of electrons, the *overall* (or net) *charge* of the ion is *positive*. However, the core electrons give a negative charge to the *external* region of the ion. At a distance, the positive ions attract the negative ions, but when they get close together, the repulsion between the negatively charged *external* regions of each ion counteracts the attraction between the net charges. The ions come closer and closer to one another until the attraction and the repulsion just balance. The distance between the nuclei of the two atoms when this balance is achieved is called the *length* of the ionic bond. The crystal structure of sodium chloride is more accurately depicted in Fig. 18.3, which shows the space filled by electrons, than it is in Fig. 18.2a, which is drawn to show the location of the atomic nuclei. The argon atom has 18 electrons, and so it is larger than neon with its 10. The Cl$^-$ ion has the argon configuration, and so it is larger than the Na$^+$ ion, which has the neon configuration.

18.9 The Relation between the Properties and Structure of Ionic Compounds

When we raise the temperature of a solid, the energy of vibration of the particles in the solid crystal increases. When the temperature gets high enough, this energy of vibration overcomes the forces of attraction between the particles; they break away from the crystal lattice and move about with the fluidity characteristic of the liquid state. We call this change "melting." The

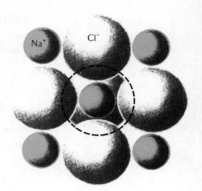

FIGURE 18.3 Cross section to scale of the crystal structure of NaCl. The dotted circle represents one Cl⁻ above and one Cl⁻ below the plane of the paper.

forces of electrical attraction between oppositely charged ions in an ionic crystal are so strong that we have to heat such compounds several hundred degrees in order to melt them. The melting point of NaCl is 801°C. In order to increase the kinetic energy of the ions in the molten liquid sufficiently to enable them to escape from the liquid (vaporize), we must raise the temperature several hundred degrees more. The boiling point of NaCl is 1413°C. The great thermal stability of ionic crystalline compounds is thus readily understood in terms of electrical forces of attraction.

The strength of the network of ionic bonds in a crystal lattice is also indicated by the hardness of the crystal. Hardness can be estimated by scratching one substance with another. If the solid substance A scratches the solid substance B, then A is said to be harder than B. Using a scale of hardness on which the hardest known naturally occurring substance is assigned the number 10 and less hard materials have lower numbers, we can assign MgO, CaO, SrO, and BaO hardnesses of 6.5, 4.5, 3.5, and 3.3, respectively. The distances between nuclei in these compounds are 2.10, 2.40, 2.57, and 2.77 Å, respectively. Clearly, the shorter the distance between these equally charged ions, the greater the force of attraction between them and the more difficult it is to remove some of the solid material by scratching it, i.e., the harder the oxide is.

If a force is applied to a part of an ionic crystal and gradually increased, the crystal is only slightly deformed. If the force is increased sufficiently, the crystal suddenly cleaves along a smooth plane parallel to one of its faces. This seems quite in accord with our concept of the ionic bonding between the charged particles in the regular lattice of the crystal. Figure 18.4 shows the initial and final stages of this process. When a sufficient force is applied to the crystal to move a set of planes one interionic distance past another, ions of like charge are brought together, and their repulsion causes the crystal to cleave between these displaced planes.

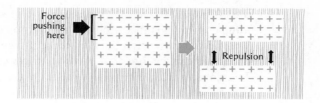

FIGURE 18.4 Cleavage of an ionic crystal

As we saw in Sec. 18.2, when a beam of x-rays is reflected from a crystal of sodium chloride, the reflected rays display a regular geometric pattern. This indicates that the crystal is composed of a three-dimensional latticework of Na^+ and Cl^- ions. When a beam of x-rays is reflected from a piece of glass, there is little regularity in the pattern of reflected rays. This indicates that the molecules in glass are present in a random arrangement.

When we heat a crystal of sodium chloride, the ions in the lattice vibrate faster and faster. Finally a temperature is reached at which the vibrations are so strong that the bonds between the ions are broken and the solid liquefies. Since all the particles in the NaCl crystal lattice are Na^+ or Cl^-, all the bonds between the particles are the same, and they all break at once; NaCl has a sharp melting point at 801°C.

When we heat a piece of glass, the vibration of the molecules in the solid first breaks the weaker bonds between molecules and the solid softens (loses some of its rigidity). As we heat to higher temperatures, more bonds are broken and the glass begins to flow. At still higher temperatures most of the bonds between molecules are broken and the glass becomes completely liquid. A lump of typical glass may begin to soften at 550°C, slowly lose its shape at 600°C, but not liquefy until heated to 850°C. Evidently, the bonding in glass is quite different from that in salt.

18.10 Compounds Composed of Ions Containing More Than One Atom

Because NaCl is the most common ionic compound in nature and because its structure is quite simple, we have considered its behavior at some length as a typical ionic compound. The halides of the alkali metals and alkaline earth metals are quite similar to NaCl, and so they were useful to us in relating chemical behavior with the electron configurations in the atoms of the elements involved.

However, we must not lose sight of the fact that most of the ionic compounds in nature are more complex than NaCl. The 29 metals in the short transition series (Table 10.3), the 30 metals in the long transition series, and the 10 metals below the metalloids in groups IIIA, IVA, VA, and VIA form ionic compounds in which the metal ions may have charges from 1+ to 4+. The nonmetallic elements in these groups and in group VIIA form many kinds of negative ions that include atoms of other elements (chiefly oxygen). The considerable differences in the sizes of metallic ions and in the sizes and shapes of negative ions lead to considerably more complicated crystal lattice patterns than the simple one for NaCl that we have studied.

Nevertheless, our understanding of the fundamental nature of the ionic compound NaCl helps us greatly in understanding the properties of hundreds of ionic compounds that form much more complicated crystal structures.

Summary

A crystalline salt is composed of a regular geometric array of positive and negative ions in a three-dimensional lattice. The ions in a crystal lattice may be monatomic as in NaCl (Na^+ and Cl^-) or polyatomic as in $(NH_4)_2SO_4$ (NH_4^+ and SO_4^{2-}).

The solid crystal is held together by the electrostatic attractions between the positive and negative ions in the crystal lattice. We call these attractions ionic bonds and the substance an ionic compound.

Because the forces of electrostatic attraction between the ions in salt crystals are very strong, these compounds have high melting points, high boiling points, and high relative hardness.

In the formation of binary salts (salts containing only two elements) of alkali or alkaline earth metals, the atoms of the metal donate electrons to the atoms of the nonmetal to form the ions in the salt crystal. By this transfer of electrons the ions formed achieve the external electron configuration of a noble gas. Except in the case of the ions Li^+, Be^{2+}, and H^-, this configuration contains eight external electrons and is called the stable octet. Li^+, Be^{2+}, and H^- have a stable duo of electrons like the He atom.

In forming alkali halides, one atom of an alkali metal (M') donates one electron to one atom of a halogen (X') to form one positive ion with a single charge (M'$^+$) and one negative ion with a single charge (X'$^-$). All alkali metal halides have the formula M'X' and have very similar properties.

In forming alkaline earth halides, one atom of an alkaline earth metal (M'') donates two electrons to two atoms of a halogen to form one positive ion with a double charge (M''$^{2+}$) and two negative ions each with a single charge (X'$^-$). All alkaline earth halides have the formula M''X$_2'$ and have very similar properties.

Alkali metals tend to react thus: $M' \longrightarrow M'^+ + e^-$
Alkaline earth metals tend to react thus: $M'' \longrightarrow M''^{2+} + 2e^-$

Halogens tend to react thus: $\qquad X_2' + 2e^- \longrightarrow 2X'^-$

Oxygen group elements tend to react thus: $\qquad X'' + 2e^- \longrightarrow X''^{2-}$

The similarities in the properties of the elements in a family may be attributed to the similarities in their electron configurations. The atoms of the alkali metals (except Li) have a single electron in a valence shell (main energy level) just above the stable octet of a noble gas. The atoms of the alkaline earth metals (except Be) have two electrons in a valence shell just above the stable octet of a noble gas. (In Li and Be the valence electrons are in the shell just above the stable duo of helium.) The atoms of the halogens have seven of the eight electrons needed for the stable octet of a noble gas. The atoms of the oxygen family elements have six of the eight electrons needed for the stable octet of a noble gas.

The difference in the properties of the elements in a family may be attributed to the differences in the distance between the electrons in the valence shell and the nucleus of the atom. The larger the atom of the metal, the greater this distance, the more easily is the valence electron donated, and the more vigorously does the element react. The smaller the atom of the nonmetal, the smaller the distance between the nucleus and the valence electrons, the greater is the atom's tendency to accept electrons, and the more vigorously does the element react.

New Terms and Concepts

BONDING: The forces of attraction between atoms in molecules.

CORE OF AN ATOM: The nucleus and all the electrons except those outer ones that can form bonds with other atoms.

CRYSTAL LATTICE: A symmetrical geometric array of atoms, ions, or molecules in a solid substance.

IONIC BOND: The bond formed between two atoms when they combine by one of them donating an electron to the other, thus forming a positive ion and a negative ion that are held together by their opposite charges.

IONIC COMPOUND: A compound in which the atoms are held together by ionic bonds.

IONIC CRYSTAL: A crystal composed of ions.

ISOELECTRONIC: Particles having the same number of electrons.

KERNEL OF AN ATOM: Another name for the core of an atom.

METAL HYDRIDE: A compound containing only a metal and hydrogen.

MONATOMIC ION: An ion containing only one atom.

POLYATOMIC ION: An ion containing more than one atom.

STABLE OCTET OF ELECTRONS: A configuration of eight electrons having the same value for n, but values of 0 or 1 for l; this is a particularly stable group, as evidenced by the inertness of the zero group gases, which possess such an octet of their outermost electrons.

VALENCE ELECTRONS: Electrons involved in forming chemical bonds.

VALENCE SHELL: The shell containing valence electrons.

Testing Yourself

18.1   What is an ionic compound? Describe the structure and properties of ionic compounds, using a simple one as an example.

18.2   What is the special nature of the electron configurations of (a) metallic elements that readily form positive ions, (b) nonmetallic elements that readily form negative ions?

18.3   What is a stable octet? A stable duo? Give examples of each in positive ions, negative ions, neutral atoms.

18.4   From your knowledge of the periodic table and the electron configurations of atoms, which of the following hypothetical compounds would you expect to exist?

Ca_2O_3 Ca_3O_2 BaS $LiCl_2$

18.5   Why is the formula for an alkali halide M'X' but that for an alkaline earth halide $M''X_2'$?

18.6   How do we explain the facts that the charge found on a sodium ion is equal to that of a proton and the charge on a calcium ion is twice that of a proton?

18.7   Why is the formula for sodium oxide Na_2O but that for magnesium oxide MgO?

18.8   Would you expect to encounter a Sr^{3+} ion in any common salt? Why or why not? What about a Br^{3-} ion?

18.9   The electron configurations of the atoms of several elements are given below. Which of these would you expect to lead to similar chemical properties?

$1s^2 2s^2 2p^6 3s^2 3p^5$ $1s^2 2s^2 2p^6 3s^2$

$1s^2 2s^2 2p^6 3s^2 3p^6 4s^2$ $1s^2 2s^2 2p^6$

$1s^2 2s^2 2p^5$ $1s^2 2s^2 2p^6 3s^2 3p^6 3d^{10} 4s^2 4p^6 5s^2$

18.10 What are the electron dot formulas for alkali metal atoms? For alkali metal ions? For oxygen group atoms? For oxygen group ions?

18.11 Using electron dot formulas for the atoms and ions involved, write chemical equations for the formation of rubidium sulfide, calcium chloride, and strontium oxide.

18.12 Of what ions are the following salts composed?

$(NH_4)_2SO_4$ K_3PO_4 $Ba(C_2H_3O_2)_2$

*The arrangement of water molecules in ice is responsible
for the hexagonal structure of snow crystals.*

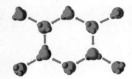

19. Covalent Bonds and Covalent Compounds

Although the concept of ionic bonding developed in Chap. 18 is very useful for interpreting the structure and properties of many simple inorganic salts, it is quite inadequate for explaining the bonding in the molecules of thousands of more complicated inorganic compounds and millions of organic compounds now known. The simplest diatomic molecule we know is that of hydrogen, H_2. Hydrogen melts at $-259°C$ and boils at $-253°C$, so that it is a gas at ordinary temperatures and pressures. When liquefied, it does not conduct an electric current. It has none of the properties of an ionic compound. How can we explain the chemical bonding in this simple molecule? If we can construct a theory to explain the bonding in H_2, can we use it to explain the bonding in other nonionic substances? These questions will be answered in this chapter.

19.1 The Electron Cloud Model for the Hydrogen Atom

When two hydrogen atoms unite to produce a hydrogen molecule, H_2, an ionic bond cannot be formed by the donation of an electron from one H atom and its acceptance by the other because each atom has the same tendency to hold on to its electron. The concept of the electron as a particle orbiting

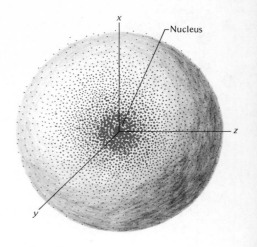

FIGURE 19.1 The $1s$ electron spherical charge cloud around the nucleus of a hydrogen atom

the atomic nucleus is very useful for explaining ionic bonding but is not of much help here. We noted in Sec. 17.13 that electrons may be thought of as particles or as waves. Picturing them as waves will help us explain the bonding in hydrogen.

If the electron in the hydrogen atom is a wave, it may be described by the equations developed by mathematical physicists for waves. These wave-mechanical equations involve complicated higher mathematics with which we need not be concerned. We will be content to accept the physical model of an electron that emerges from the solutions of these theoretical equations.

The wave-mechanical theoreticians find that the $1s$ electron in the hydrogen atom consists of a spherical region in space surrounding the atomic nucleus. The charge of the electron is dispersed throughout this space in what is called a *charge cloud*. The electron is thought of not as a concentrated tiny particle of negative charge, but as a diffuse cloud of negative charge (Fig. 19.1).

charge cloud

19.2 How Do Two Hydrogen Atoms Join to Form a Molecule?

When two H atoms approach one another, each of the positive nuclei will attract both of the negative electron clouds. At the same time the nuclei will repel each other, and the electron clouds will repel each other (Fig. 19.2). As the atoms approach one another, the space between the two positive nuclei (protons) will have the greatest attraction for negative charges. It seems likely that the two electron clouds will be drawn into this region until the electrostatic attractions between the positive nuclei and the negative electrons will be just balanced by the electrostatic repulsion between the nuclei and be-

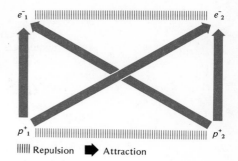

FIGURE 19.2 Electric interactions between the two protons and the two electrons in two H atoms

e^-_1 ||||||||||||||||||||||||||||||||||||||||||||||| e^-_2

p^+_1 ||||||||||||||||||||||||||||||||||||||||||||||| p^+_2

|||||| Repulsion ➡ Attraction

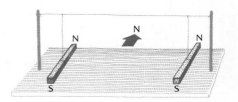

FIGURE 19.3 Two bar magnets suspended some distance apart above a table

covalent bond

tween the electron clouds. The overlapping and merging of the two electron clouds between the two nuclei, and the sharing of this pair of electrons between the two atoms, produce the bond that holds the two H atoms in the H_2 molecule. We call this shared pair a *covalent bond*.

How can the two electron clouds merge in the space between the nuclei? This is related to the question we asked about the filling of energy levels in polyelectronic atoms (Sec. 17.8): How can two electrons occupy the same 1s orbital in a helium atom? In the latter case, we said that two electrons could occupy the same energy level (orbital) if they had opposite spins. We use the same principle to explain the merging of two electron clouds to produce a shared pair of electrons (covalent bond) in the H_2 molecule. If the electrons have opposite spins, they can merge to form the covalent bond. But what does electron spin have to do with these cases? To answer this question, we need briefly to consider the behavior of magnets.

If two small bar magnets are suspended horizontally on threads some distance apart (Fig. 19.3), they swing around until one end of each points north and the other south. We call the end pointing north the northseeking or north pole of the magnet and the opposite end the south pole. If we bring the north pole of one magnet near to the north pole of another, suspended horizontally, the north pole of the suspended magnet swings away from the

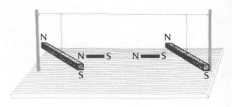

FIGURE 19.4 Repulsion between like poles and attraction between unlike poles in magnets

north pole of the magnet held near it (Fig. 19.4). The same repulsion is seen between two south poles. On the other hand, the north pole of one magnet *attracts* the south pole of the other.

If we place a magnetic compass (a small bar magnet on a pivot) on a horizontal table top under a copper wire running north and south, the compass needle points north (Fig. 19.5*a*). If we now pass a stream of electrons from right to left through the wire by attaching a battery to it, the north pole of the compass needle swings toward the west (Fig. 19.5*b*). We conclude that the flow of electrons through the wire has created a magnetic field around the wire that has been detected by the compass. If we pass a stream of electrons from left to right through the wire, the north pole of the compass needle swings toward the east (Fig. 19.5*c*). We conclude that the reversal of the direction of flow of the electrons has reversed the magnetic field and its effect on the compass.

FIGURE 19.5 Behavior of a magnetic compass needle placed below a north-south wire, (*a*) with no electrons flowing through the wire, (*b*) when electrons are flowing from right to left, (*c*) when electrons are flowing from left to right

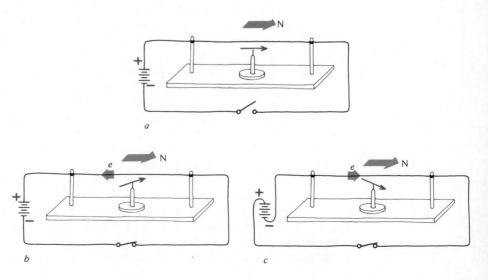

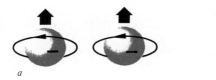 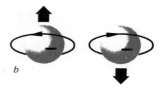

FIGURE 19.6 (a) Two electrons spinning in the same direction behave like two magnets with their north poles pointed in the same direction. (b) Two electrons spinning in opposite directions behave like two magnets with their north poles pointed in opposite directions.

By analogy we assume that the spinning motion of an electron creates a magnetic field, giving the electron some of the properties of a tiny bar magnet. Two electrons spinning in the same direction behave like bar magnets with their north poles pointing in the same direction (Fig. 19.6a). The magnetic interaction of these two is repulsion, and they cannot occupy the same orbital in an atom. But two electrons spinning in opposite directions behave like bar magnets with their north poles pointing in opposite directions (Fig. 19.6b). The magnetic interaction of these two is attraction, and they can occupy the same orbital in an atom.

Reasoning in a similar fashion, we conclude that if a hydrogen atom with an electron having its spin in one direction approaches another hydrogen atom with an electron having the *same* spin, the electron clouds cannot merge, and *no bond forms*. But if a hydrogen atom with an electron having its spin in one direction approaches another hydrogen atom with an electron having the *opposite* spin, the electron clouds can merge and *form a covalent bond*.

We can visualize the merging of the two electron clouds to form a bond between the two H nuclei as shown in Fig. 19.7a. When the bond is formed, each hydrogen nucleus has a duo of electrons associated with it. Simpler symbols for showing the formation of the covalent bond between two H atoms are given in Fig. 19.7b and c. The merging of two *s* electron orbitals is called an *s—s bond*.

s—s bond

19.3 The Chlorine Molecule

From Gay-Lussac's studies on combining volumes of gases (Sec. 6.8) we learned that the gaseous element chlorine consists of diatomic molecules, Cl_2. How can two atoms of this element combine to form a molecule? If two atoms can combine, why don't many more combine to form larger molecules? Our understanding of bond formation in H_2 helps us answer these questions.

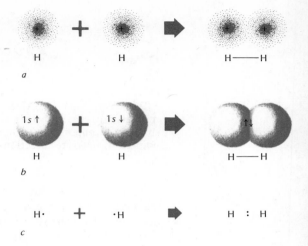

FIGURE 19.7 Three ways of symbolizing the formation of an *s—s* bond between two H atoms to form an H$_2$ molecule

The electron configuration of chlorine atoms is (Ne) $3s \uparrow\downarrow 3p \uparrow\downarrow \uparrow\downarrow \uparrow$. Since the neon configuration is stable (nonreactive), the formation of the Cl—Cl bond in Cl$_2$ probably involves the electrons in the $n = 3$ shell. When two chlorine atoms approach one another closely, a bond may form between them by the merging of the half-filled $3p$ orbital of one atom with the similarly half-filled $3p$ orbital of the other—*if the spins of the $3p$ electrons in these orbitals are opposite one another.* This union is symbolized in Fig. 19.8, which shows two $3p$ electron clouds paired in the region between the two chlorine nuclei. The merging of these two electron clouds is called a *p—p bond*.

Since there is only one half-filled orbital in a Cl atom, the formation of the *p—p* bond in Cl$_2$ achieves a stable octet of electrons around each of the atoms. Hence there is no tendency to form molecules of chlorine containing more than two atoms.

In the gaseous state the other halogens—fluorine, bromine, and iodine— also exist as diatomic molecules with a stable octet for each atom. The covalent bonds holding these atoms in such molecules are *p—p* bonds also—

p—p bond

FIGURE 19.8 The formation of a *p—p* covalent bond between two Cl atoms

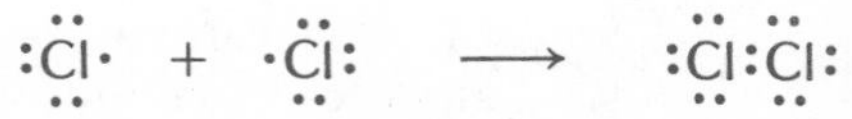

$$:\overset{..}{\underset{..}{Cl}}\cdot \; + \; \cdot \overset{..}{\underset{..}{Cl}}: \; \longrightarrow \; :\overset{..}{\underset{..}{Cl}}:\overset{..}{\underset{..}{Cl}}:$$

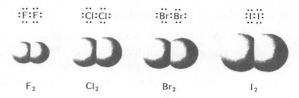

FIGURE 19.9 Scale models of the halogen molecules

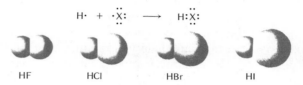

FIGURE 19.10 Formation of $s-p$ covalent bonds in hydrogen halides; scale models of hydrogen halide molecules

$2p-2p$ in F_2, $4p-4p$ in Br_2, and $5p-5p$ in I_2. Because the $2p$ orbital in the fluorine atom is smaller than the $3p$ orbital in the chlorine atom, the F_2 molecule is smaller than the Cl_2 molecule. The $4p$ orbital in bromine is larger than the $3p$ orbital in chlorine, and so Br_2 is larger than Cl_2. The $5p$ orbital in the iodine atom is still larger, and so I_2 is larger than Br_2. Drawings of scale models in Fig. 19.9 show the relative sizes of the halogen molecules.

19.4 Hydrogen Halides

Hydrogen fluoride (HF), hydrogen chloride (HCl), hydrogen bromide (HBr), and hydrogen iodide (HI) are gases at ordinary temperatures. When they are liquefied, they do not conduct an electric current, and so the bonds in the compounds must be covalent rather than ionic. Each halogen atom has seven electrons in its valence shell and can achieve a stable octet by sharing a pair of electrons with another atom. Hydrogen atoms have only one electron in the valence shell and can achieve a stable duo by sharing a pair with another atom. The combining of hydrogen atoms with halogen atoms to form hydrogen halides may be symbolized as in Fig. 19.10. The relative sizes of the hydrogen halide molecules are also shown. In these molecules, the shared pair forms an $s-p$ *bond.*

$s-p$ bond

19.5 The Shape of the Water Molecule

As we saw in Sec. 15.1, water has unusual properties that make it of primary importance to all living things. These properties arise from the structure and bonding in the water molecule. Since water does not readily conduct electricity, we believe that the hydrogen and oxygen atoms are held together by covalent bonds.

X-ray studies of ice crystals and electron scattering by liquid and gaseous water show that the water molecule is not straight like this: H—O—H; instead it is bent (Fig. 19.11). There is an angle of about 105° between the two lines drawn from the hydrogen nuclei to the oxygen nucleus. We call this the *bond angle*. How can we account for the bond angle? Does this shape of the molecule help us explain the properties of water?

bond angle

The electron configuration of the oxygen atom is:

$$\text{(He) } 2s \uparrow\downarrow \; 2p \uparrow\downarrow \uparrow \uparrow$$

When a hydrogen atom approaches an oxygen atom, the H atom's half-filled 1s orbital may merge with one of the half-filled p orbitals of the oxygen atom to form an s—p bond. The half-filled 1s orbital of a second hydrogen atom may merge with the other half-filled p orbital of the oxygen atom to form a second s—p bond. The oxygen kernel now has a stable octet of electrons around it.

Since each of the orbitals of the oxygen atom now consists of a region in space occupied by *two* electrons, the four pairs of electrons in the octet repel one another and get as far from each other as they can and still be adjacent to the core (kernel) of the oxygen atom. They can achieve this if they locate at the corners of a regular *tetrahedron* with the oxygen atom at its center. This is a pyramid whose base and three sides are equilateral triangles (Fig. 19.12*a*). The angle between lines drawn from the center of the tetrahedron to its corners is 109° (Fig. 19.12*b*).

tetrahedron

From the tetrahedral arrangement of the electron pair clouds around the oxygen kernel, we would *predict* a bond angle of 109° for the bent water molecule. How can we explain the different value of 105° that is *observed*?

FIGURE 19.11 The water molecule

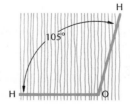

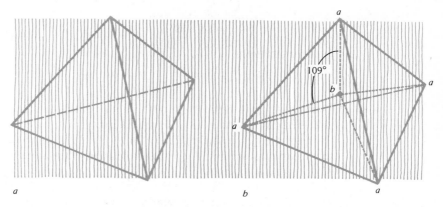

FIGURE 19.12 Tetrahedral geometry. (*a*) A tetrahedron. (*b*) The points at *a* are tetrahedrally oriented about the point *b* which is at the center of the tetrahedron.

Figure 19.13 shows the tetrahedral arrangement with the oxygen kernel at the center of the figure, an H atom at each of two corners, and an unshared pair of electrons at each of the other two corners. This accounts for all the valence electrons—six from the oxygen atom and two from the hydrogen atoms. The presence of the positive H nucleus near a shared pair of electrons will partly reduce the negative charge of the cloud; two shared pairs will not repel each other as strongly as two unshared pairs, and so the angle between the two bonds to the H atoms (with their shared pairs) will be slightly less than 109°. The angle between two lines drawn from the oxygen kernel to the two unshared pairs will be slightly greater than 109°. The tetrahedron will be slightly irregular.

FIGURE 19.13 The tetrahedral arrangement of electron pairs around an oxygen atom in a water molecule

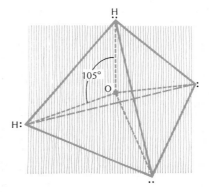

FIGURE 19.14 Formation of two
covalent bonds between hydrogen
atoms and an oxygen atom to form a
molecule of water

$$2H\cdot \ + \ \cdot \ddot{\underset{\cdot\cdot}{O}} \colon \ \longrightarrow \ H \colon \overset{H}{\underset{\cdot\cdot}{\ddot{O}}} \colon$$

Because it is difficult to picture the three-dimensional tetrahedron in a two-dimensional drawing, we frequently show the octet of bonding electrons in a square planar arrangement as in Fig. 19.14. But we must remember that the geometry in this symbolic array is not that actually present in the three-dimensional molecule.

We would predict that the bonding and structure of H_2S, H_2Se, and H_2Te would be quite similar to that of H_2O. The orbitals involved in forming the hydrides of S, Se, and Te would be $3p$, $4p$, and $5p$, respectively. The p orbitals increase in size with increasing value of the quantum number n, and so we would expect the distance between the nucleus of the hydrogen atom and the atom to which it is bonded to increase in the order H—O, H—S, H—Se, H—Te. This distance between the nuclei of two atoms bonded together is called the *bond length*. Scattering of x-rays and electrons is used to determine bond lengths. Table 19.1 summarizes some data for the hydrides of the oxygen family. These data bear out our prediction that the bond lengths in the hydrides will increase as the size of the oxygen group atom increases. The relative sizes of the molecules of the hydrides of the oxygen family are shown in Fig. 19.15.

bond length

19.6 Polarity in the Water Molecule

One of the very important properties of water in the environment is its ability to dissolve ionic compounds. Indeed, water is the best of all solvents for such compounds. Somehow or other, water must reduce the electrostatic attractions between the ions in crystals so that they can break away from the lattice and enter the solution. What is there about the water molecule that gives

TABLE 19.1 *Some Characteristics of the Oxygen Family Hydrides*

Characteristic	H_2O	H_2S	H_2Se	H_2Te
Bond length, Å	1.0	1.3	1.5	1.6 (estimate)
Boiling point, °C	100	−60.7	−41.5	−1.8
Melting point, °C	0	−85.6	−60.4	−51

FIGURE 19.15 Scale
models of the molecules
of the hydrides of the
oxygen family

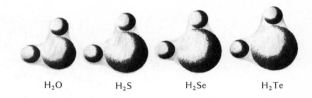

H$_2$O H$_2$S H$_2$Se H$_2$Te

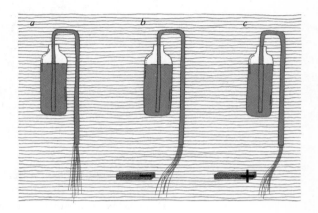

FIGURE 19.16 The
effect of electric
charges on a stream
of water

it this great solvent power? Obviously it must have something to do with electrostatic attractions, so let's see how water is affected by electrostatic charges.

If we set up a bottle with a glass tube like that shown in Fig. 19.16, a stream
of water will run downward as in part *a*. If we rub a plastic rod with wool and
bring the negatively charged rod close to the stream of water, it is attracted to
the rod as shown in *b*. If we rub a glass rod with silk and bring the positively
charged rod close to the stream of water, it is attracted to the rod as shown
in *c*. How can water be attracted to both positive and negative charges?

Let us assume that the positive atomic nuclei and the negative electrons in
the water molecule are not symmetrically distributed, but that one part of the
molecule has a slight excess of positive charge and another a slight excess of
negative charge. An oversimplified symbol for the molecule would be that in
Fig. 19.17. This unsymmetrical molecule is called a *dipole*.

dipole

Because water molecules are in constant motion at ordinary temperatures,
the orientation of the dipoles in the stream of water in Fig. 19.16*a* will be

FIGURE 19.17 A simple symbol for a
molecule in which the positive and
negative charges are not symmetrically
distributed: a dipole

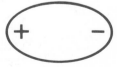

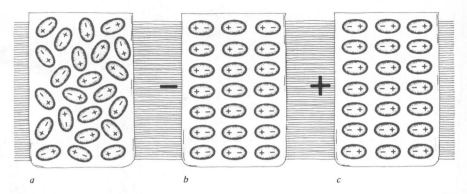

FIGURE 19.18 (*a*) Random orientation of water dipoles in absence of an electrostatic charge; (*b*) orientation of water dipoles with positive ends pointing toward a negative external charge; (*c*) orientation of water dipoles with negative ends pointing toward a positive external charge

random (Fig. 19.18*a*). When we bring a negatively charged rod up close to the stream of water, the dipoles will orient themselves as in Fig. 19.18*b*. Since the positive ends of the dipoles are closer to the rod than the negative ends, the attraction between the rod and the nearby positive charges will be greater than the repulsion between the rod and the more distant negative charges, so that the molecules will be drawn toward the rod, as in Fig. 19.16*b*. When we bring a positively charged rod up close to the stream of water, the dipoles will orient themselves so that their negative ends will point toward the positive rod, as in Fig. 19.18*c*. Again, the attraction between the rod and the nearby negative charges of the dipoles will exceed the repulsion between the rod and the more distant positive charges of the dipoles, so that the molecules will be drawn toward the rod as in Fig. 19.16*c*.

polar compound

Any compound that behaves toward positive and negative charges as water does must be made up of molecules that are dipoles. Such substances are called *polar compounds*, and the electric charges in their molecules are unsymmetrically distributed. How can this come about?

19.7   Polar Covalent Bonds and the Concept of Electronegativity

In Secs. 18.3 and 18.4 we interpreted differences in the chemical reactivity of various metallic elements by noting that atoms of different metals have different tendencies to *donate* electrons and thus to become positive ions. In Sec. 18.5 we saw that the differences in chemical reactivity of various nonmetallic elements could be explained in terms of the differences of atoms in their tendencies to *accept* electrons and thus to become negative ions. How convenient it would be if we could account for chemical reactivity more generally by developing a numerical scale of electron donation-acceptance for all elements! Exactly this was done by Linus Pauling, an American chemist who won a Nobel Prize for his contributions to our understanding of chemical bonds between atoms. Pauling proposed a scale of *electronegativity* (Table 19.2) to describe the tendency of an atom of an element in a compound to attract electrons. It is important to note that the electronegativity refers to the combined—not the free—element.

electronegativity

When two elements combine to form a compound, energy in the form of heat is often released. To decompose the compound into its elements, we have to supply heat or an equivalent amount of some other form of energy. If the bonds between the atoms in the compound are strong—i.e., if it takes a great deal of energy to break them and recover the elements—we say that the *bond energy* in the compound is high. Pauling reasoned that when bonds of high energy are formed, the electrons involved in the bonding must be much more strongly attracted to one of the atoms in the compound than to the other. He assumed that the atom with the greater attraction for electrons in the compound has a higher electronegativity than the other. Pauling designed a mathematical equation to calculate differences in electronegativity among the atoms of various elements in terms of the bond energies of the compounds they formed. Knowing that the element fluorine has a greater attraction for electrons than any other element, he arbitrarily assigned it an electronegativity of 4 as a basis for his calculation of the electronegativities of other elements.

bond energy

TABLE 19.2 *Electronegativities of the Elements*

												H 2.1					
Li 1.0	Be 1.5	B 2.0													C 2.5	N 3.0	O 3.5	F 4.0
Na 0.9	Mg 1.2	Al 1.5													Si 1.8	P 2.1	S 2.5	Cl 3.0
K 0.8	Ca 1.0	Sc 1.3	Ti 1.5	V 1.6	Cr 1.6	Mn 1.5	Fe 1.8	Co 1.8	Ni 1.8	Cu 1.9	Zn 1.6	Ga 1.6	Ge 1.8	As 2.0	Se 2.4	Br 2.8
Rb 0.8	Sr 1.0	Y 1.2	Zr 1.4	Nb 1.6	Mo 1.8	Tc 1.9	Ru 2.2	Rh 2.2	Pd 2.2	Ag 1.9	Cd 1.7	In 1.7	Sn 1.8	Sb 1.9	Te 2.1	I 2.5
Cs 0.7	Ba 0.9	La–Lu 1.1–1.2	Hf 1.3	Ta 1.5	W 1.7	Re 1.9	Os 2.2	Ir 2.2	Pt 2.2	Au 2.4	Hg 1.9	Tl 1.8	Pb 1.8	Bi 1.9	Po 2.0	At 2.2
Fr 0.7	Ra 0.9	Ac 1.1	Th 1.3	Pa 1.5	U 1.7	Np–No 1.3										

From Linus Pauling, *The Nature of the Chemical Bond*, 2d ed., The Cornell University Press, Ithaca, 1960.

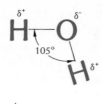 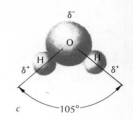

a b c

FIGURE 19.19 Symbols for the water molecule: (a) shared and unshared pairs of electrons shown by dots; (b) unsymmetrical distribution of electric charge; (c) rough approximation of the space filled by the electron clouds

When an atom of low electronegativity combines with an atom of high electronegativity, the first atom donates an electron to the second. The electron donor becomes a positive ion and the electron acceptor a negative ion. These are held together in a crystal by ionic bonds. When two atoms have the same electronegativity, they may combine by forming a covalent bond in which they share a pair of electrons equally. If the difference in electronegativity of two atoms is considerable but not great enough to produce ions by electron transfer, a covalent bond can form, but the pair of electrons will not be shared equally between the two atoms. The atom having the higher electronegativity will attract the electrons more strongly, and this will lead to a slight imbalance of electric charge. The space occupied by the atom of higher electronegativity will become slightly negative and that occupied by the atom of lower electronegativity will become slightly positive. The attraction between

polar covalent bond two atoms sharing a pair of electrons unequally is called a *polar covalent bond*.

The covalent bond between two atoms of the same element is nonpolar. The covalent bond between two atoms of different elements having the same electronegativity is nonpolar. The covalent bond between any two atoms having different electronegativities will be polar; the greater the difference in electronegativity, the more polar the bond.

EXAMPLE 19.1 Will the following covalent bonds be polar or nonpolar: F—H, O—H, N—H, P—H, O—C?

ANSWER

Bond	F—H	O—H	N—H	P—H	O—C
Difference in electronegativity	4.0 − 2.1 1.9	3.5 − 2.1 1.4	3.0 − 2.1 1.1	2.1 − 2.1 0	3.5 − 2.5 1.0
Polar or nonpolar	polar	polar	polar	nonpolar	polar

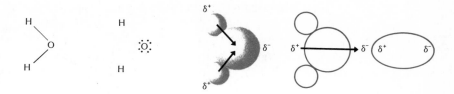

FIGURE 19.20 Polarity in the water molecule

19.8   The Distribution of Charge in the Water Molecule

The structure of the water molecule is shown in Fig. 19.19. Because of the difference in electronegativity of H atoms and O atoms, the H—O bond is polar; the O atom pulls the electron clouds of the shared pairs away from the H nucleus (the proton) and toward itself. This leaves the space around each H nucleus slightly positive and that around the O atom slightly negative. To symbolize a slight excess of charge, we use the Greek letter delta, δ; by $\delta+$ we mean a slight excess of positive charge, by $\delta-$ a slight excess of negative charge. The symbol of polarity in a bond is an arrow pointing from the region of excess positive charge to that of excess negative charge. Various ways of indicating the structure and polarity of the water molecule are shown in Fig. 19.20. The polarities of the two bonds added together give polarity to the molecule as a whole.

19.9   Why Is Water Such a Good Solvent for Ionic Compounds?

We now have the background required to answer the questions posed in Sec. 19.6. In a water solution of NaCl the ions will be surrounded by water molecules. The positive sodium ions, Na^+, will attract the negative region (the region around the oxygen atom) of each water molecule near it, and the negative chloride ion, Cl^-, will attract the positive regions (the regions around the H atoms) of each water molecule near it (Fig. 19.21). The ions in a water solution are *hydrated*—surrounded by an envelope of water molecules.

hydrated ion

When a crystal of solid NaCl is placed in water, the corner of the crystal where a sodium ion is located will attract H_2O molecules around it, with the negative regions pointed towards the Na^+ ion. Similarly, the corner of the crystal where a chloride ion is located will attract H_2O molecules around

FIGURE 19.21 The attraction of water molecules to Na$^+$ and Cl$^-$

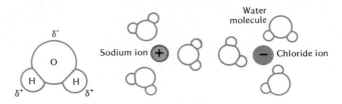

it, with the positive regions pointed towards the Cl$^-$ ion (Fig. 19.22).

The water molecules close to the sodium ion partially neutralize its positive charge and weaken its attraction to the negative ions in the crystal lattice; it can then break away and enter the solution. The water molecules close to the chloride ion have an analogous effect so that it, too, can break away from the crystal and enter the solution. This process continues with the ions that become exposed as those originally at the corners dissolve. In solution, the envelopes of water molecules around the ions partially neutralize the charges and thus serve as insulators between the ions. No other common liquid

FIGURE 19.22   The dissolving of NaCl in water

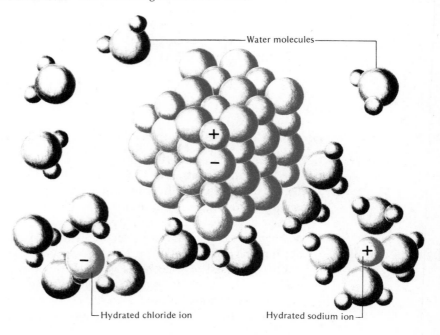

is as good an insulator as water, and so water is the best solvent for ionic compounds.

If the forces of attraction between the positive and negative ions in the crystal lattice of a given ionic compound are very high, they cannot be overcome by the hydration of the ions, and the substance is insoluble.

19.10   Bonds between Molecules of Water: Hydrogen Bonds

When we plot the boiling and melting points of the hydrides of the oxygen family of elements against the period in which the element lies (the horizontal row in the periodic table), we get the upper two graphs shown in Fig. 19.23. The boiling and melting points of water are astonishingly high. From our experience with the gradual variation of properties within a family of elements we would expect that we could extrapolate the graphs by means of the dotted lines and find the boiling point of water to be near $-90°C$ and its melting point near $-115°C$. When we plot the boiling points of the hydrides of the carbon family in Fig. 19.23, we find the regular variation we expect.

FIGURE 19.23   Melting and boiling points of some hydrides and noble gas elements

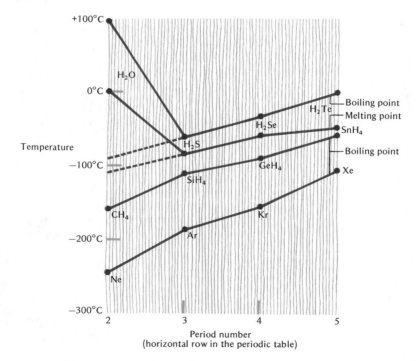

Likewise, the graph of the boiling points of the noble gas family shows the expected regularity. What is so special about water that it should boil and freeze at such unexpectedly high temperatures?

Some special attraction must be holding the water molecules tightly together in the ice crystal if we have to heat it to 0°C—far above the expected −115°C—in order to jostle the molecules out of the crystal lattice to form liquid water. This special attraction must also exist at the boiling point, for we have to heat water to 100°C—far above the expected value of −90°C—to give the molecules in the liquid rapid enough motion so that they can escape from the liquid and fly off into the gaseous state. What kind of attraction might there be between water molecules to hold them so tightly together?

Each proton (H atom nucleus) in a water molecule is rather exposed, because much of the electron cloud that was around it in the hydrogen atom before the H_2O molecule was formed has been pulled into the space between the H atom and the O atom when the H—O bond was formed. Might an exposed proton (H atom nucleus) in one molecule of water be attracted to an unshared electron pair in the oxygen atom of another molecule? The structure of ice crystals indicates that this is the case. The crystal consists of oxygen atoms arranged in a tetrahedral pattern with hydrogen atoms between them. This arrangement is shown for six water molecules in Fig. 19.24. The attraction between a proton of one molecule and an unshared electron pair of an oxygen atom of another molecule is shown by an arrow in Fig. 19.24. This attraction is called a *hydrogen bond* or a *hydrogen bridge*. The tetrahedral pattern in Fig. 19.24 is three-dimensional; thus water molecules 1, 3, and 5 have additional hydrogen bonds to water molecules *below* the plane of the paper, and molecules 2, 4, and 6 have additional hydrogen bonds to water molecules *above* the plane of the paper.

In Fig. 19.25 the structure of the ice crystal is shown more nearly to scale. Here the larger electron clouds are those of oxygen, and the smaller ones are

hydrogen bond or
hydrogen bridge

FIGURE 19.24 The arrangement of water molecules in ice

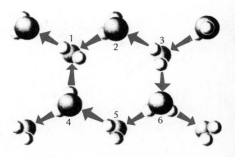

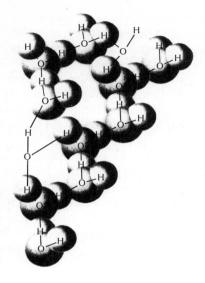

FIGURE 19.25 The structure of ice

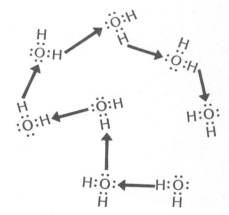

FIGURE 19.26 Hydrogen bonding in liquid water. Aggregates of water molecules contain varying numbers of molecules in various patterns.

those of hydrogen. This structure is quite open; it is honeycombed with empty space in hexagonal channels running through the layers of water molecules perpendicular to the plane of the paper in Fig. 19.25. When ice melts, some of the hydrogen bonds are broken, and these channels collapse, allowing the water molecules to draw closer together, so that the density of liquid water is greater than that of ice.

When ice melts, the rigid structure breaks up into fragments like that shown in Fig. 19.26. Such fragments have been detected by electron diffraction. Because the ice structure is quite strong, a crystal has to be heated to its unexpectedly high melting point before the structure is shaken apart to produce liquid water.

Some hydrogen bonds between water molecules must persist right up to the boiling point, thus decreasing the tendency of individual molecules to fly off into the gaseous state. The water has to be heated to an unexpectedly high temperature to make it boil.

Because the electronegativities of the other members of the oxygen family are so much less than that of oxygen, there is no tendency to form hydrogen bonds between the molecules of H_2S, H_2Se, or H_2Te. For this reason, the melting and boiling points of these compounds are quite normal, as shown in Fig. 19.23.

Fluorine has an electronegativity of 4.0, so it will strongly attract the shared pair of electrons in the HF molecule, leaving the proton (H atom nucleus) somewhat exposed. Hydrogen bonds form between the exposed proton of one HF molecule and the F of another HF molecule thus: H—F---H—F (the solid line indicating the polar covalent bond and the dotted line the hydrogen bond). This produces aggregates of HF molecules somewhat like the aggregates of H_2O molecules in water. The hydrides of the other halogens do not form hydrogen bonds.

19.11   The Ammonia Molecule

Many substances, such as proteins, which are found in the human environment and are associated with chemical processes in living organisms, are related to ammonia, NH_3. Knowledge of structure and bonding in the ammonia molecule is helpful in understanding these substances and the processes they undergo.

As we noted in Sec. 14.10, when hydrogen and nitrogen are heated to a high temperature under high pressure, they combine to form ammonia. Since ammonia has none of the properties of ionic compounds, we believe the atoms within it are covalently bonded. Studies of x-ray reflection from crystals of solid NH_3 and of electron scattering by liquid and gaseous NH_3 show that the angles between the N—H bonds are 107°. How can we account for this geometry?

The electron configuration for nitrogen is (Ne) $2s$ ↑↓ $2p$ ↑ ↑ ↑. When a hydrogen atom approaches a nitrogen atom, the H atom's half-filled $1s$ orbital can merge with one of the half-filled p orbitals of the nitrogen atom to form an s—p bond. A second and a third H atom can be bound to the nitrogen by the same kind of bond. The nitrogen kernel now has a stable octet of electrons about it (Fig. 19.27).

Since each of the orbitals of the nitrogen atom now consists of a region in space occupied by *two* electrons, these electron pairs repel one another and take up positions at the corners of a tetrahedron, as was the case with the wa-

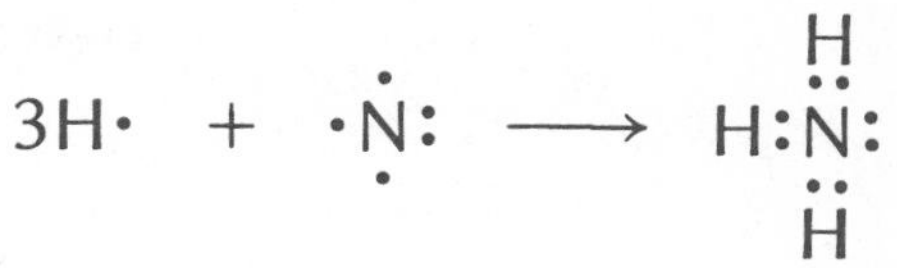

FIGURE 19.27 Formation of three covalent bonds between hydrogen atoms and a nitrogen atom to give an ammonia molecule

$$3H\cdot \ + \ \cdot\ddot{N}\!: \ \longrightarrow \ H\!:\!\overset{\displaystyle H}{\underset{\displaystyle H}{\ddot{N}}}\!:$$

FIGURE 19.28 The tetrahedral geometry of the ammonia molecule

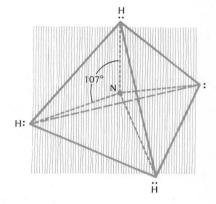

ter molecule. This arrangement is pictured three-dimensionally in Fig. 19.28.

We would predict that the bonding and structure of PH_3, AsH_3, and SbH_3 would be quite similar to that of NH_3 but that the bond lengths would increase because the bonding p orbitals in these compounds are larger ($3p$, $4p$, and $5p$, respectively). These predictions are confirmed by studies of these molecules: there is a regular increase in size, as shown in Fig. 19.29.

Nitrogen has an electronegativity of 3.0, and so it will strongly attract the shared pairs of electrons in the NH_3 molecule, leaving the protons (H atom nuclei) somewhat exposed. Hydrogen bonds form between the exposed proton of one NH_3 molecule and the N of another NH_3 molecule thus:

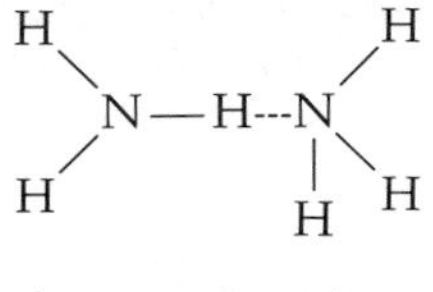

$$\begin{array}{c} \text{H} \qquad\qquad \text{H} \\ \diagdown \qquad\qquad \diagup \\ \text{N}\!-\!\text{H}\text{-}\text{-}\text{-}\text{N} \\ \diagup \qquad\qquad \diagdown \\ \text{H} \qquad\qquad\ \text{H} \\ \qquad\quad \big|\quad\ \ \diagup \\ \qquad\quad \text{H} \end{array}$$

The solid lines indicate the polar covalent bonds. The dotted line is the hydrogen bond. The hydrides of the other elements in the nitrogen family do not form hydrogen bonds.

When an ammonia molecule reacts with a hydrogen chloride molecule, the crystalline ionic compound ammonium chloride is formed. The interaction of these two covalent molecules produces two ions: the ammonium

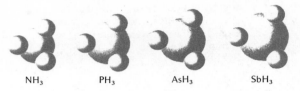

FIGURE 19.29 Scale models of the hydrides of the nitrogen family of elements

ion and the chloride ion. In the ammonium ion, the four H atoms are bonded covalently at the corners of a tetrahedron with the N atom at the center. The nitrogen atom is thus surrounded by a stable octet of electrons. In the chloride ion, the chlorine atom is likewise surrounded by a stable octet. This bonding is symbolized in Fig. 19.30.

19.12 Structure and Bonding in Methane and Carbon Tetrachloride

Huge amounts of natural gas, which consists largely of methane, CH_4, are brought from underground to the surface of the earth daily to fuel the world's industries and to warm the homes and working places of people in cold climates. Methane is also the principal component of "marsh gas," which bubbles up from the muddy bottoms of shallow bodies of water. Where the bottom mud is rich in the detritus from plants and animals that have died and sunk down through the water, there is not enough oxygen dissolved in the water for ordinary microorganisms to metabolize the organic material and produce CO_2. Other anaerobic organisms (organisms living or active in the absence of uncombined oxygen) metabolize the organic substances present and produce CH_4 and other substances. Methane and other hydrocarbons with more than one atom of carbon per molecule are the chief components of all natural gas and petroleum, as well as the organic part of oil-bearing sands and shales.

Methane is also produced by microorganisms involved in the digestive processes in the rumen (part of the stomach) of some animals that eat plants.

FIGURE 19.30 Formation of ionic NH_4Cl from covalent NH_3 and HCl

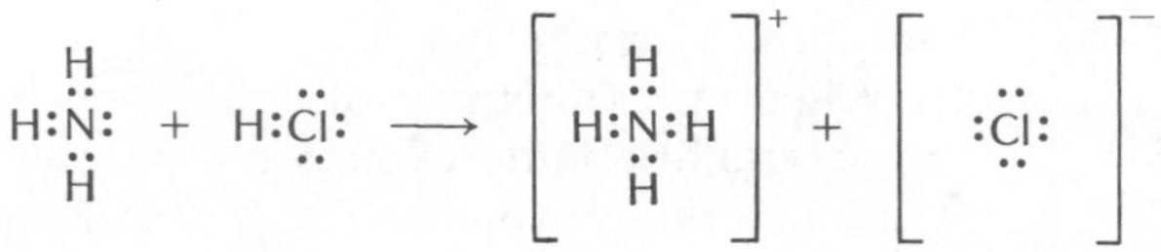

$$\text{H:N:} + \text{H:Cl:} \longrightarrow \left[\text{H:N:H}\right]^+ + \left[\text{:Cl:}\right]^-$$

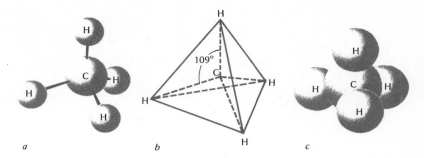

FIGURE 19.31 The geometry of the methane molecule: (*a*) a ball-and-stick model; (*b*) tetrahedral geometry of the molecule; (*c*) model showing relative sizes of the atoms

Microorganisms in the manure of pigs also produce methane. In Taiwan, where pigs are of prime importance in the meat supply and natural gas is not readily available, the manure is put in a covered steel box in the pigpen, and the gas produced (of which about 50 percent is methane) is piped to the kitchen and burned for cooking. The manure from 20 pigs produces enough combustible gas to do the cooking for one family.

X-ray and electron scattering studies of CH_4 indicate that the molecule consists of a central carbon atom surrounded tetrahedrally by the four hydrogen atoms (Fig. 19.31). All members of the carbon family in the periodic table form tetrahedral hydrides (Fig. 19.32).

Carbon tetrachloride, CCl_4, can be made from methane. This compound is a liquid at room temperature. When a stream of carbon tetrachloride is tested with negative and positive charges as water was tested in Fig. 19.16, the flow of the liquid is undeflected by either kind of charge. Apparently, the CCl_4 molecule as a whole is nonpolar even though each of the four C—Cl bonds is polar (because the electronegativity of Cl is 3.0 and that of C is 2.5). Somehow, the effects of the four polar bonds must cancel out. One geometric ar-

FIGURE 19.32 Scale models of the molecules of the hydrides of the carbon family

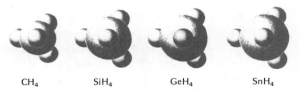

CH₄   SiH₄   GeH₄   SnH₄

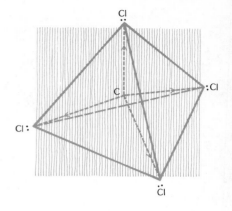

FIGURE 19.33 Polarity in the CCl₄ molecule

rangement of bond angles that will achieve this is the location of the C atom at the center of a tetrahedron and each Cl atom at a corner. This indirect evidence thus supports tetrahedral orientation of the four bonds formed by the carbon atom. The polarity of the bonds in CCl₄ is shown in Fig. 19.33.

Summary

Millions of the compounds we know are nonelectrolytes; that is, the atoms in them are not held together by ionic bonds. The compounds are held together because the atoms share pairs of electrons. A shared pair of electrons is called a covalent bond, and compounds held together by such bonds are called *covalent compounds.*

covalent compound

The concept of bond formation by the sharing of pairs of electrons is made more understandable if we modify our model of the electron and consider it to be a diffuse cloud of negative charge rather than a tiny point charge in orbit around the atomic nucleus. This concept evolved from the mathematical equations for describing an electron as a wave (wave mechanics).

Two H atoms join together to form an H_2 molecule by sharing the two $1s$ electrons between the two nuclei. We describe this bond by saying that the $1s$ electron cloud of one atom has merged or overlapped with the $1s$ electron cloud of the other. The duo of electrons is a stable configuration called an s—s bond. Two electrons with opposite spins can overlap to form a covalent bond; two electrons with the same spin cannot form a covalent bond.

Covalent bonds can be formed by the overlapping of two half-filled p orbitals as in Cl_2 and the other halogens, X_2. The formation of this bond achieves a stable octet of electrons around each atom in X_2. Again, the electrons to be shared must have opposite spins. Hydrogen can form halides by the overlapping of its $1s$ electron cloud with the half-filled p orbital cloud of a halogen atom to form an s—p bond in HX. A stable duo is thus achieved for H and a stable octet for X.

The water molecule is bent, with an angle of 105° between the two s—p bonds. The oxygen is at the center of a tetrahedron, and two H atoms occupy two corners of the tetrahedron; the other two corners are occupied by unshared pairs of electrons. The shapes of the hydrides of the other elements in the oxygen family are very similar.

The positive and negative charges in the water molecule are not symmetrically arranged, and so the molecule is polar. The oxygen atom, being more electronegative than the H atom, tends to pull the shared pairs in the bonds away from the H atoms. This makes the region around the oxygen atom slightly negative and that around each hydrogen slightly positive. This polarity accounts for the excellence of water as a solvent for ionic compounds. The polarity also causes hydrogen bonding between water molecules in ice and in liquid water. This hydrogen bonding accounts for the low density of ice and the unexpectedly high melting and boiling points of H_2O compared to the hydrides of the other elements in the oxygen family.

Nitrogen forms NH_3 with three s—p bonds pointing toward the corners of a tetrahedron occupied by three H atoms; there is an unshared pair of electrons at the fourth corner. The other members of the nitrogen family form similar tetrahedral hydrides.

Carbon forms CH_4 with four s—p bonds pointing toward the corners of a tetrahedron occupied by four H atoms. The other members of the carbon family form similar tetrahedra. CCl_4 is also tetrahedral but is nonpolar because the polarities of the four C—Cl bonds cancel out.

New Terms and Concepts

BOND ANGLE: The angle between the lines drawn through the nuclei of atoms joined together in a compound.

BOND ENERGY: The energy liberated when a bond forms between two atoms; this is the same as the energy absorbed when the bond is broken.

BOND LENGTH: The distance between the nuclei of atoms joined together in a compound.

CHARGE CLOUD: The smeared out or diffuse charge of an electron.

COVALENT BOND: The bond holding two atoms together by the sharing of a pair of electrons.

COVALENT COMPOUND: A compound in which the atoms are held together by covalent bonds.

DIPOLE: A pair of electric charges of equal magnitude but with opposite signs, separated by a small distance.

ELECTRONEGATIVITY: The tendency of an atom in a compound to attract electrons.

HYDRATED ION: An ion surrounded by water molecules.

HYDROGEN BOND OR HYDROGEN BRIDGE: The attraction between a hydrogen atom in a molecule and an unshared pair of electrons on a highly electronegative atom in another molecule of the same species.

p — p BOND: A covalent bond formed by the merging of a half-filled p orbital of one atom with a half-filled p orbital of another atom.

POLAR COMPOUND: A compound in which the positive and negative charges are arranged unsymmetrically.

POLAR COVALENT BOND: A covalent bond in which the electron pair is not shared equally by the two atoms bonded together.

s — p BOND: A covalent bond formed by the merging of a half-filled s orbital of one atom with a half-filled p orbital of another atom.

s — s BOND: A covalent bond formed by the merging of a half-filled s orbital of one atom with a half-filled s orbital of another atom.

TETRAHEDRON: A solid figure bounded by four faces.

Testing Yourself

19.1 What is the essential difference between an ionic and a covalent bond? Can a sharp line be drawn between these two kinds of bonds?

19.2 Describe the distribution of electric charge in the H atom and in the H_2 molecule.

19.3 If two Br atoms combine to form one Br_2 molecule, why does the combination process stop there? Why don't Br atoms combine to form Br_3, Br_4, etc., molecules?

19.4 How does a bond form between the atoms in the HI molecule?

19.5 What is a tetrahedron? How is this geometric form useful to us in explaining the bond angles in water molecules? In ammonia molecules? In methane molecules?

19.6 How do we account for the fact that the boiling point and freezing point of water are so far out of line with those of the other hydrides of group VI elements (Fig. 19.23)?

19.7 Account for the fact that water expands when it freezes. Is this the usual behavior of a liquid when it freezes? Of what importance is this behavior of water to living organisms?

19.8 Describe the process by which water dissolves a crystal of an ionic compound.

19.9 Would you expect the following substances to be polar or nonpolar? Why? Chloroform, $CHCl_3$; methyl alcohol, CH_3OH; hydrogen sulfide, H_2S; iodine molecules, I_2; carbon tetrabromide, CBr_4.

19.10 Carbon dioxide is nonpolar. What does this tell you about the shape of the molecule? Would you expect carbon disulfide to be polar or nonpolar? Why?

Crystals of cholesterol are shown in the photomicrograph; its structure is illustrated in the drawing.

20. Hydrocarbons and Their Derivatives Related to Foods

Millions of covalent compounds are known to the chemist today. Thousands of them are used in our highly technological industries. Hundreds of thousands more are used by living organisms to build their tissues and carry out the chemical processes involved in their metabolism. Most of these covalent substances are carbon compounds. We call them organic compounds because many of them were first found in chemical systems associated with living organisms. Today most of these compounds have been synthesized in non-living systems in the laboratory, and so we think of organic chemistry as the chemistry of carbon and its compounds.

In this book we have space to consider only a few of the multitude of organic chemicals important to us today. We start with the simplest organic compounds, the hydrocarbons. We then consider more complex compounds derived from hydrocarbons. Finally, we discuss the organic compounds which comprise our basic foods—carbohydrates, fats, and proteins.

20.1 Simplification by Classification

Confronted by an array of millions of organic compounds, we can simplify learning about them by classifying them into various categories. It is convenient to think of all organic compounds as being derivatives of the simplest ones, called hydrocarbons because they contain only hydrogen and carbon. These compounds illustrate a most important property of carbon—its ability to combine with itself to form molecules containing carbon-to-carbon bonds. Another important property of the carbon atom is its ability to form a stable octet by sharing its four valence electrons with four other atoms to form four covalent bonds. This makes it possible for carbon to combine with many different elements to form a wide variety of compounds.

functional group
characteristic group

We shall consider classes of compounds in which one or more hydrogen atoms have been replaced by an oxygen atom, an —OH group, an —NH$_2$ group, and various combinations of these substituents. These are called *functional* or *characteristic groups*, because their presence in a molecule makes it react in a characteristic way. The classes of compounds containing the functional groups known as alcohols, aldehydes, ketones, acids, esters, and amines are important in the structures of carbohydrates, fats, and proteins.

20.2 Alkanes

Natural gas and petroleum are mixtures containing many different hydrocarbons. The simplest is methane, CH_4. Other compounds have the formulas C_2H_6, C_3H_8, C_4H_{10}, etc., with as many as 30 C atoms per molecule. From the formulas of the four compounds given here we can derive the general formula C_nH_{2n+2}. Hydrocarbons of this type are called *alkanes*. Some information about the first 10 alkanes is given in Table 20.1. (The isomers in Table 20.1 will be discussed in Sec. 20.3.)

alkanes

How are the atoms held together in these molecules and what are the shapes

TABLE 20.1 *Names, Formulas, and Isomers of the First 10 Alkanes*

Name	Molecular formula	Condensed structural formula	Number of possible isomers
Methane	CH_4	CH_4	1
Ethane	C_2H_6	CH_3CH_3	1
Propane	C_3H_8	$CH_3CH_2CH_3$	1
Butane	C_4H_{10}	$CH_3(CH_2)_2CH_3$	2
Pentane	C_5H_{12}	$CH_3(CH_2)_3CH_3$	3
Hexane	C_6H_{14}	$CH_3(CH_2)_4CH_3$	5
Heptane	C_7H_{16}	$CH_3(CH_2)_5CH_3$	9
Octane	C_8H_{18}	$CH_3(CH_2)_6CH_3$	18
Nonane	C_9H_{20}	$CH_3(CH_2)_7CH_3$	35
Decane	$C_{10}H_{22}$	$CH_3(CH_2)_8CH_3$	75

of the molecules? We have already noted the four covalent bonds between C and H atoms in CH_4 and the tetrahedral shape of the molecule. The bond angles in all alkanes are found to be 109°. What might be the shape of the ethane molecule, C_2H_6? If the two C atoms have a covalent bond between them, each has three unpaired valence electrons to share with H atoms. From the observed bond angles we would expect tetrahedral orientation of the four shared pairs around each C atom. We might think of the shape of the C_2H_6 molecule as that of two tetrahedra sharing a corner (Fig. 20.1). Because of the difficulty of making a two-dimensional drawing of a three-dimensional tetrahedron, we noted previously that we might symbolize the presence of the four shared pairs of electrons in CH_4 as in Fig. 20.2. We can use the same simplification to indicate the structure of C_2H_6 in two dimensions (Fig. 20.3). Similar reasoning leads us to conclude that the propane molecule, C_3H_8, might have the shape of a chain of three tetrahedra. There would be three C atoms in a chain with covalent bonds between them, and the remaining valence electrons would be paired with those of the H atoms.

FIGURE 20.1 Two C atoms bonded in ethane by sharing a pair of electrons. The six H atoms are at the other corners of the tetrahedra.

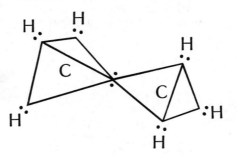

FIGURE 20.2 Two ways of symbolizing the structure of the methane molecule

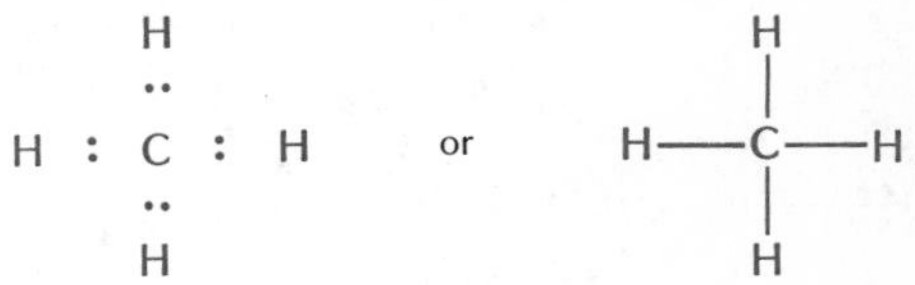

$$H : \overset{\cdot\cdot}{\underset{\cdot\cdot}{C}} : H \quad \text{or} \quad H-\overset{H}{\underset{H}{C}}-H$$

FIGURE 20.3 Two ways of symbolizing the structure of the ethane molecule

$$H : \overset{\cdot\cdot}{\underset{\cdot\cdot}{C}} : \overset{\cdot\cdot}{\underset{\cdot\cdot}{C}} : H \quad \text{or} \quad H-\overset{H}{\underset{H}{C}}-\overset{H}{\underset{H}{C}}-H$$

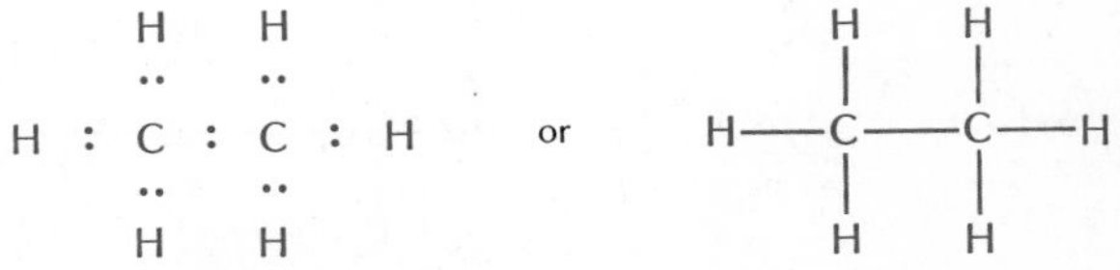

The arrangement of the four C atoms in the butane molecule, C_4H_{10}, is suggested in different ways in Fig. 20.4. To help visualize the three-dimensional relationships in complicated organic molecules, we find ball-and-stick models very useful. The holes in the balls representing C atoms are bored so that when sticks are inserted there will be angles of 109° between the sticks (they will point toward the corners of a tetrahedron). To show how the electron clouds fill the space in a molecule we use plastic or wooden spheres with one side cut off so that when the flat faces of the atomic units are put together, the model will show the shape of the space occupied by the electron clouds, that is, the shape of the molecule.

20.3 Structural Isomers of Alkanes

There are two kinds of butane, C_4H_{10}, one of which boils at $-0.5°C$, the other at $-10.2°C$. The difference in boiling points indicates some difference in structure. Studies of structure show that the compound boiling at $-0.5°C$ consists of molecules with the four C atoms in a chain. In the molecules of the compound boiling at $-10.2°C$ there are three C atoms in a chain and the

FIGURE 20.4 Four ways of symbolizing the butane molecule: (*a*) electron dot formula; (*b*) structural formula; (*c*) ball-and-stick model; (*d*) space-filling model.

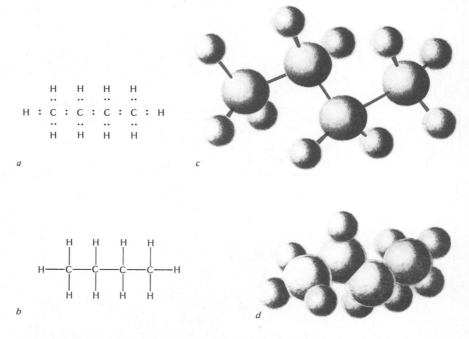

isomer

fourth on a branch from the chain. We call the compound with the un-branched chain *normal butane* and the one with the branched chain *iso-butane*. Two or more compounds having the same general formula but different structures and properties are called *isomers*. Various ways of indicating the structures of the two isomers of butane are shown in Fig. 20.5.

When a half-filled orbital of one carbon atom merges with that of another to form a C—C bond by sharing a pair of electrons, there is nothing about the bond to prevent rotation of one atom with respect to the other in the molecule. Therefore, a chain may easily twist in various directions. Studies of the shapes of hydrocarbon molecules show that this twisting is common. Figure 20.6*a* shows space-filling models of the normal butane molecule in various configurations. Figure 20.6*b* shows the shape of isobutane.

EXAMPLE 20.1 The structural formula of pentane, C_5H_{12}, can be written in several different ways. How many different isomers of pentane are shown below?

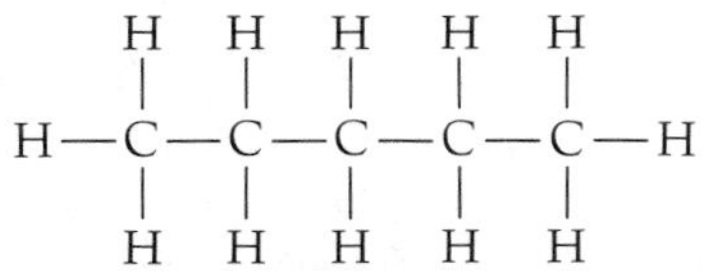

a

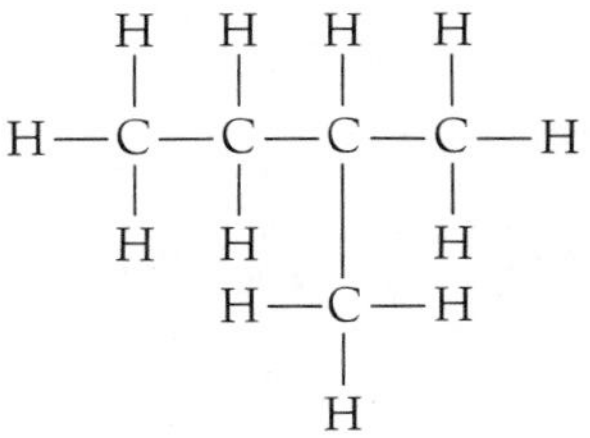

b

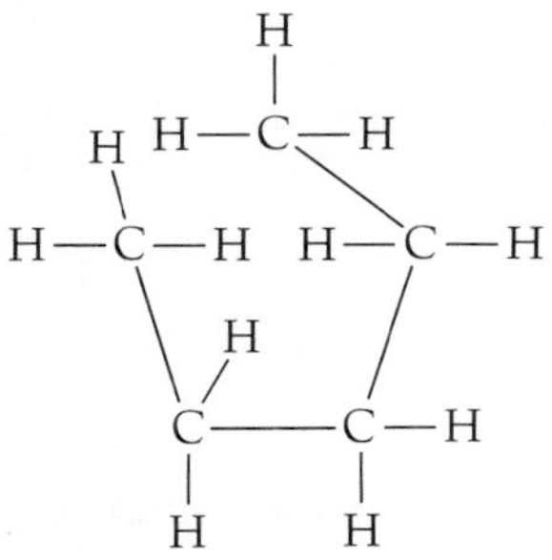

c

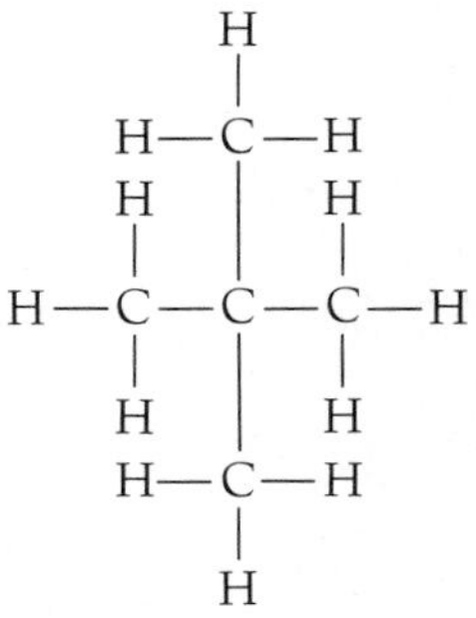

d

ANSWER Forms *a* and *c* are the same because each consists of five C atoms in an unbranched chain. The chain is just twisted more in *c*. Forms *b* and *d* have branched chains, but *b* has a single branch and *d* has two branches, and so these two are different. Three different isomers of pentane are shown in the four structural formulas.

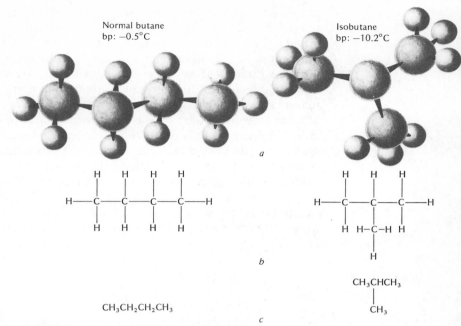

FIGURE 20.5 Structural isomers of butane: (*a*) ball-and-stick models; (*b*) structural formulas in two dimensions; (*c*) condensed structural formulas.

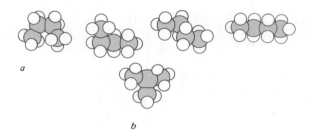

FIGURE 20.6 Various configurations of the butane molecule; open circles are H atoms; shaded circles are C atoms: (*a*) normal butane and (*b*) isobutane.

20.4 Alkanes from Petroleum

The main source of alkanes is petroleum, which is a mixture of hydrocarbons. They are separated by distillation, the main process in refining petroleum. Because petroleum contains so many alkanes with different numbers of carbon atoms and different isomeric structures, many of which have boiling points very close together, the separation of specific compounds from it is very difficult. For this reason, refining rarely produces individual hydrocarbons, but usually a series of hydrocarbon mixtures which have been vaporized and collected within various temperature ranges. The uses of various petroleum fractions are given in Table 20.2. As the alkanes increase in molecular weight, their boiling points rise. Apparently, larger molecules exert considerable forces of attraction upon one another, and these intermolecular attractions and increasing molecular weights make it necessary to heat to higher and higher temperatures in order to disengage the molecules from the liquid and drive them into the gaseous state. We believe that these intermolecular attractions arise from the electrical interactions of the various nuclei and electron charge clouds in polyatomic molecules; the greater the number of positive nuclei and negative charge clouds, the greater the interaction and the higher the boiling point. These intermolecular attractions are

TABLE 20.2 *The Boiling Ranges and Uses of Hydrocarbons from Petroleum*

Product	Carbon-chain size	Boiling range, °C	Use
Natural gas	C_1–C_5	−164 – 30	Fuel; synthesis of carbon, hydrogen, gasoline
Petroleum ether (ligroin)	C_4–C_{10}	30–90	Solvent; dry cleaning
Gasoline	C_4–C_{13}	40–225	Motor fuel
Kerosene	C_{10}–C_{16}	175–275	Fuel; lighting
Gas oil, fuel oil, diesel oil	C_{15}–C_{18}	250–400	Diesel and furnace fuel
Lubricating oils, greases, petroleum jelly	C_{20} up	350 up	Lubrication
Paraffin (wax)	C_{23}–C_{29}	50–60 (melting point)	Candles; waterproofing coatings for cartons; matches; home canning
Pitch (tar)	Viscous liquid residues		Artificial asphalt; paving, roofing
Petroleum coke	Solid residues		Electrodes; fuel

called *van der Waals bonds*, named after the Dutch chemist Johannes van der Waals.

20.5   Unsaturated Hydrocarbons

When petroleum is refined to yield gasoline and other fuels, three hydrocarbon gases with two C atoms per molecule are formed as by-products. These are ethane, C_2H_6, ethylene, C_2H_4, and acetylene, C_2H_2. In the ethane molecule each C atom can achieve a stable octet of electrons by forming *one* C—C bond and three C—H bonds (Fig. 20.7a). In ethylene there are only two H atoms per C atom, and so each C can achieve a stable octet only by forming *two* bonds (a *double bond*) between the C atoms and two C—H bonds (Fig. 20.7b). In acetylene there is only one H atom per C atom, and so each C can achieve a stable octet only by forming *three* bonds (a *triple bond*) between C atoms and one C—H bond (Fig. 20.7c).

Our concept of double and triple bonding between C atoms can be simplified by thinking about the tetrahedral orientation of electrons around a C atom. When a C—C (single) bond is formed, the two tetrahedra share a corner (one pair of electrons, Fig. 20.8a). When a C=C (double) bond is formed, the two tetrahedra share an edge (two pairs of electrons, Fig. 20.8b). When a C≡C (triple) bond is formed, the two tetrahedra share a face (three pairs of electrons, Fig. 20.8c).

When a mixture of ethylene and hydrogen is heated, no observable reaction takes place. But if the hot mixture is passed through a fine screen made of platinum wire, the double bond in the ethylene molecule opens up, and two atoms of hydrogen add on to form ethane:

$$C_2H_4 + H_2 \xrightarrow[\text{of platinum}]{\text{in the presence}} C_2H_6 \qquad\qquad (20.1)$$

FIGURE 20.7 Single, double, and triple bonds between C atoms

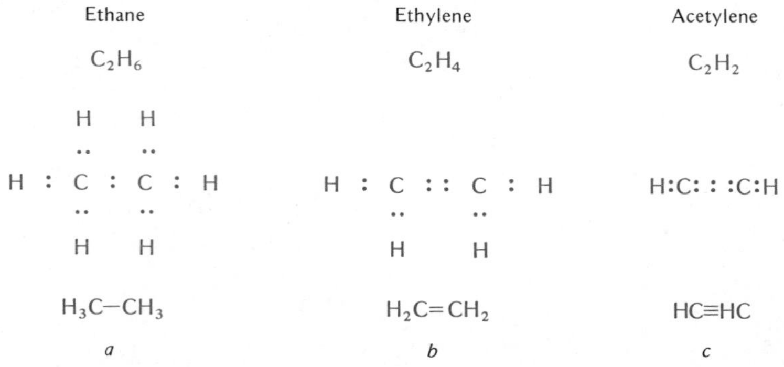

Ethane	Ethylene	Acetylene
C_2H_6	C_2H_4	C_2H_2

<pre>
      H   H
      ·· ··
  H : C : C : H   H : C :: C : H   H:C : :C:H
      ·· ··   ··   ··
      H   H   H   H
</pre>

H_3C—CH_3	H_2C=CH_2	HC≡HC
a	*b*	*c*

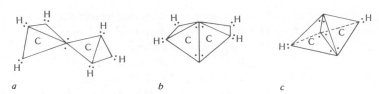

FIGURE 20.8 Tetrahedra joined (*a*) corner to corner; (*b*) edge to edge; (*c*) face to face.

unsaturated
hydrocarbon

Because ethylene can add on more hydrogen, we call it an *unsaturated hydrocarbon*, and we describe Eq. (20.1) as hydrogenation. Acetylene is also an unsaturated hydrocarbon; it can be hydrogenated to ethylene:

$$C_2H_2 + H_2 \xrightarrow[\text{of platinum}]{\text{in the presence}} C_2H_4 \qquad (20.2)$$

If acetylene is hydrogenated with a limited supply of hydrogen, it forms ethylene. In the presence of excess hydrogen, it forms ethane.

catalyst

catalysis

If two or more substances are mixed and no observable reaction takes place, but a reaction can be promoted by the addition of another substance which is not consumed in the reaction, we call the latter a *catalyst*. We say that the reaction has been catalyzed by the added substance; the process is called *catalysis*. In the catalytic hydrogenation of ethylene and acetylene, the platinum is unchanged. A catalyst is present after the reaction in the same amount as before the reaction; it is not used up by the reaction. Catalysis is very important in most reactions in living organisms and in the chemical industry, but the mechanism of catalysis is too complicated to be included here.

20.6   Structure and Bonding in Diamond and Graphite

We have established one of the unique properties of carbon—ease of forming C-to-C bonds—as illustrated with the hydrocarbons. Now let us investigate the structure of pure carbon, which exists as diamond and also as graphite. Diamond is a clear, colorless, transparent substance and the hardest known naturally occurring solid; it will scratch any other natural solid and all but one or two synthetic ones. Diamond dust is widely used as an abrasive. Diamond has a melting point of about 3500°C—higher than that of any other element!

In contrast, graphite is a black, shiny, opaque solid which is so soft that it will rub off on paper; the "lead" in a pencil contains graphite. Its softness also makes it very useful as a lubricant for machinery. It does not melt but sublimes (evaporates from the solid state directly to the gaseous state without becoming a liquid) at about 3650°C. How can the same element have such different properties in its different pure forms? Can we relate these properties to the structures of diamond and graphite?

X-ray studies of diamond crystals show that every C atom is surrounded by four others at equal distances; the bonds around each carbon atom are tetrahedrally oriented (Fig. 20.9a). This produces a structure which is extremely resistant to distortion. It is very difficult to scratch any atoms out of the lattice. To do so involves breaking many strong bonds. The crystal has to be heated to a very high temperature in order to shake the atoms out of their strongly bonded pattern and so liquefy the solid.

X-ray studies of graphite show that the solid is composed of sheets of hexagonal rings of carbon atoms extending in two dimensions in a plane (Fig. 20.9b). The distance between these sheets (3.40 Å) is greater than that between atoms (1.42 Å) in the rings. Three of the valence electrons of each C atom are paired with electrons from three adjacent C atoms in the same plane. This forms a very stable planar network of atoms. The fourth valence electron in each C atom is paired with the fourth electron of an atom in a different layer. Since each atom in the ring is held to three others in its plane by three bonds but is held to only one atom in an adjacent plane, rubbing graphite does not disrupt the sheets of atoms, but breaks the bonds between sheets. One plane can easily slide along another. This gives graphite a flaky quality, making it soft and slippery, a good lubricant.

20.7 Alcohols

When wood is heated to a high temperature in a closed vessel so that it cannot burn by union with oxygen, various liquids and tarry substances are driven

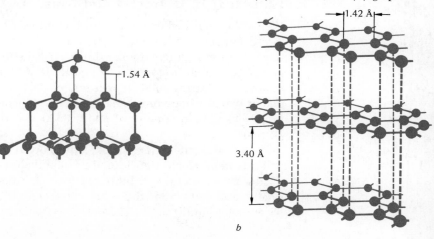

FIGURE 20.9 The arrangement of carbon atoms in (a) diamond and (b) graphite

a

b

off, and almost pure carbon in the form of charcoal is left in the vessel. One of the principal components of the liquid which is distilled off is wood alcohol, with the formula CH_3OH. The scientific name for this liquid is methyl alcohol or methanol.

When yeast organisms grow in solutions containing sugar (fruit juices, molasses, honey) or starch (potatoes, corn, barley, rye), they produce alcohol and carbon dioxide. This process is called *fermentation*. The alcohol produced is C_2H_5OH, called ethyl alcohol or ethanol. Mankind has been drinking ethyl alcohol in fermented beverages for thousands of years, but *methyl* alcohol is a dangerous *poison*. It is often added to ethyl alcohol to "denature" the latter, to make it unfit for drinking. There are many other *alcohols*—propanol, butanol, etc.—but they are of little importance in man's natural environment.

Methanol is a derivative of methane in which one H atom has been replaced by —OH (a *hydroxyl group*). A hydroxyl group is bound to the C atom by sharing a pair of electrons with it. Ethanol is a derivative of ethane. The formulas for these two alcohols are

alcohol

hydroxyl group

Methanol

Ethanol

alkyl group

active site

The type formula for an alcohol thus derived from an alkane is ROH, where R stands for any *alkyl group*, C_nH_{2n+1}. The methyl group is —CH_3; the ethyl group is —C_2H_5.

In the molecule ROH, the R is comparatively unreactive. Hydroxyl, —OH, is called an *active site*, or a functional group, in a molecule. It is a part of the molecule's structure that can be active, or it can function in a chemical reaction. Its presence confers on the molecule a set of properties characteristic of the functional group. As will be shown in Sec. 20.8, the —OH in a molecule can react with oxygen, so that C—O—H becomes C=O, another functional group. In Sec. 20.13 we see how the —OH can react with an acid. The chemical properties of an organic compound depend on the functional groups that are present in its molecules.

If an H atom on each of the two carbons in ethane is replaced by an — OH, the product is called ethylene glycol. This *dihydric alcohol* readily mixes in all proportions with water; it is widely used as an antifreeze (Sec. 12.5). When an H atom on each of the three carbons in propane is replaced by an —OH the product is glycerine, a *trihydric alcohol* used in cough drops, as a lubricant, and in many hand lotions and similar cosmetics. The structural formulas of ethylene glycol (glycol) and glycerine (glycerol) are

$$
\begin{array}{cc}
\quad\ \ \text{H}\ \ \text{H} & \quad\ \text{H}\ \ \text{H}\ \ \text{H} \\
\quad\ \ |\ \ \ | & \quad\ |\ \ \ |\ \ \ | \\
\text{H}-\text{C}-\text{C}-\text{H} & \text{H}-\text{C}-\text{C}-\text{C}-\text{H} \\
\quad\ \ |\ \ \ | & \quad\ |\ \ \ |\ \ \ | \\
\quad\ \text{OH OH} & \quad\text{OH OH OH}
\end{array}
$$

Glycol   Glycerol

(a dihydric alcohol)   (a trihydric alcohol)

20.8 Aldehydes

When a mixture of methyl alcohol vapor and air is passed over a catalyst such as finely divided platinum, silver, or copper at a slightly elevated temperature, formaldehyde is formed. When a mixture of ethyl alcohol vapor and air is so treated, acetaldehyde is formed. The equations for the reactions are

$$
2\left[\ \text{H}-\overset{\displaystyle \text{H}}{\underset{\displaystyle \text{H}}{\text{C}}}-\text{O}-\text{H}\ \right] + \text{O}_2 \longrightarrow 2\left[\ \text{H}-\overset{\displaystyle \text{H}}{\text{C}}=\text{O}\ \right] + 2\text{H}_2\text{O}
$$

Methyl alcohol   Formaldehyde

$$
2\left[\ \text{CH}_3-\overset{\displaystyle \text{H}}{\underset{\displaystyle \text{H}}{\text{C}}}-\text{O}-\text{H}\ \right] + \text{O}_2 \longrightarrow 2\left[\ \text{CH}_3-\overset{\displaystyle \text{H}}{\text{C}}=\text{O}\ \right] + 2\text{H}_2\text{O}
$$

Ethyl alcohol   Acetaldehyde

In general, the mild oxidation of an alcohol, RCH_2OH, yields an *aldehyde*, RCHO. The functional group

$$
-\text{C}\!\!\begin{array}{c}\diagup\text{H} \\ \diagdown\text{O}\end{array}
$$

is called the aldehyde group. The oxidation of an alcohol to an aldehyde must be done at only slightly elevated temperatures or the alcohol will be burned up. When any organic compound is heated to a high temperature in the presence of oxygen, the carbon and hydrogen present are strongly oxidized (burned) to CO_2 and H_2O.

Solutions of formaldehyde are used to "pickle" biological specimens. Formaldehyde readily combines with animal tissues and stiffens them so that their form and structure are preserved. Acetaldehyde is a sedative. Many of the aldehydes of larger molecular weight have pleasant odors or flavors: vanilla, cinnamon, and lilac, for example.

20.9 Ketones

ketone

When an alcohol with two alkyl groups attached to the carbon bonded to the —OH group is oxidized by oxygen at a slightly elevated temperature, a *ketone* is produced. Thus, when isopropyl alcohol is oxidized, acetone is produced:

$$2\begin{bmatrix} CH_3 \;\; H \\ \;\;\diagdown\;\diagup \\ \;\;\;\;C-O-H \\ \diagup \\ CH_3 \end{bmatrix} + O_2 \longrightarrow 2\begin{bmatrix} CH_3 \\ \diagdown \\ \;\;\;C=O \\ \diagup \\ CH_3 \end{bmatrix} + 2H_2O$$

$$\text{Isopropyl alcohol} \qquad\qquad\qquad \text{Acetone}$$

The general formula for a ketone is

$$\begin{array}{c} R-C-R' \\ \parallel \\ O \end{array}$$

R and R' may be the same or different alkyl groups.

Ketones are useful solvents for organic compounds and are widely used in making plastics, paints, adhesives, etc. Acetone is one of the intermediate products in the metabolism of sugar in the human body. Diabetics sometimes have acetone in their breath, an indication that their metabolism of sugar is faulty. Some ketones of larger molecular weight are fragrant; one has the odor of violets. Some are used in perfumes that contain a combination of odors, e.g., civet from the African civet cat, ambergris from whales, and musk from the musk deer.

20.10 Compounds with More Than One Functional Group

We have already noted that glycol and glycerol are alcohols with two and three hydroxyl groups, respectively. Many molecules are known which have more than one functional group, either several of the same kind or of several different kinds. Among the most important of these are foods—carbohydrates, fats, and proteins. In such compounds each functional group contributes its properties to the characteristics of the molecule as a whole.

20.11 Carbohydrates

polyhydroxy-
aldehyde
polyhydroxyketone

Most carbohydrates are compounds with the general formula $C_xH_{2y}O_y$. Molecules of these substances, called *polyhydroxyaldehydes* or *polyhydroxyketones*, contain several hydroxyl groups and either an aldehyde or a ketone group.

Of all the foods consumed by man, carbohydrates are by far the most abundant and the cheapest. They are eaten by many animals and contribute to the production of fats and of proteins like meat, milk, and eggs. Ultimately, all the energy utilized by living matter comes from carbohydrates. Their photosynthesis by green plants from CO_2 and H_2O was mentioned in Sec. 15.2. The carbohydrates with the general formula $C_6H_{12}O_6$ have a sweet taste and are called *hexose* sugars (Fig. 20.10). Because of the presence of the five —OH groups, they are soluble in water, easily crystallizable from water solutions, and diffuse through the membranes of living cells. If not attacked by bacteria in the digestive tract, they are absorbed from the small intestine into the blood without chemical change.

hexose

The three most important hexoses are glucose, fructose, and galactose.

FIGURE 20.10 Structural formulas of three hexoses

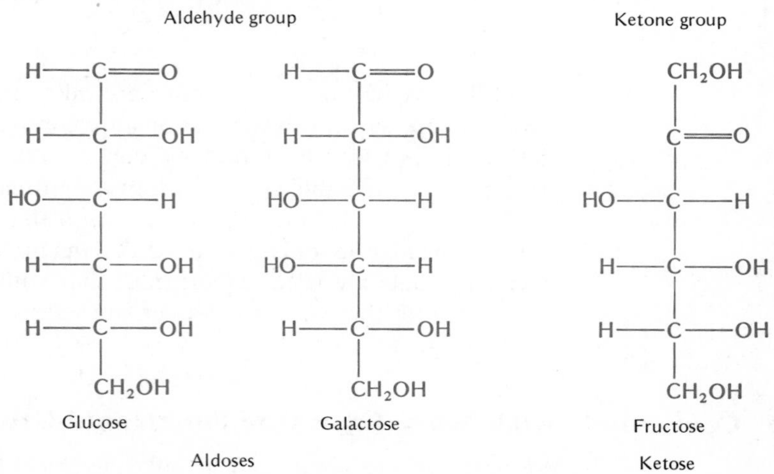

Glucose

Galactose

Fructose

Aldoses

Ketose

Glucose (also called dextrose, grape sugar, corn sugar, blood sugar, starch sugar) is widely distributed in nature. It occurs in the blood of all animals to the extent of about 0.1 percent and is abundant in fruits and the juices of many plants. More than 50 percent of the solid matter in grapes is glucose; sweet corn and onions contain considerable amounts of it. Fructose (also called fruit sugar and levulose) occurs along with glucose in plant juices and fruits; it constitutes about half of the solid matter in honey. Galactose does not occur in natural products but arises from the partial digestion of lactose (milk sugar), so it is of some importance for human nutrition. Some infants are unable to metabolize galactose; it accumulates in the body's tissues, including the central nervous system, and damages the cells. If milk is excluded from the diet of such an infant and other sources of carbohydrates are supplied, the child grows normally. Glucose and galactose contain an aldehyde group and are called *aldoses*; they are isomers differing only in the geometrical arrangement of the H and OH along the chain of six C atoms. Fructose differs from glucose in that it has a ketone structure instead of an aldehyde group; it is called a *ketose*.

aldose

ketose

The "sugar" we are familiar with as a sweetener is sucrose; it is found in all land plants—in their juices, seeds, flowers, stems, and roots. It is especially abundant in the juices of sugar cane, sugar beets, sorghum, and the saps of some maple and palm trees. One molecule of sucrose, $C_{12}H_{22}O_{11}$, contains two hexose units—one glucose and one fructose—joined by an oxygen bridge arising from the removal of a molecule of water from two — OH groups, one on glucose and one on fructose. Because it contains two hexose units, sucrose is called a *disaccharide*. Lactose present in milk is a disaccharide consisting of one unit of glucose and one of galactose. Maltose is produced in grain seeds that have been "malted" (allowed to sprout); it is a disaccharide whose molecule consists of two units of glucose.

disaccharide

Starches are *polysaccharides* with the general formula $(C_6H_{10}O_5)_n$, where n is greater than 1000; the *monosaccharide* units in starches are glucose. Over half the solid matter in cereal grains is starch; an even larger proportion of the total solids of potatoes, bananas, and chestnuts is starch. When starch is chewed, the saliva converts it into maltose; this is broken down (digested) to glucose in the small intestine, from which it is absorbed into the blood. Glucose may be injected directly into the blood to nourish people whose digestive systems are disrupted by disease or surgery.

polysaccharide
monosaccharide

The fibrous tissue in the cell walls of plants and trees contains considerable cellulose, a polysaccharide consisting of long chains of glucose units. Wood is about 50 percent cellulose; cotton is almost pure cellulose. The digestive juices present in man's alimentary canal do not contain the biological catalysts (called *enzymes*) that can break down cellulose, and so it must be

enzyme

structurally different from starch. Termites and plant-eating animals can metabolize cellulose because they have bacteria in their alimentary canals which produce enzymes that can break this polysaccharide down to glucose.

20.12 Organic Acids

If fermenting fruit juice is not protected from the air, the oxygen in the air will encourage a kind of fermentation that produces acid instead of alcohol. Thus wine or apple cider will "turn sour" if not sealed from the air. The acid in these sour solutions (vinegars) is acetic acid, $HC_2H_3O_2$. If acetic acid is mixed with a base like sodium hydroxide, NaOH, only one of the hydrogens reacts, and the product is sodium acetate, $NaC_2H_3O_2$. Obviously, one of the hydrogen atoms in acetic acid is bonded differently from the other three. Structural studies show that molecules of acetic acid are arranged thus:

$$H : \overset{\displaystyle H}{\underset{\displaystyle H}{C}} : \overset{\displaystyle :O:}{C} : \overset{..}{O} : H \qquad \text{Replaceable H}$$

Frequently abbreviated as
CH_3COOH

carboxyl group

Acetic acid is conveniently thought of as a derivative of ethane in which one CH_3 has been oxidized to —COOH, called a *carboxyl group*. The H atom in the —COOH is the one that dissociates to give acetic acid its sour taste and that reacts with NaOH thus:

$$CH_3COOH \quad + \quad HONa \longrightarrow H_2O + NaOOCCH_3$$

Acetic   or   hydrogen   sodium   water   sodium
acid   acetate   hydroxide   acetate

Formic acid, which is secreted by many ants and gives the sting to an ant's bite, has the formula HCOOH. Some primitive people catch ants and boil them up to produce a sour sauce which they use much as we use vinegar. Like methane, formic acid has only one C atom per molecule; it is given the scientific name methanoic acid. The scientific name for acetic acid is ethanoic acid. Propanoic acid is a derivative of propane and butanoic of butane. Butanoic acid has the common name butyric acid because it is formed when butter becomes rancid (spoils).

dicarboxylic acid

If an H atom on each of the two C atoms in ethane is replaced by a —COOH group, a *dicarboxylic acid* is produced; its common name is oxalic acid, and it is present in considerable amounts in rhubarb, giving this fruit its very sour taste. Citric acid is a *tricarboxylic acid* abundantly present in citrus fruit: oranges, lemons, limes, grapefruit. This acid is also found in many other fruits in lesser amounts. Citric acid is called a *hydroxyacid* because

tricarboxylic acid

hydroxyacid

it contains the —OH group as well as three —COOH groups. The formulas of formic, oxalic, and citric acids are

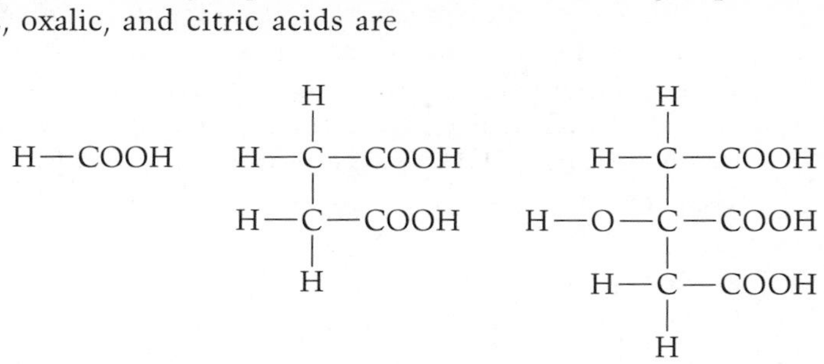

$$\text{H—COOH} \qquad \begin{array}{c} \text{H} \\ | \\ \text{H—C—COOH} \\ | \\ \text{H—C—COOH} \\ | \\ \text{H} \end{array} \qquad \begin{array}{c} \text{H} \\ | \\ \text{H—C—COOH} \\ | \\ \text{H—O—C—COOH} \\ | \\ \text{H—C—COOH} \\ | \\ \text{H} \end{array}$$

Formic acid   Oxalic acid   Citric acid

20.13 Esters

carboxylic acid
dehydrating agent
ester

When an alcohol and a *carboxylic acid* are heated together in the presence of a *dehydrating agent* (a substance that combines with water), they produce an *ester*. The reaction is called esterification:

$$\text{R—O—H} + \underset{\underset{O}{\|}}{\text{H—O—C—R}'} \longrightarrow \underset{\underset{O}{\|}}{\text{R—O—C—R}'} + H_2O$$

Alcohol + carboxylic acid $\longrightarrow$ ester   + water

The alkyl groups R and R′ may be the same or different.

Methyl alcohol + acetic acid $\longrightarrow$ methyl acetate + water
Ethyl alcohol + butyric acid $\longrightarrow$ ethyl butyrate + water

Esters give many fruits and flowers their characteristic flavors and odors: methyl butyrate, apples; ethyl butyrate, pineapples; ethyl formate, rum; octyl acetate, oranges; normal (straight-chain) pentyl acetate, apricots; isopentyl (branched-chain) acetate, bananas.

20.14 Fats and Soaps

When an ester is heated with sodium hydroxide, it yields an alcohol and a sodium salt, thus:

Ethyl   +   sodium $\longrightarrow$ ethyl   +   sodium
acetate   hydroxide   alcohol   acetate

$$C_2H_5OOCCH_3 + \quad NaOH \quad \longrightarrow C_2H_5OH + NaOOCCH_3$$

soap

When animal or vegetable fats are boiled with a strong solution of NaOH, glycerol and *soap* are produced. Soap is a mixture of the sodium salts of several different long-chain carboxylic acids; these are called *fatty acids*. Since the reaction of fats with NaOH produces glycerol and sodium salts of the fatty acids, we conclude that fats are the glycerol esters of the fatty acids. Natural fats are mixed esters; they contain more than one fatty acid component combined with the glycerol. The equation for the *saponification* (soaping) of a fat by NaOH may be written in general form as follows:

$$
\begin{array}{ccc}
\underset{\text{Fat}}{
\begin{array}{l}
CH_2-O-\overset{\displaystyle O}{\overset{\|}{C}}-R \\[4pt]
CH\ -O-\overset{\displaystyle O}{\overset{\|}{C}}-R' \\[4pt]
CH_2-O-\overset{\displaystyle O}{\overset{\|}{C}}-R''
\end{array}}
\; + NaOH \longrightarrow &
\underset{\text{Glycerol}}{
\begin{array}{l}
CH_2-OH \\[4pt]
CH\ -OH \\[4pt]
CH_2-OH
\end{array}}
\; &
\underset{\substack{\text{Sodium salts}\\ \text{of fatty acids}\\ \text{(soaps)}}}{
\begin{array}{l}
+\ R-COONa \\[4pt]
+\ R'-COONa \\[4pt]
+\ R''-COONa
\end{array}}
\end{array}
$$

The R, R′, and R″ in the fat may be the same or different, and so the resultant fatty acid salts (soaps) may be the same or different.

Some of the hydrocarbon chains of fatty acids contain no double bonds; others have one, two, or three (Table 20.3). When hydrogen is bubbled through fats at elevated temperatures in the presence of finely divided nickel as a catalyst, these double bonds open up, and a hydrogen atom adds on to

TABLE 20.3 *Some Common Fatty Acids*

Number of carbon atoms	Structure	Common name	Melting point, °C	Source
Saturated fatty acids				
16	$CH_3(CH_2)_{14}COOH$*	Palmitic	63	Palm oil
18	$CH_3(CH_2)_{16}COOH$	Stearic	70	Mutton, beef
Unsaturated fatty acids				
16	$CH_3(CH_2)_5CH{=}CH(CH_2)_7COOH$	Palmitoleic	−1	Milk
18	$CH_3(CH_2)_7CH{=}CH(CH_2)_7COOH$	Oleic	13	Olives, pork
18	$CH_3(CH_2)_4CH{=}CHCH_2CH{=}CH(CH_2)_7COOH$	Linoleic	−5	Soybeans
18	$CH_3CH_2(CH{=}CHCH_2)_3(CH_2)_6COOH$	Linolenic	−11	Linseed oil

*$CH_3CH_2CH_2CH_2CH_2CH_2CH_2CH_2CH_2CH_2CH_2CH_2CH_2CH_2CH_2COOH$

unsaturated fat

saturated fat

each of the two carbon atoms formerly doubly bonded to one another. Because hydrogen can be added to these fats, they are called *unsaturated fats*, and the acids containing double bonds are called unsaturated fatty acids. Since hydrogen cannot be added to fats made up of fatty acids containing no double bonds, these are called *saturated fats*, and the acids present are saturated fatty acids.

The melting points of the fats derived from various fatty acids show the same trends as those of the acids themselves. Fats formed from saturated fatty acids have higher melting points than those derived from unsaturated fatty acids. Glyceryl stearate and palmitate are solids at body temperature; glyceryl palmitoleate, oleate, linoleate, and linolenate are liquids. Animal fats are mostly saturated fats and are solids at body temperature. Vegetable fats are mostly unsaturated fats and are liquids (oils) at body temperature.

glyceride

Cholesterol is a compound found in varying amounts in almost all living organisms. In man, there is some indication that above-normal concentrations of cholesterol in the blood may be associated with heart failure arising from atherosclerosis (hardening of the arteries), which accompanies the deposition of fatty substances in the inner layer of the arteries. The concentration of cholesterol in the blood may be reduced if the fats in the diet are the *glycerides* (glyceryl esters) of polyunsaturated fatty acids. Vegetable fats consist mainly of the glycerides of the unsaturated fatty acids oleic acid, linoleic acid, and linolenic acid. Animal fats consist principally of the glycerides of the saturated stearic and palmitic acids. For this reason some dieticians recommend that animal fats be replaced in the diet by polyunsaturated vegetable fats to reduce blood cholesterol.

20.15   The Cleansing Action of Soap

One end of a soap molecule consists of a long hydrocarbon chain and the other of a carboxylate ion and a sodium ion (Fig. 20.11*a*). Ions are soluble in water, whereas nonpolar molecules, such as hydrocarbons, are not. When a globule of soap is put into water, the soluble ionic portions of the molecules tend to dissolve, but the insoluble hydrocarbon chains tend to cluster toward the center of the globule (Fig. 20.11*b*).

Soap cleans because the long hydrocarbon chain of the fatty acid unit is soluble in grease (fats, oils, or petroleum hydrocarbons) and the carboxylate ion and sodium ion dissolve in water. Most dirt on clothes or on the surface of the skin consists of foreign matter held in place by a film of grease. The hydrocarbon tails of the soap molecules enter the grease, and the carboxylate and sodium ions are held in the surrounding water. When the water is washed out, it carries the hydrocarbon tail and the greasy dirt with it. The "dissolving" of the dirt is indicated schematically in Fig. 20.12.

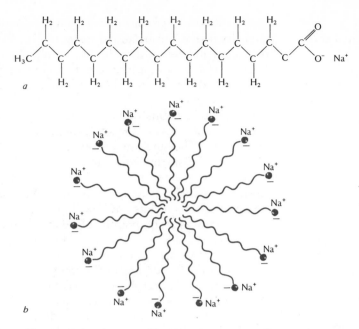

FIGURE 20.11 (*a*) The sodium stearate molecule. (*b*) Globule of soap dissolved in water.

FIGURE 20.12 The cleansing action of soap. Hydrocarbon tails of soap molecules penetrate greasy dirt; COO⁻Na⁺ ends dissolve in water.

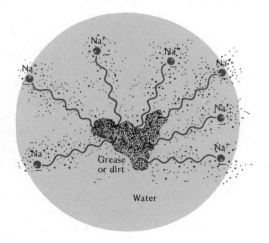

Water that has trickled through limestone dissolves small amounts of calcium and magnesium ions, Ca^{2+} and Mg^{2+}. When soap is added to this water, insoluble soaps (such as calcium and magnesium stearate) precipitate in the form of gummy curds. Such water is said to be hard. Because soap is used up to precipitate the curd, extra soap has to be added before it can dissolve grease (Fig. 20.12). Because it is hard to remove the gummy curds of calcium and magnesium soap from fabrics, laundering in hard water is difficult. Water may be softened by removing the dissolved calcium and magnesium ions or by forming compounds with them so that they cannot react with soap to produce the curds. Various kinds of sodium phosphates can serve this latter purpose, but these salts are undesirable because they are excellent nutrients for algae and accelerate eutrophication of rivers and lakes (Sec. 13.6).

20.16 Amino Acids

amine

All living organisms contain considerable amounts of nitrogen in various compounds, most of which may conveniently be thought of as related to ammonia. For this reason they are called *amines*. When ammonia is bubbled through an alcohol solution of methyl chloride, the following reactions take place:

$$NH_3 \quad + \quad CH_3Cl \longrightarrow \quad CH_3NH_3Cl$$
$$CH_3NH_3Cl + \quad NH_3 \quad \longrightarrow \quad CH_3NH_2 + NH_4Cl$$
$$\text{Methyl}$$
$$\text{amine}$$

amino group

The $-NH_2$ *group* is called *amino*, and so methyl amine might equally well be called aminomethane.

All amines—like their parent substance, ammonia—are basic [Eq. (12.17)]. Ammonia added to a water solution of HCl neutralizes the acid, forming ammonium chloride; an amine behaves in the same way.

$$NH_3 \quad + \quad HCl \quad \longrightarrow \quad NH_4Cl \quad \text{or} \quad NH_4{}^+ \quad + Cl^-$$
$$\text{Ammonia} + \text{hydrogen} \longrightarrow \quad \text{ammonium}$$
$$\text{chloride} \qquad \text{chloride}$$

$$CH_3NH_2 + \quad HCl \quad \longrightarrow \quad CH_3NH_3Cl \quad \text{or} \quad CH_3NH_3{}^+ + Cl^-$$
$$\text{Methyl} \quad + \quad \text{hydrogen} \quad \longrightarrow \quad \text{methylammonium}$$
$$\text{amine} \qquad \text{chloride} \qquad \text{chloride}$$

Many nitrogen compounds occurring in living organisms contain both the basic $-NH_2$ group and the acidic $-COOH$ group in the same molecule. These compounds are called amino acids; in most of them, the $-NH_2$ and the —COOH are attached to the same C atom. The simplest amino acid is aminoacetic acid; its common name is glycine. Aminopropanoic acid is called alanine. Many different groups of atoms may replace one of the H atoms attached to the C atom binding the —NH_2 and —COOH in an amino acid. The formulas for glycine, alanine, and amino acids in general are

$$
\begin{array}{ccc}
\text{NH}_2 & \text{NH}_2 & \text{NH}_2 \\
| & | & | \\
\text{H}-\text{C}-\text{COOH} & \text{CH}_3-\text{C}-\text{COOH} & \text{R}-\text{C}-\text{COOH} \\
| & | & | \\
\text{H} & \text{H} & \text{H} \\
\text{Glycine} & \text{Alanine} & \text{Amino acid}
\end{array}
$$

where R stands for any group of atoms which may be attached to the C.

20.17 Proteins

Proteins are the very stuff of life; all living organisms contain many proteins in many forms utilized in many ways. Fibrous proteins called keratins are responsible for the structures of skin, hair, claws, and nails. Collagen is found in muscles, tendons, and connective tissues in flesh. Other proteins are required as enzymes (biological catalysts) for chemical reactions essential to the growth and maintenance of living organisms.

When a protein is boiled with a strong acid, it breaks down into a mixture of amino acids; about 20 different amino acids have been isolated from the tissues of the human body. Most of these can be synthesized in the body from carbohydrates and fats and a nitrogen compound, but some must be essential amino acid provided ready-made in the diet. These are called *essential amino acids* and are contained in animal protein (meat, fish, cheese, eggs). A diet adequate to maintain the adult human in good condition should contain a minimum of 2.5 oz/day of such protein material. Although cereal grains (rice, wheat, corn, rye, barley) contain considerable amounts of different proteins, some of the essential amino acids are missing. A diet based solely on cereals is thus inadequate; it must be supplemented by some animal protein.

The molecular weights of proteins are very high, varying from 6000 for insulin to 20,000,000 for keratins. Apparently, hundreds of amino acid molecules are linked to form these very complex molecules. Since amino acids contain both the acidic carboxyl group —COOH, and the basic amino group —NH_2, it seems reasonable to assume that in a protein molecule the —COOH of one amino acid molecule has reacted with the —NH_2 of an-other (Fig. 20.13). The bond between the C of the —COOH and the N of peptide bond the —NH_2 is called a *peptide bond*. The C in the —COOH of a glycine molecule may be attached to the N in the —NH_2 of a molecule of alanine to form glycylalanine:

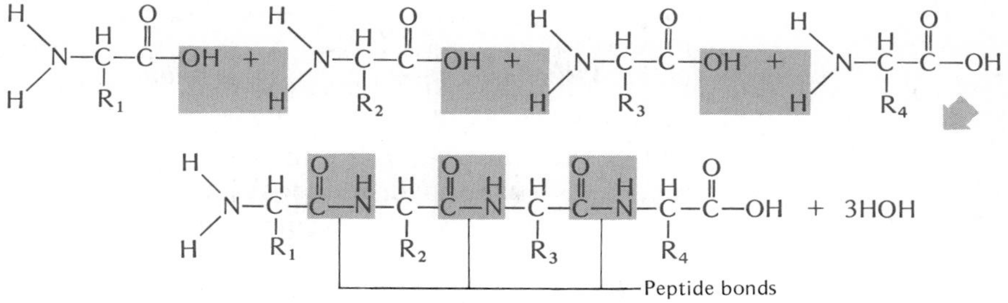

Four amino acids → a four-unit peptide + 3 water molecules

FIGURE 20.13 Four amino acids combine to form a polypeptide chain of four units held together by three peptide bonds.

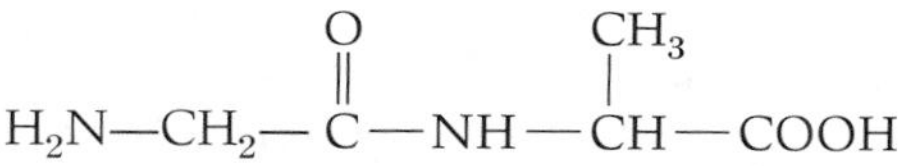

$$H_2N-CH_2-\overset{\overset{\displaystyle O}{\|}}{C}-NH-\overset{\overset{\displaystyle CH_3}{|}}{CH}-COOH$$

Glycylalanine

peptide

When amino acid units are united, the compound is called a *peptide*; when many units are united, a polypeptide.

Polypeptides with large numbers of amino acids are formed in nature. Polypeptides with as many as 18 amino acid units have been synthesized by chemists in the laboratory. These compounds have the properties of small protein molecules. The characteristics of polypeptides depend on the sequence of amino acids within the chain as well as on the kinds of amino acids included. Using only one of each of the 20 known amino acids in a 20-unit polypeptide, changes in sequence could produce about 2.5×10^{18} different kinds of molecules! Since proteins contain many more than 20 amino acid units per molecule, there is practically no limit to the number of different proteins which may be involved in the structures and metabolism of living plants and animals.

Summary

The element carbon is unique in its capacity for combining with itself to form the framework for very large molecules. This framework carries unshared electrons that can form covalent bonds with atoms of hydrogen, oxygen, nitrogen, and many other elements. Millions of these carbon compounds are known; they are usually called organic compounds.

Natural gas and petroleum contain hundreds of hydrocarbons—compounds composed of hydrogen and carbon only. Alkanes are hydrocarbons with the general formula C_nH_{2n+2}. The carbon atoms in these compounds are linked by single bonds to form chains, branched and unbranched. Various patterns of branching lead to different structures for molecules with the same number of C and H atoms; these are called isomers. The properties of molecules depend on the kind *and arrangement* of atoms present; isomers have different properties. In some hydrocarbons there are double or triple bonds between carbons. In the presence of a catalyst these hydrocarbons open their multiple bonds and add on more hydrogen, and so they are called unsaturated hydrocarbons. A catalyst is a substance that speeds up a chemical reaction but is not used up in the reaction. The differences in the properties of diamond and graphite can be explained by the difference in the bonding between the C atoms in these two forms of the element; properties are determined by structure.

There are various combinations of C atoms with atoms of other elements to form *functional groups*. These groups confer certain properties on substances composed of molecules containing the groups. Molecules of alcohols contain one or more —OH groups. Molecules of aldehydes, ketones, carboxylic acids, and amines have the following type formulas:

$$R-C\begin{smallmatrix}H\\\\O\end{smallmatrix} \qquad \begin{smallmatrix}R'\\\\R\end{smallmatrix}C{=}O \qquad R-C\begin{smallmatrix}O\\\\OH\end{smallmatrix} \qquad R-NH_2$$

Aldehydes   Ketones   Carboxylic acids   Amines

A carboxylic acid can react with an alcohol to form an ester. Esters contribute to the odors and flavors of many flowers and fruits.

Glucose and fructose are carbohydrates with six carbons in a chain (hexoses), five hydroxyl groups, and an aldehyde or ketone group. Sucrose is formed by the combination of two glucose molecules and starch is formed from many such molecules.

Fats are glycerol esters of carboxylic acids with long hydrocarbon chains, some of which contain double bonds. Unsaturated fats (those containing unsaturated chains) are liquid at body temperature and are called oils. There is some evidence that foods containing unsaturated fats are less likely to produce undesirable amounts of cholesterol in the blood than saturated fats.

Soaps are sodium salts of fatty acids. They owe their cleansing action to the solubility of the sodium part of the compound in water and the hydrocarbon chain in grease.

Amino acids contain both the acid carboxyl and the basic amine groups. Long chains of amino acid units can form by the interaction of the carboxyl group on one amino acid with the amino group of another (*peptide* bond).

Protein molecules are composed of such long chains of these amino acid groups that they have molecular weights of several thousand.

New Terms and Concepts

ACTIVE SITE: A group of atoms which allows an organic molecule to react in a characteristic way.

ALCOHOL: An organic compound containing the —OH (hydroxyl) group.

ALDEHYDE: An organic compound with the formula $R-C\diagup^{H}_{\diagdown\!\!\!\diagdown O}$

ALDOSE: A sugar containing one or more aldehyde groups.

ALKANE: A hydrocarbon with the formula C_nH_{2n+2}.

ALKYL GROUP: A hydrocarbon group with the formula $-C_nH_{2n+1}$.

AMINE: A compound with the formula RNH_2.

AMINO GROUP: $-NH_2$.

CARBOXYL GROUP: $-COOH$.

CARBOXYLIC ACID: An acid containing the —COOH group.

CATALYSIS: The action of a catalyst; the modification of the rate of a chemical reaction by a catalyst.

CATALYST: A substance that modifies the rate of a chemical reaction without being consumed by the reaction.

CHARACTERISTIC GROUP: A group of atoms which allows an organic molecule to react in a characteristic way.

DEHYDRATING AGENT: A substance that unites with water.

DICARBOXYLIC ACID: An organic acid containing two —COOH groups.

DIHYDRIC ALCOHOL: An alcohol containing two —OH groups.

DISACCHARIDE: A carbohydrate that reacts with water to form two monosaccharides.

DOUBLE BOND: Two pairs of electrons shared between two atoms.

ENZYME: A protein produced by a living organism and functioning as a biological catalyst in a living organism.

ESSENTIAL AMINO ACID: An amino acid that cannot be synthesized in the human body.

ESTER: The product of the reaction of an organic acid and an alcohol.

FATTY ACID: An acid produced by the interaction of water with a fat.

FUNCTIONAL GROUP: A group of atoms that allows an organic molecule to react in a characteristic way.

GLYCERIDE: A glyceryl ester.

HEXOSE: A sugar containing six atoms of carbon.

HYDROXYACID: An organic compound containing both the —OH group and the —COOH group.

HYDROXYL GROUP: —OH.

ISOMER: A compound having the same percentage composition and molecular weight as another compound but differing in chemical or physical properties.

KETONE: An organic compound with the formula $\begin{smallmatrix} R' \\ \diagdown \\ R \diagup \end{smallmatrix} C{=}O$.

KETOSE: A sugar containing a ketone group.

MONOSACCHARIDE: A simple sugar that cannot be decomposed by water.

PEPTIDE: A compound composed of amino acid units.

PEPTIDE BOND: The chemical bond between the —COOH of one amino acid and the —NH$_2$ of another.

POLYHYDROXYALDEHYDE: An organic compound containing several —OH groups and an aldehyde group.

POLYHYDROXYKETONE: An organic compound containing several —OH groups and a ketone group.

POLYSACCHARIDE: A carbohydrate composed of many monosaccharides linked together.

SAPONIFICATION: The reaction of an ester with a strong base to produce an alcohol and the salt of an organic acid, especially the reaction of a fat with NaOH or KOH to produce glycerol and salts of the fatty acids which are called soaps.

SATURATED FAT: A fat derived from saturated (completely hydrogenated) fatty acids.

SOAP: The sodium or potassium salt of a fatty acid.

TRICARBOXYLIC ACID: An acid containing three —COOH groups.

TRIHYDRIC ALCOHOL: An alcohol containing three —OH groups.

TRIPLE BOND: Three pairs of electrons shared between two atoms.

UNSATURATED FAT: A fat derived from unsaturated (incompletely hydrogenated) fatty acids.

UNSATURATED HYDROCARBON: A hydrocarbon containing one or more double or triple bonds.

VAN DER WAALS BOND: The attraction between the positive nuclei of atoms in one molecule and the electron clouds of atoms in other molecules leading to very weak bonds between molecules.

Testing Yourself

20.1   What are hydrocarbons? Saturated hydrocarbons? Unsaturated hydrocarbons?

20.2   What are the names, formulas, and structural formulas for the following substances: The first four alkanes with unbranched chains, ethylene, acetylene, isobutane?

20.3 What is an alcohol? What are the structural formulas for ethyl alcohol and glycerol?

20.4 What is a carbohydrate? An aldose? A ketose? Sucrose? Starch?

20.5 What is an organic acid? Name some organic acids commonly found in foods.

20.6 What is a fat? What is the chief difference between fats that are liquid and those that are solid at body temperature?

20.7 What is a soap? Explain its cleansing action.

20.8 What is an amino acid? A polypeptide? A protein?

20.9 Of what importance are amino acids nutritionally?

20.10 Why must we have proteins in our diets?

*Vitamin B1, seen in the photomicrograph, needed only in
minute quantities, is nonetheless essential to complete nutrition.*

21. Chemistry and Mankind's Need for Food

The most pressing problem faced by man today is to control his rate of re-production. Next in urgency is to control his environment so that he can produce enough food to eliminate starvation. Modern science and technology have given man the power to stabilize his birth rate. Whether or not he will do so depends primarily on political and social action. Modern science and technology have also enabled man to grow prodigious amounts of food, but at considerable cost in the pollution of the environment. More science and technology as well as political and social action are needed if we are to minimize pollution without cutting food production. The "green revolution" that has multiplied food production around the world grew out of the applications of biological and chemical knowledge to agriculture. We have almost realized man's age-old dream of freedom from hunger by our tremendous success in developing new food crops and multiplying their productivity by applying chemical fertilizers, herbicides, insecticides, fungicides, and hormones.

We may consider the human body one of nature's chemical factories, a factory more complex in its operations than any in industry. Inorganic acids, bases, and salts, organic compounds by the thousands react with one another in the preservation of human life. The raw materials for this factory are the air we breathe, the water we drink, and the food we eat. In our study of the interactions between man and his environment, it is important to find answers to the following questions, which we shall consider in this chapter. What kinds of foods and how much of them do we need to be well nourished? How do we go about producing enough food to meet these needs? How is the environment affected by the chemicals used in modern food production? How can we minimize their bad effects?

21.1 The Kinds of Food We Need

vitamin

Our bodies use foods for three different purposes: (1) to produce the energy needed to keep us alive and active; (2) to build and renew body tissues; (3) to regulate chemical reactions within the body and maintain the internal conditions required for them to take place. Our energy requirements can be met simply by eating the requisite amounts of fats and carbohydrates. But the tissues in our bodies contain nitrogen, so that building and renewing them requires that we eat proteins. And our body chemistry is regulated by various biological catalysts which include enzymes, *vitamins*, and hormones (substances which control growth). These are included in our diet or are synthesized from our food. A properly balanced diet, therefore, must contain fats, carbohydrates, proteins, vitamins, and minerals. The required daily amounts of these substances vary immensely from hundreds of grams of carbohydrates to millionths of a gram of minerals.

21.2 Measuring Energy Content of Foods by Calorimeters

Almost 200 years ago the French scientist Antoine Lavoisier and the English scientist Joseph Priestley showed that when a mouse or a guinea pig is kept in a closed jar, it soon sickens, but that it can be restored quickly by admitting fresh air to the jar. A burning candle in a jar behaves in the same way: when it starts to go out, fresh air will quickly restore the flame. If the small animal is allowed to die in the jar, the air remaining will not support a candle flame. A small animal placed in a jar in which a candle has burned out will soon die. Priestley noted that if a mouse and a lighted candle are placed in the same jar, when the flame goes out, the mouse dies at the same time. Both Lavoisier and Priestley concluded that the oxidation (burning or union with oxygen) of the candle wax is similar to the oxidation (respiration or using up of oxygen) by the animal. They reasoned that an animal burns up food in his body as the flame of the candle burns up the wax. We now know that, although the overall effect is the same, the mechanism of burning in the candle is different from burning in an animal. In the cells of the body, the reaction occurs in many steps which are catalyzed by various enzymes.

Present knowledge of energy metabolism has grown out of Priestley's reasoning. The output of heat from oxidizing a food can be measured by (1) burning it in oxygen or (2) feeding it to a person. The apparatus for the former is quite simple (Fig. 21.1). A heavy steel bomb is lined with gold (or some other nonreactive metal) and contains a cup in which a weighed sample of food is placed. The bomb is filled with gaseous oxygen under a pressure of 20 atm (atmospheres) or more and is immersed in a known weight of water into which a thermometer dips. The sample is ignited by passing an electric current through the coil of wire in contact with the sample. The wire gets red-hot and starts burning the sample. The heat released by the combustion raises the temperature of the bomb, the water, and the water container. The total amount of heat released by the burning of the food can be calculated from the rise in temperature (correcting for the heat generated by the ignition wire).

The output of energy when a person metabolizes a weighed amount of food can be determined by putting him in a small, thermally insulated room and measuring the heat he produces. The design of such a "whole-body" calorimeter is shown in Fig. 21.2. The calorimeter chamber is 7 ft long, 4 ft wide, and 6.5 ft high. The walls are kept at the same temperature as the room

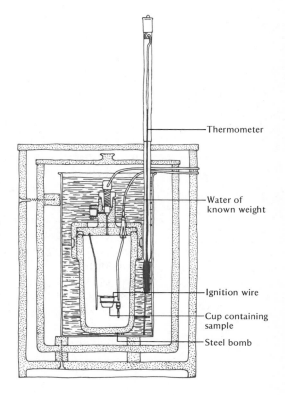

FIGURE 21.1 An oxygen bomb calorimeter for measuring heat produced from burning foods. (Redrawn with permission; from H. C. Sherman, *Chemistry of Food and Nutrition*, The Macmillan Company, New York, 1952.)

Thermometer

Water of known weight

Ignition wire

Cup containing sample

Steel bomb

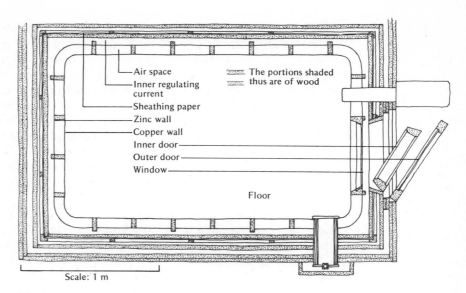

Air space
Inner regulating current
Sheathing paper
Zinc wall
Copper wall
Inner door
Outer door
Window

The portions shaded thus are of wood

Floor

Scale: 1 m

FIGURE 21.2 A whole-body calorimeter. (Redrawn with permission; from H. C. Sherman, *Chemistry of Food and Nutrition*, The Macmillan Company; New York, 1952.)

so that no heat will leak through them. The temperature of the room is kept constant by circulating cooling water through a coil of pipe near the ceiling. The amount of heat generated by the occupant's metabolism is the sum of the heat of vaporization of the water generated by his respiration plus the amount of heat extracted from the chamber by the cooling water. Air at the constant temperature maintained in the calorimeter is circulated through external absorbers to collect the water and carbon dioxide produced by the occupant. Enough oxygen is added to the circulating air to replace that removed as H_2O and CO_2. The kinds and weights of substances excreted in the occupant's urine and feces are also determined. Food and excreta are passed through a special compartment. The human body generates considerable amounts of heat, and the unit used in measuring energy metabolism is the kilocalorie, symbolized by kcal. This is the amount of heat required to raise the temperature of one *kilogram* of water one degree Celsius: 1 kcal = 1000 cal. The term "calorie" used in publications about nutrition is really the kilocalorie.

The number of kilocalories produced by a subject who metabolizes a weighed amount of a carbohydrate agrees closely with the number of kilocalories produced by burning the same weight of the same carbohydrate in a bomb calorimeter. The same is true for fats. Evidently, the end products of the combustion of carbohydrates and fats in the body are the same as those in the bomb calorimeter. The carbon in carbohydrates and fats is metabolized to CO_2 in the body and burned to CO_2 in the bomb. The hydrogen in these foods is metabolized to H_2O in the body and burned to H_2O in the bomb.

When the number of kilocalories produced by the metabolism of a weighed amount of a protein is compared to the number produced by burning an equal weight of it in a bomb calorimeter, there is a significant difference. This arises from the fact that the nitrogen in proteins is liberated as the uncombined element N_2 in the bomb calorimeter but is metabolized in the body into urea,

$$NH_2CONH_2 \qquad \overset{H}{\underset{H}{\diagdown}} N - \overset{\overset{\textstyle O}{\|}}{C} - N \overset{H}{\underset{H}{\diagup}}$$

and some more complicated nitrogen compounds. These end products of oxidation in the cell can be burned to produce more heat. This accounts for the fact that the body produces about 1.3 kcal less heat per gram of protein in food than is produced per gram of protein in the bomb calorimeter. On the average, we metabolize 98 percent of the carbohydrates, 95 percent of the fats, and 92 percent of the proteins we eat. We excrete the remainder in urine and feces.

21.3 Measuring Energy Output by Indirect Calorimetry

The amount of heat produced by the burning or metabolism of a food is directly related to the amount of oxygen consumed. For sucrose, the relation is given by the equation

$$C_{12}H_{22}O_{11}(s) + 12O_2(g) \longrightarrow 12CO_2(g) + 11H_2O(l) + 1350 \text{ kcal}$$

$$\text{1 mole} \quad + \text{12 moles} \longrightarrow \text{12 moles} + \text{11 moles}$$

$$12(22.4 \text{ liters at STP}) = 269 \text{ liters}$$

The consumption of 1 liter of oxygen measured at STP for the metabolism of sucrose leads to the production of $1350/269$ or 5.04 kcal.

The equation for the oxidation of glucose is

$$C_6H_{12}O_6(s) + 6O_2(g) \longrightarrow 6CO_2(g) + 6H_2O(l) + 673 \text{ kcal}$$

$$\text{1 mole} \quad + \text{6 moles} \longrightarrow \text{6 moles} + \text{6 moles}$$

$$6 (22.4 \text{ liters at STP}) = 134 \text{ liters}$$

The consumption of 1 liter of oxygen measured at STP for the metabolism of glucose leads to the production of $673/134$ or 5.01 kcal.

These two values for the number of calories of heat associated with the using up of 1 liter of oxygen in the human body are approximately—though not exactly—equal. Studies of the amounts of heat produced when 1 liter of oxygen is used for the combustion of other carbohydrates yield values of about 5 kcal/liter measured at STP.

Carbohydrates have the general formula $C_x(H_2O)_y$. Each molecule contains enough oxygen (Fig. 20.9) to oxidize all the hydrogen to H_2O, and so the oxygen used up in combustion or metabolism is needed only to oxidize the C present to CO_2. The fatty acids (Table 20.3) in fats contain much more hydrogen than can be converted into water by the oxygen present in the molecule. Therefore, part of the oxygen used to burn fat goes to form water. The fats commonly present in our diet produce about 4.7 kcal of heat per liter of oxygen (measured at STP) that is consumed. The production of heat from the proteins common in our diet is about 4.8 kcal/liter of oxygen at STP. The amount of heat produced from the consumption of oxygen to metabolize a food composed of the usual mixture of carbohydrates, fats, and proteins is approximately 4.8 kcal/liter of oxygen at STP.

FIGURE 21.3 Apparatus for determining heat output of a man by measuring the volume of oxygen consumed by his respiration

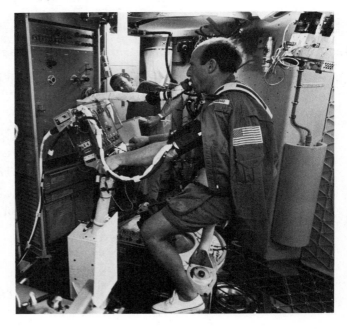

FIGURE 21.4 Determining energy production while doing heavy physical work

The energy produced by a person may be determined by providing a known volume of air to the subject, collecting the air he exhales, absorbing the CO_2 and H_2O from it, and measuring the volume of oxygen consumed. Apparatus for performing such a test with an astronaut pedaling an exercise machine is shown in Fig. 21.3. Air is delivered to the man from one tank, and his exhalations are returned to another.

If we wish to determine the energy production of a person doing various kinds of work which involves moving about, we may equip him with a mask (Fig. 21.4) that takes in air from the atmosphere but forces his exhalations into a collecting bag. Analysis of the air in the bag then shows how much oxygen was consumed during the test period and thus the subject's output of energy.

21.4 Man's Energy Output

basic metabolic
rate (BMR)

When an average-sized (154-lb) male adult lies quietly in bed without food for 24 hr, his output of heat is about 1680 kcal. This is called his *basic metabolism* or *basic metabolic rate (BMR)*. This heat is produced by the chemical reactions associated with breathing, heart action, maintenance of body temperature and muscular tension, and excretion. If the man is given food but remains at rest, his 24-hr output rises to 1840 kcal because the processes of digesting and absorbing food produce heat. With rest in bed for 8 hr, sitting in a chair for 16 hr, and eating a normal amount of food, he produces 2168 kcal/day.

One's basic metabolism depends upon his weight, age, and sex. A normal person usually has a BMR equal to 24 times his body weight in kilograms. Children have a larger BMR than adults, and the BMR for men is about 5 percent greater than that for women. The BMR is closely related to the activity of the thyroid gland. Overactivity of this gland may double the BMR, and underactivity may cut it in half.

The performance of physical exercise increases the output of heat (Table 21.1).

The amount of heat a person puts out in a day can be estimated by considering the number of hours he sleeps, sits, stands, walks, and does muscular work. If he eats amounts of food that would oxidize to produce more than this amount of heat, the unneeded food is stored in his body as fat and glycogen (a carbohydrate), and he gains weight. If he eats less food than that required to produce this amount of heat, some of his body tissues will be oxidized to produce the necessary heat, and he will lose weight. When we stop growing and reach adulthood, we should balance our intake of food with our output of heat so that we keep our weight nearly constant. The average number of kilocalories of heat produced by an individual (and, therefore, the number of kilocalories of heat that must be liberated by the combustion of the food he eats) is given in Table 21.2.

TABLE 21.1 *Energy Production by an Average-sized (154-lb) Man under Different Conditions of Muscular Activity*

Activity	kcal/hr
Sitting at rest	15
Standing relaxed	20
House painting or carpentry	150–200
Walking at 3 mi/hr	270
Sawing wood	400–600
Cycling at 10 mi/hr	450
Playing tennis	400–500
Swimming, breaststroke or backstroke	300–650
Swimming, crawlstroke	700–900
Running	800–1000

TABLE 21.2 *Daily Dietary Allowances in Kilocalories Recommended by the Food and Nutrition Board of the U.S. National Research Council*

	Sedentary	*Physically active*	*With heavy work*
154-lb man	2400	3000	4500
123-lb woman	2000	2400	3000

"Physically active" in Table 21.2 means "using the larger muscles." A typist and a worker sitting at an assembly line for radios do physical work, but they do not use their larger muscles and should be classed as "sedentary." Executives and office supervisors who work under nervous tension get tired, but their work does not involve "physical activity" in the biological sense; they, too, are sedentary. The dietary allowances suggested in Table 21.2 should not be applied too rigidly; modifications of 15 to 20 percent up or down to fit particular circumstances should be considered. The figures given are group averages rather than individual prescriptions. If an adult performs his usual tasks and eats his usual diet and neither gains nor loses weight, his caloric intake is balanced with his needs.

21.5 Heat Furnished by the Metabolism of Foods

The amount of heat produced by the combustion of various kinds of foods has been studied extensively. While there are slight differences in the heats of combustion of different sugars and starches, the *average* number of kilocalories released from eating *carbohydrates* (Sec. 20.11) as they occur in our diets is *4 kcal/g*. The *average* amount of heat released from *fats* (Sec. 20.14) is *9 kcal/g*, and from *proteins* (Sec. 20.17) it is *4 kcal/g*.

The calorific value of foods which are mixtures of nutrients can be estimated reasonably well from an analysis giving the percentage of carbohydrates, fats, and proteins present. For example, the average composition of whole milk is 4.9 percent carbohydrate, 3.9 percent fat, and 3.5 percent protein; the rest is water. One hundred grams of milk will furnish

$$
\begin{array}{lll}
4.9 \text{ g carbohydrate} & \times\ 4 \text{ kcal/g or} & 19.6 \text{ kcal} \\
3.9 \text{ g fat} & \times\ 9 \text{ kcal/g or} & 35.1 \text{ kcal} \\
3.5 \text{ g protein} & \times\ 4 \text{ kcal/g or} & \underline{14.0 \text{ kcal}} \\
\text{Total} & & 68.7 \text{ kcal}
\end{array}
$$

Eggs contain about 0.7 percent carbohydrate, 11.5 percent fat, 12.8 percent protein. One hundred grams of egg will furnish approximately 158 kcal.

The percentages of protein, fat, and carbohydrate in some common foods and their total calorific values are given in Table 21.3.

TABLE 21.3 *Composition and Calorific Values of Some Typical Foods*

Food	Percent of protein	Percent of fat	Percent of carbohydrate	kcal/100 g
Almonds, dried	18.6	54.1	19.6	597
Apples, raw	0.3	0.4	14.9	58
Apple pie	2.1	9.5	39.5	246
Bacon	9.1	65.0	1.1	630
Bananas	1.2	0.2	23.0	88
Beans, dried	21.4	1.6	61.6	338
Beef, lean	19.2	12.5	0.0	195
Bread, white	8.5	2.0	52.3	261
whole wheat	9.5	3.5	48.0	262
Butter	0.6	81.0	0.4	733
Cheese, cheddar	25.0	32.2	2.1	398
Corn, whole	10.0	4.3	73.4	372
Cornflakes	8.1	0.4	83.0	385
Halibut	18.6	5.2	0.0	126
Ice cream	4.0	12.5	20.6	207
Peanut butter	26.1	47.8	21.0	576
Peas, green	6.7	0.4	17.7	98
Pork, lean cuts	14.5	32.7	0.0	357
Potatoes, raw	2.0	0.1	19.1	83
Soybeans, dried	34.9	18.1	34.8	331
Sugar	0.0	0.0	99.5	385
Watermelon	0.5	0.2	6.9	28
Yeast, brewer's, dried	36.9	1.6	37.4	273

21.6 Amino Acids Essential to Man and Their Sources in Protein

Of the 20 different amino acids (Sec. 20.16) isolated from the tissues of the human body, there are 10 (called the essential 10) that must be included in the diet. Apparently our bodies cannot synthesize the essential 10 from others we take in, but from the essential 10 they can synthesize the other 10 within our tissues. The essential 10 were identified by feeding rats on diets containing no protein but with measured amounts of different pure amino acids as dietary supplements. Normal rates of growth and reproduction were obtained only when the essential 10 were supplied. They are arginine, histidine, isoleucine, leucine, lysine, methionine, phenylalanine, threonine, tryptophane, and valine. The essential 10 are present in greatest abundance in foods containing "animal" protein (milk, eggs, meat, fish), but considerable amounts are found in various species of peas and beans, especially soybeans.

Some of the enzymes (catalysts for biochemical reactions) required for the normal operation of the body are derivatives of amino acids. Some of these are: glutathione, which plays an important role in some oxidation reactions taking place in the cell; thyroxine, a substance secreted by the thyroid gland, which regulates the basal metabolism of the body; adrenaline, which gives a quick rise in basal metabolic rate when the body needs to pour out energy to meet an emergency; and insulin, which is required for the metabolism of sugar.

When our diet includes a sufficient amount of food to keep body weight about constant, the amount of protein required is only about 65 g/day* for an average (154-lb or 70-kg) man and 55 g/day for an average (128-lb or 58-kg) woman. This latter figure should be increased to 65 g/day for a pregnant woman and to 75 g/day for a nursing mother. Extensive studies of the amount of protein required to keep the body in good condition indicate that increasing the muscular work done by a person does not increase his need for protein if he is given extra rations of carbohydrates and fats to provide the kilocalories of energy necessary for the extra work.

21.7 Vitamins

In the eighteenth century, when sailing vessels made journeys lasting several months, sailors lived on very restricted diets because only the simplest foods could be preserved for such a long time. European sailors were plagued with scurvy and Japanese sailors with beriberi. When they returned to shore and ate their usual diets, they quickly recovered. Although a doctor in the British navy discovered in 1747 that scurvy could be prevented by adding citrus fruit to the sailors' diet, it took 50 years for the navy to accept and make proper use of this discovery. When lemon and lime juices were added to the sailors' rations, scurvy disappeared. Soon thereafter, British sailors became known the world over as "limeys."

In 1883 the director of the naval medical bureau of Japan discovered that Japanese sailors would not contract beriberi if their usual diet of polished rice was supplemented with generous amounts of meat, fish, and vegetables. It took the Japanese 25 years to accept this discovery and to eliminate beriberi by providing adequate diets for their sailors.

*Recommended Dietary Allowances, Publication 1694, National Academy of Sciences, Washington, D.C. (1968).

In 1912 Dr. Casimir Funk, a Polish scientist working in London, published a book entitled *Die Vitamine* in which he suggested that the diseases of scurvy, beriberi, rickets, pellagra, and sprue were all due to dietary deficiencies. His book made a considerable impression on the scientific world, and the term "vitamin" became widely adopted as the name of a substance present in only a limited number of foods but very necessary for complete nutrition.

Chemists soon began to search for pure substances that could be isolated from foods and used to prevent diseases of dietary deficiency. In 1926 a substance called thiamine, isolated from the hulls of rice grains, was found to alleviate beriberi when included in the diet. Since it takes 100 *kg* (kilograms) of rice hulls to yield 100 *mg* (milligrams) of thiamine, it would appear that thiamine is present in rice hulls to the extent of only *1 ppm* (one part per million). At the time of the original research on thiamine, there were no chemical tests sufficiently delicate to detect such small amounts, and so the weight of thiamine required to cure beriberi was determined by feeding the pure compound to a subject suffering from the disease. Small animals like rats, guinea pigs, and chickens often suffer from the same dietary deficiency diseases as man. Hence they are widely used to determine the minimum dosage of pure vitamin needed to prevent a given disease. Similar *bioassays* (feeding experiments) are run to determine the vitamin content of various foods. In recent years, vitamins have been prepared by direct chemical synthesis instead of by extraction from foodstuffs.

bioassay

The vitamins about which we need to be concerned in our daily diets are vitamin A (which promotes healthy growth in young animals and children and is necessary for good vision in adults), thiamine, ascorbic acid (also called vitamin C, which prevents scurvy), and riboflavin (which is essential to the healthy growth of children, lengthens the "prime of life," slows the development of senility in older persons, and decreases the incidence of infectious diseases at all ages). Vitamin D is needed for proper growth in young children but is present in milk in sufficient amounts to meet their needs. If we include some animal proteins, milk, citrus fruits, greens, and yellow vegetables and fruits (like carrots, squash, and cantaloupes) in our diets, the few milligrams of the various vitamins we need will be supplied.

21.8 Mineral Elements in Food and Nutrition

The chemical elements found in significant quantities in the human organism are given in Table 21.4.

Nutritionists refer to the elements other than oxygen, carbon, hydrogen, nitrogen, and sulfur (which are present in the organic compounds found in food) as the "mineral elements" or "inorganic elements." These elements

TABLE 21.4 *Approximate Composition by Weight of the Adult Human Body*

Element	Percent	Element	Percent	Element	Percent
Oxygen	65	Potassium	0.35	Copper	0.00015
Carbon	18	Sulfur	0.25	Iodine	0.00004
Hydrogen	10	Sodium	0.15	Other elements of	
Nitrogen	3	Chlorine	0.15	doubtful function	
Calcium	2	Magnesium	0.05		
Phosphorus	1	Iron	0.004		

are required as constituents of the bones and teeth, as essential parts of the structures of certain complex organic compounds present in soft tissues (muscles, blood cells, and so on), and as soluble salts in tissue fluids such as blood, digestive juices, and other secretions. Under average conditions of diet, activity, and health a man excretes 20 to 30 g of mineral salts daily. These consist largely of chlorides, sulfates, and phosphates of sodium, potassium, magnesium, and calcium.

Calcium and phosphorus in the form of the slightly soluble solid calcium phosphate are the main mineral (inorganic) constituents of bone. The slightly soluble solids calcium fluoride and calcium carbonate are also present. Practically all sodium compounds are soluble; they are present in the body largely as salts in solution in blood plasma and other body fluids. Most potassium compounds are also soluble; they are found principally as salts in blood corpuscles, the protoplasm of muscles and other organs, and in milk. Of the total magnesium in the body, 71 percent is present in slightly soluble solid compounds in the bones; the remainder is present in muscles and blood cells. Phosphorus in soluble phosphates is heavily involved in enzymes regulating the liberation of energy from the oxidation of fats, carbohydrates, and proteins. Iron is an essential atom in the hemoglobin molecule, and copper catalyzes the synthesis of this important component of red blood cells. Iodine is required by the body for the synthesis of thyroxine, which contains 65 percent iodine.

Few people in the United States suffer from lack of necessary minerals. A well-balanced diet which will supply the needed vitamins will also contain the essential minerals. The addition of sodium iodide to table salt and the fluoridation of municipal water supplies are the only measures necessary for supplementing the diet with adequate amounts of minerals.

For good nutrition what we need is a balanced diet, one which does not depend too much on any one kind of food. The danger of many so-called "crash" reducing diets lies in their undue limitation on the variety of foods necessary for good nutrition.

21.9 Hunger in the World

Increasing birth rates, decreasing infant mortality, and a lengthening of man's average lifespan through better medical care have caused the population explosion of recent years. In the April 12, 1968, issue of *Medical World News* there appeared this statement:

> By year's end, estimates the Population Reference Bureau, world population will pass the 3.5 billion mark. And if current trends continue, seven billion persons are projected for the year 2000. The bureau points out there are no indications that food production can keep pace with such population growth, and more ominously, most of the increases occur in those countries least able to provide even for their present population.

On the other hand, since 1952 the *worldwide* increase in the rate of food production per capita has been 3 percent a year, while the population has increased an average of 1.7 percent a year. This slight gain is part of the general trend that the increase in total food production has exceeded that in total population since 1850. The fight against hunger today must focus on the underdeveloped countries with high population densities.

Yet hunger still rages in parts of the world, though no one knows precisely how many people are hungry today. Demographers estimate that from 40 to 60 percent of the world's people live on diets varying from substandard to insufficient to maintain life. Though total food production increased 40 percent in the period 1952 to 1966, the population in many underdeveloped countries increased faster than this, and the production per capita in these countries declined.

In the United States, where there is about 1 acre of cultivated land per person, we produce food equivalent to about 11,000 kcal of edible crops per day. We eat about 2000 kcal of this, export another 2000, and feed 7000 to the livestock and poultry from which we obtain the remaining 1000 kcal in our average daily diets. If the productivity of the world's 3.3 billion acres now under cultivation could be raised to that per acre in this country, the 7 billion people anticipated by the year 2000 could be fed, but the diets of many would include only about half the animal protein we are used to in the United States.

In the countries where the population is most dense, there is little unused arable land, and so increased production of food must come from increasing the yield per acre. These areas need an agricultural revolution similar to the one in Mexico, where the application of modern agricultural technology trebled the production of food between the mid-forties and the early sixties. In 1968 the Philippines became self-sufficient in rice production for the first time in this century as the direct result of planned agricultural modernization. In the same year modern methods of production raised India's food harvest 12 percent above its best previous year. Modern agricultural methods can produce 3 times the yield per acre of traditional practices. Furthermore, plants can be made to mature fast enough so that in some places two and even three crops can be grown in 1 year. For the first time in history, it is *technologically possible* to produce enough food for all mankind. There are two ways of bringing about the necessary agricultural revolutions to achieve this goal: (1) by improving the yields of traditional foods, and (2) by developing new foods.

21.10 Improving Yields of Traditional Foods

In the United States we are in the midst of an agricultural revolution. There has been a 500 percent increase in food production per farmer in the 100 years since colleges of agriculture were established in state universities. A farmer in the Orient raises enough food for 2 people; one in the United States raises enough for 20. In 1920, the United States farmer cultivated 2 acres per person; in 1962, 1. Five factors have dominated this revolution: (1) mechanization of cultural and management practices for crops and livestock (70 million acres of United States cropland formerly used to feed horses was released for the production of human food when tractors replaced horses); (2) use of chemicals as fertilizers, micronutrients, and controllers of insects, fungi, nematodes, and weeds, as antibiotic protective agents, and as regulators of plant and animal growth and health; (3) genetic improvements through (*a*) hybridization of plants to increase both total production and the protein content in grains, and (*b*) selective breeding of animals to increase milk, egg, and meat production; (4) extension of irrigation to arid and semiarid lands (yields on such lands in the western United States have been increased between 50 and 100 percent); (5) application of corporate management systems to large acreages of farmland.

For 140 years (1800–1940) yields of corn (the number-one crop in the United States) remained essentially constant at 22 to 26 bushels/acre. Since 1940, yields of corn have more than tripled; yields of potatoes have likewise tripled; and those of wheat and soybeans have doubled. Each year we see new production records. Milk production per cow has increased each year for the last 20 years, and the curve of production is still rising steeply. New records are being achieved in production efficiency for eggs, broilers, turkeys, beef, and pork.

Increased production of corn is indicative of future developments for other grains. The average yield in the United States is 80 bushels/acre, but the record is 304, and yields of 400 bushels/acre or more are predicted. Of all the major food crops, corn is the most efficient in using solar energy in photosynthesis. Approached only by sugarcane and some tropical grasses, corn consistently produces more total digestible nutrients than any other crop grown in the Northern states. Still further increases in productivity can be achieved by earlier planting, more plants per unit area, precision planting in equidistant array, uniformity in height of plants, starter fertilizer, irrigation, fertilization during growth, and use of weedkillers. In one experiment, the *complete* elimination of weeds resulted in a 700 percent increase in the yield of corn! The effectiveness of one man working in a factory producing weedkillers is equivalent to that of 800 men hoeing weeds out by hand!

To make the most efficient use of sunlight, new shapes of corn plants are being bred that will hold their uppermost leaves vertically and their lower leaves horizontally. Tassels will be small and the plants cone-shaped. Density of planting will be increased from the 25,000 plants per acre now common to 35,000. Such close planting will minimize the impact of rainfall on soil structure, reduce weed growth, and increase efficiency in the use of water.

Wheat is the most important human food crop of the world, and of first importance in the United States (corn is grown as food for both animals and humans). It is grown over a larger area of the world and in greater quantities than any other crop. The revolution in the culture of corn that got under way 25 years ago is just now beginning with wheat. New varieties of this grain are resistant to mildews, smuts, root rots, virus diseases, Hessian fly, cereal leaf beetle, sawfly, and winter injury; they will soon be available for planting. New hybrids are being created which not only produce more food grain per acre but also bear a 20 to 25 percent increase in total protein, with a distribution of amino acids that makes them more desirable for human food, approaching the biological quality of skim milk.

Rice is second only to wheat in amount consumed by humans. A billion people in the Orient depend on rice as their principal food. Research at the International Rice Research Institute in the Philippines has led to increases in rice production in Oriental lands as dramatic as those in corn and wheat in our own country. Millions of acres of croplands in Asia are now producing the new "miracle" rices which are filling millions of hungry stomachs in many lands.

21.11 New Research on Increasing Food Production

The American farmer's main enemies—weeds, insects, and plant diseases—reduce the annual crop by about 20 percent. Stated in another way, the equivalent of 75 million acres is being used to feed insects, weeds, and organisms that cause plant diseases. Estimates indicate that 600 species of weeds cause an annual loss of $2.5 billion in the United States, and that another $2.5 billion is spent on weed control. Insects cause a loss of $4 billion in crops and livestock.

The recent discovery of the chemical structure of the juvenile hormone which prevents an insect from developing beyond the pupal stage promises to open a new chapter in controlling the depredations of insects on our food supply. The hormone affects all insects and is effective in dosages of only 1 g/acre! Care will have to be used in applying the hormone lest we kill our insect friends as well as our enemies.

Millions of acres of formerly unproductive semiarid sandy soils may be brought into production by placing strips of asphalt that are $\frac{1}{8}$ in. thick about 2 ft below the surface. These strips, expected to last 15 or 20 years, prevent seepage of groundwater into layers of soil below the depth of plant roots. Yields of high-value crops like tomatoes, potatoes, cabbages, beans, and cucumbers have been increased 70 percent in some Michigan soils by the use of such water barriers. With this treatment crops on sandy irrigated soils in Arizona have required substantially less water. An elevenfold increase in rice production with one-eighth the usual amount of irrigation was achieved in asphalt-lined paddies in Taiwan. And sugarcane crops in that country increased 60 to 100 percent and used less irrigation water.

High-value crops grown in greenhouses with atmospheres enriched in CO_2 to stimulate photosynthesis (Sec. 15.2) have responded with increased yields of 25 to 100 percent. Corn, soybeans, sugar beets, barley, and rice react dramatically to atmospheric levels of CO_2 higher than normal. Some research is being done in open fields to test the response of crops to fertilization by CO_2.

The development of new glasses and transparent or translucent plastics has extended production of foods in greenhouses. Japan leads the world in this area with increases of more than 10 percent a year. Annual yields of tomatoes in greenhouses have exceeded 120 tons/acre. Growing plants in nutrient solutions instead of in soil, heating growth media, and electric lighting of greenhouses are receiving renewed attention. The use of waste heat from nuclear and other types of power plants for stimulating food production is under intensive study.

A "food factory" has been built in the desert near Puerto Penasco on the Gulf of California (Fig. 21.5). Inflated double-unit greenhouses measuring 50 by 100 ft are constructed of very thin plastic on a frame of wooden strips. The atmosphere in the greenhouses has almost 100 percent relative humidity

FIGURE 21.5 Agricultural Research Station at Puerto Penasco, Mexico

and 1000 ppm CO_2. The water requirement in these conditions is but 1 to 5 percent of that in open fields. Seawater is used for cooling the greenhouses in summer and heating them in winter. Electricity and CO_2 are generated by diesel engines, and plans are under way for using atomic energy to generate electricity and to desalinate seawater with CO_2 as a by-product. A similar food factory has been constructed in Abu Dhabi, a small sheikdom south of Kuwait on the Persian Gulf. There are 18,500 mi of uninhabited, unproductive desert shoreline on the earth for the future development of such factories.

Certain chemicals improve the growth characteristics of plants. Wheat treated with 2-chloroethyltrimethyl ammonium chloride grows with shorter, thicker stems that are less susceptible to flattening by wind before harvest, and its heads are more fully filled with seed. One or 2 lb of the chemical per acre is effective, and all varieties of wheat respond. Other chemicals also alter the growth habits of food plants and increase their effectiveness in photosynthesis. Fruit trees treated with plant growth regulators begin to bear in 4 years after planting instead of the usual 7. Still other substances enhance the desirable qualities of fruits and improve production. Gibberellin is such a substance, especially helpful in maximizing the quality of white grapes and citrus fruits.

Some chemicals that have been developed as weedkillers also stimulate the production of protein. Rye treated with subtoxic doses of these chemicals produces 20 to 80 percent more protein than untreated plants. Pea plants so treated produce peas with a 40 percent increase in protein and with the same kinds of amino acids as before. Protein content can also be increased this way in rice foliage, alfalfa hay, and oats, thus increasing the value of these crops as animal feed.

Urea,

$$H_2N-\underset{\underset{O}{\|}}{C}-NH_2$$

is a nitrogen-rich compound that is produced cheaply by the reaction of ammonia with carbon dioxide:

$$2NH_3 + CO_2 \longrightarrow NH_2CONH_2 + H_2O$$

Studies of animal nutrition show that a dietary supplement of urea is an excellent source of nitrogen for the production of protein. More than 125,000 tons of urea are used annually in the United States for animal feeds. The 47 percent efficiency with which a dairy cow will convert protein in her feed to protein suitable for human food is increased to 70 percent when urea is added to her diet. The efficiency by which feed is converted to beef in growing steers is increased 2.5 times by adding urea to the feed.

Efficiency in the production of eggs and chickens has been improved more in recent years than that of any other food. As a result, the price of eggs fell steadily during the period 1947–1972, while the prices of most other foods rose sharply. Consequently, increasing numbers of poor people have been able to afford this protein-rich food. The ratio of food consumed by the hen to the number of eggs she produces has been reduced during this same period from 7 lb of feed per dozen eggs to 4 lb per dozen. Efficiency in the production of chicken meat has likewise increased. Twenty-five years ago it took 4 lb of feed to produce 1 lb of chicken. The figure today is about half that. Chicken was formerly a luxury item in the diet. Now it is one of the cheapest of meats. Similar increases in the efficiency of growing turkeys have been achieved and this former once-a-year Thanksgiving dinner treat is widely available all year round to purchasers in the lower-income brackets.

Research in the physiology and nutrition of other animals will in the next few years undoubtedly bring about increased efficiency in the production of beef and pork which will be comparable to the results already achieved with poultry. It is hoped that these developments will enable more and more low-income consumers to increase their intake of vital proteins. Since all animals have a "comfort zone" of temperature and humidity within which they flourish, raising animals in controlled climates will undoubtedly increase the efficiency of producing them for food.

21.12   The Development of New Foods

From recent increases in the production of food grains in countries like India, the Philippines, and Pakistan, where starvation has been commonplace for hundreds of years, it seems likely that soon no man need suffer from an insufficient number of dietary calories per day. The most pressing problem now is the production of enough food protein to provide the essential amino acids required for the normal growth of children and the maintenance of good health in adults. Increasing the protein content of conventional foods has already been mentioned, and current research is developing several new foodstuffs that may be valuable as protein supplements.

Increasing the amount of seafood in man's diet seems to be the likeliest means to an immediate increase in the intake of essential amino acids. Officials of the United Nations Food and Agriculture Organization believe that the development of more efficient fisheries could increase the present harvest from 55 million tons/yr to 200 million tons/yr without endangering any species of fish. About half the present catch is consumed directly, and half is ground into fish meal, which is widely used as animal feed. "Fish flour" containing 80 percent protein can be prepared by grinding up whole fish and extracting the fats and water with the solvent isopropyl alcohol. What is left is an almost completely tasteless and odorless fine grayish powder so rich in *animal* protein that 10 g/day could fill the *animal* protein needs of the average human being. Supplementing this with 55 g of vegetable protein would complete the daily requirement of 65 g of protein for the average man. Fish flour is stable, does not require refrigeration, and can be added to many foods without affecting their palatability. It has been incorporated into tasty breads, pastas, cereals, and cookies in amounts from 5 to 25 percent. After extensive deliberation, the U.S. Food and Drug Administration has pronounced fish flour fit for human consumption.

Ocean "farming" is also increasing the world's supply of food protein. Oysters, shrimps, and clams can be grown in brackish coastal water. Japan now leads the world in farming the seas.

All human food comes ultimately from photosynthesis performed in green plants, but our best crop plants use barely 2 percent of the sunlight that falls on them, and most use only 0.5 percent. Furthermore, farm crops cover only part of the ground for only part of the year, and half or more of a crop plant is inedible structural material consisting of cellulose or other indigestible substances. For 20 years food scientists have been fascinated with the possibility of growing algae as a food crop. These one-celled organisms, composed primarily of proteins, carbohydrates, and fats, utilize solar energy much more efficiently than our present crop plants, can be grown the year round, and carry little structural material. Compared with algae, animals are very inefficient producers of protein. A 1000-lb steer makes about 1 lb of useful protein in one day's growth, but 1000 lb of microorganisms can make 4000 lb of protein in a day.

The genus of algae known as *Chlorella* grows rapidly in laboratory cultures and, by changes in the conditions under which it grows, can be made to produce a dried crop containing from 7 to 88 percent protein, 6 to 38 percent carbohydrate, and 1 to 75 percent fat. Under conditions of maximum growth rate the product is about 50 percent protein, 43 percent carbohydrate, and 7 percent fat. *Chlorella* crops are rich in vitamins, seem to contain all the essential amino acids, and produce fats not very different from those in common foods. *Chlorella* flour has been used as a supplement in preparing bread, noodles, soups, and ice cream. In Central Africa a type of blue-green algae growing on highly alkaline shallow ponds, harvested and eaten by local tribes, contains 62 to 68 percent protein and several vitamins. Japanese food scientists have developed a process for cultivating *Chlorella* in sewage, producing an animal feed containing 50 percent protein. This is used as a feed supplement, an important commodity since approximately 10 percent of the monetary value of Japanese imports goes for animal feeds. *Chlorella* supplements in animal diets have increased the rate of human food production markedly.

Other microorganisms are being cultivated to produce food. The yeast *Torula* is grown in the solutions of tree sugars produced by cooking woodpulp for making paper. When the yeast is separated from the liquid culture and dried, it serves as a valuable food supplement for animals. Cellulose waste —including newspaper, computer printout paper, and cellulosic agricultural wastes (straw, bagasse from sugar cane, cornstalks)—can be used as food for a mixed culture of microorganisms called *Cellulomonas* and *Alcaligenes* to produce an animal food containing about 75 percent protein.

The *Candida* strain of yeast can produce food when grown in a culture consisting of alkanes from petroleum with chains of 10 to 18 carbons mixed with a water solution of ammonium salts to furnish nitrogen; phosphorus and potassium are supplied as general fertilizers to stimulate the growth of the organisms; and some trace elements and growth vitamins are also added. Oxygen is bubbled through the culture medium, which is stirred vigorously to mix the oil and water phases intimately. When the organisms are removed from the culture medium and dried, a solid is produced which looks like light-brown sugar and has no pronounced odor or taste. A kilogram of hydro-carbon can produce a kilogram of yeast, and the product contains about 50 percent protein. A similar protein material can be made by the action of microorganisms on methane.

Today half the world's people live on a poorly balanced diet that retards normal growth and prevents buoyant good health. And Gandhi's statement that "If God should appear to an Indian villager, it would be in the shape of a loaf of bread" still holds for untold millions. Yet we now have the scientific and technological ability to end starvation, once and for all. The most formidable barriers to the realization of this utopia are not scientific. They are political, economic, and cultural. And they are very real. Less formidable but still great are the problems of maintaining the quality of the environment as we expand food production.

21.13   The Use of Insecticides

For obvious reasons, dichlorodiphenyltrichloroethane is called DDT. This insecticide has saved more human lives than any other single substance. It has killed billions of the insects, spiders, mites, and ticks that carry malaria, typhus fever, plague, and 27 other diseases that have killed millions of human beings in years past. The World Health Organization (WHO) of the United Nations points out that DDT has an amazing record in eradicating malaria for 550 million people, having saved about 5 million lives and prevented 100 million illnesses in the period between 1944 and 1953, and having served at least 2 billion people without causing the loss of a single life by poisoning from DDT. The WHO concludes that to discontinue the use of DDT immediately would be a disaster to world health. At present there is no substitute in sight for controlling malaria, typhus, and some other insect-borne diseases.

DDT has also killed billions of insects that attack food crops, including codling moths, potato beetles, corn borers, thrips, aphids, and cutworms. Without insecticides we cannot grow foods free from insect damage and insect remains. If we give up insecticides, we shall have to get used to eating wormy and weevilly foods. Substitutes for DDT are being studied, but the 1972 summer issue of the *Massachusetts Institute of Technology Review* states, "The degree of safety and effectiveness that DDT at first apparently gave the world may not be attained before the end of this century. Meanwhile, the people of the U.S. will have to put up with conditions that many will find barely acceptable."

In the first annual report of the U.S. Environmental Protection Agency summarizing seven months of hearings with 125 witnesses concerning the dangers and effects of pesticides on humans, it was stated that DDT holds no threat of cancer or birth defects to man, that the DDT uses in question did not harm fish or wild birds, that the benefits of DDT's continued use outweigh any possible adverse effect, and that "there is a present need for the essential uses of DDT."

DDT kills all kinds of insects; new chemicals are being sought that will kill our insect enemies but not our friends like bees and butterflies. DDT is very persistent; it takes a very long time to disappear from the environment by natural processes of biological degradation. Aldrin, a chemical with a structure quite similar to that of DDT, is a fairly good insecticide that disappears from the environment in 2 months. Malathion, a highly toxic organic compound containing phosphorus, is an insecticide which quickly degrades; 97 percent of it disappears by natural processes in 8 days. Thousands of organic compounds containing phosphorus are known: it seems likely that we can eventually find one which is both specific in killing a particular kind of insect and quickly biodegradable—but it will take a long search.

Methods of killing insects other than by the use of chemical agents are being developed. More than 100 biological parasites and predators have been imported into the United States to control insect infestations. For example, 80 years ago "cottony-cushion scale" was killing California's citrus trees. The *Vedalia* beetle which was imported to kill the scale-causing organism saved the citrus industry from collapse. Bacteria and viruses which attack insects but are harmless to higher animals have also been cultivated as effective controls for some pests.

pheromone

Another method of insect control uses various insect attractors, called *pheromones*. Methylbutanol attracts and kills male Oriental fruit flies and is widely used in control programs. Two hundred sex attractants were discovered in the decade 1960–1970, and 20 of these have been useful in attracting and trapping cotton bollworms, cabbage loopers, armyworms, and other pests. Various hormones that play an important role in the growth and development of insects can be sprayed on plants. The absorption of these substances by insects then interferes with their normal development, and the organisms die before maturing.

Genetics also plays a part in controlling insects. Male insects may be grown in the laboratory and sterilized by gamma-ray irradiation or by being fed certain hormones. When these males are released in the field, they mate with females in the area, but the eggs the females then lay are infertile. The Mexican fruit fly and the screwworm fly are kept in check in the southern part of the United States by this technique. The breeding of genetically imperfect and therefore sterile strains of insects is also being developed. In some cases, lethal genes can be bred into a strain of insects which are then released to spread the gene through the natural population. Physics also contributes a method of insect control. Radio-frequency irradiation of stored grains kills many insects without damaging the grain.

The greatest obstacles to the improvement of insect control are political and economic. It costs about $10 million and 8 to 10 years to develop one of the new types of pesticides. One successful insecticide may emerge from the testing of 10,000 possible compounds. Who will supply the time and money needed? The present economic relations between the government, the farmer, and the chemical industry are not favorable to the introduction of new pest controls. New political institutions must be created to handle the development of the new pesticides which have the potential of saving millions of dollars worth of foods and other agricultural products without polluting the environment.

21.14   Herbicides and Fungicides

Throughout the world the most important foods for man are grains. Wheat, rice, corn, and other cereal grains grow on plants which have narrow leaves and are grasslike in their anatomy and biochemistry. Weeds, on the other hand, are chiefly broad-leaved plants (like dandelions and plantains), and their biochemistry is different. The chemical called 2,4-D is a plant hormone which stimulates root growth in broad-leaved plants but has little effect on those with narrow leaves. When 2,4-D is sprayed on a mixture of growing grasses and weeds, the roots of the weeds grow so fast that these plants die from imbalance in their physiological chemistry. The hormone 2,4-D has been used for years as a weedkiller in lawns. If it is used carefully on food grain crops, it can greatly increase production by cutting down on the competition of weeds for nutrients and water.

The substance called 2,4,5-T is a much more lethal compound which kills practically all higher plants with which it comes in contact. It was widely used in warfare in Viet Nam to kill the vegetation that sheltered enemy fighters and their supply lines and depots. In places, the entire landscape has been denuded of plants by heavy applications of 2,4,5-T. It has also been used to kill desert vegetation in an effort to conserve water in arid regions.

teratogenic

Recent studies of the effects of 2,4-D and 2,4,5-T on mice and rats indicates that these herbicides may be *teratogenic*—that is, they may cause malformations of the embryos of these animals, leading to abnormal young at birth. Though there is no evidence that these compounds would have the same effects on humans, considerable caution is called for in the use of these herbicides. The United States Surgeon General has recommended that 2,4,5-T not be used around the home, for killing vegetation around water supplies, or on food crops. Such limitations have not been suggested for 2,4-D.

The use of compounds containing mercury to kill fungus spores on seed grains is also fraught with danger. Farm animals may eat grains treated with fungicide and absorb enough mercury to kill them or render their tissues unfit for human food. Farmers' children who have eaten meat from animals fed on treated seed grains have been made very ill.

Just as methods of insect control are being developed to substitute for the use of undesirable insecticides, so are other methods of weed control to take the place of dangerous herbicides. Four million acres of land in California which had been taken over by Klamath weed was recovered by breeding and setting free a special kind of beetle that feeds on this weed. Waterways can be kept free of undesirable plants by special breeds of snails and fish that feed on them. Careful attention to the best times for plowing, seeding, fertilizing, cultivating, and harvesting can maximize the natural immunity of food crops to pests.

Summary

Our bodies require food to produce energy, build tissues, and regulate metabolism. The most convenient way for measuring amounts of food required under various conditions is in terms of the kilocalories of energy produced by the body, and thus the number of kilocalories that must be furnished by the metabolism of food. All kinds of foods furnish energy, but proteins are necessary for building body tissue. Because the body cannot synthesize all the different kinds of amino acids needed for building tissue, some animal protein must be included in the diet to furnish them. Vitamins must also be supplied in the diet, since they cannot be synthesized by the body. Mineral elements are required in small amounts for the synthesis of some body catalysts.

For the first time in the history of mankind, we can eliminate starvation from the world if we can achieve the political and social action required to apply modern agricultural science and technology worldwide. The breeding of new plants and the use of fertilizers, insecticides, fungicides, weedkillers, and plant hormones have dramatically increased world food production, but at the cost of considerable environmental pollution. Not immediately, but within a few years, modern science and technology can produce new, less polluting techniques for fertilizing crops and protecting them from insects, weeds, and plant diseases. In the meantime we must use the available chemicals with great care to maximize their value and minimize their damage. Man will go hungry if we eliminate chemical methods for increasing food supplies.

In addition to improving production of conventional foods by conventional methods, we should learn how to farm the seas, how to grow more foods by the activities of microorganisms, and how to control the environment in greenhouses to speed the growth of especially valuable crops.

New Terms and Concepts

BASIC METABOLIC RATE (BMR): The rate at which energy is used by an organism at complete rest.

BIOASSAY: Evaluation of the effect of a substance on the health of an organism by administering it in measured amounts to the organism.

PHEROMONE: A substance secreted by an organism which evokes a specific response in another member of the same species.

TERATOGENIC: Producing abnormal offspring.

VITAMIN: A complex organic substance occurring naturally in plant and animal tissues and essential in small amounts for the control of metabolic processes.

Testing Yourself

21.1   For what purposes does an animal use food?

21.2   Why do nutritionists find it so convenient to discuss diets in terms of "calories" (really kilocalories)? Do "calories" tell the whole story?

21.3   What is meant by metabolic rate? How is it measured? How does it vary from person to person and with occupation?

21.4   What is the best way for you to judge whether your diet is adequate for you as an adult?

21.5   Why must we have some protein in our diet? Some animal protein?

21.6   What are vitamins and why must we include them in our diet? Is it necessary to take vitamins as pills?

21.7 What is the "green revolution"? What are its benefits and its hazards?

21.8 How can we reduce pollution by insecticides? Fertilizers? Weedkillers?

21.9 When most of the cropland in a given area is planted to one crop, we refer to the practice as "monoculture." What are some of the advantages and disadvantages of monoculture? Why is it hard to get farmers to practice diversified agriculture?

21.10 Suggest some ways in which the diets of people in underdeveloped countries can be made more adequate.

A lunar rock is shown in the photomicrograph. One theory about the moon's future is outlined: the moon's orbit gradually enlarges; then the faster rotation of the earth (due to the sun's gravitational pull) pulls the moon close enough to shatter into many fragments that would orbit the earth, forming a "ring."

22. The Evolution of the Environment

In our study of chemistry and man's environment we have concentrated on the short-range changes growing from the impact of his demands upon earth's natural resources. As we become aware of how we have ravaged our earthly home in the past 100 years, we are rightly worried about whether it will be fit for our children and their children's children in the next 100 years. The short-range outlook is gloomy indeed, if the groups of thinkers who made the predictions noted in Figs. 1.1, 1.2, and 1.3 are right.

Perhaps we can achieve some confidence in our ultimate ability to cope with our environment by taking a look at earth's long past and how man emerged in it. This look may even make us optimistic about our future—may give us hope that the human race will not only survive but will rise to greater glories on planet earth. In this final chapter we shall view the sweep of time from the beginning of the universe to the present day.

In the words of Warren Weaver, consultant on scientific affairs to the Alfred P. Sloan Foundation,

> We are just in the process of gaining a scientific picture of the total ascent of life. By far more vast and significant than the Darwinian view, this modern evolutionary doctrine begins with the elementary particles of the nuclear physicist and moves through the whole range of the atomic and molecular world up to the nucleic acids which, in their capacity to reproduce pattern and to pass on coded information, seem capable of forming

the primitive basis for a living organism. From this point it is conceivable to move on to the gene, the chromosome, the cell, and ultimately to human life. Whether or not man is the present climax of this ascent is itself now under question, for we have radar-listening devices, directed at inconceivably distant parts of the cosmos, seeking to determine whether there are other and possibly more advanced beings there, trying to communicate with earth-bound man.

22.1 The Stuff of Which Elements Were Made

On the basis of evidence now at hand, many cosmologists (scientists who study the creation and evolution of the universe) believe that about 15 billion years ago the evolution of the universe began with a big bang—the tremendous explosion of what some have called the great "ur-atom," or "original" atom, an extremely dense ball of neutrons which were the progenitors of all the atoms in the universe today. At the moment of explosion the temperature of the ur-atom was a few billion degrees, but as the neutrons flew off into empty space, they slowed down (which is another way of saying that their temperature fell), and each neutron disintegrated into a proton-electron pair—a hydrogen atom, as we know it today. Studies of the stars indicate that they have served as "atomic furnaces" in which hydrogen has been transformed into the trillions of tons of other elements found in all the galaxies of the universe. Even today, 93 percent of all the atoms in the universe are hydrogen! Some cosmologists believe that hydrogen is being produced by continuous creation, but this difference of opinion does not change our theories of how hydrogen is transformed into other elements.

22.2 The Relative Abundances of the Elements

How did the hundred or so elements we now find in nature evolve from the primordial hydrogen? Believing that knowledge of the relative abundances of the elements now in the universe may be the key to understanding the processes by which they were synthesized, cosmologists gather data on abundances from every available source. Geologists and geochemists determine the relative amounts of the elements in the solid earth, the liquid seas, and the gaseous atmosphere. Astrochemists analyze the meteors that fall onto the earth from outer space and the rocks brought back from the moon. Astrophysicists measure the colors and intensities of light coming to us from planets, stars, galaxies, and nebulas. Radioastronomers study the radio waves that emerge from interstellar matter in the great black holes in the luminous night sky above us. Nuclear physicists determine the kinds and intensities of cosmic rays that bombard us from outer space.

By combining these quantitative analyses of various kinds of matter with measurements of the masses of the various parts of the universe, cosmologists calculate the abundances of each kind of element in the universe as a whole. For purposes of comparison, it is convenient to think about *relative* abundances rather than absolute numbers of atoms of a given element in the universe. Cosmologists use 1 million (10^6) atoms of silicon, Si, as the basis for comparison. A diagram of the relative abundances of the elements is shown in Fig. 22.1. The abundance of Si is 10^6 on the vertical scale. Carbon is 10

FIGURE 22.1 Relative abundances of the elements in the universe. (Figure redrawn from L. H. Ahrens, *Distribution of the Elements in Our Planet*, McGraw-Hill Book Company, New York, 1965.)

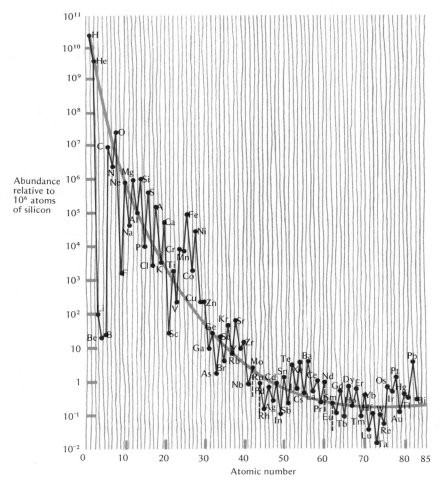

times more abundant than Si, so its abundance is at 10^7 on the scale. The abundance of ruthenium, Ru, is about one-millionth the abundance of Si, so its abundance is at 1 on the scale, etc. The curved line shows the general trend of change of abundance with atomic number.

From Fig. 22.1 we note the following relationships.

1 Hydrogen and helium are by far the most abundant elements.
2 Elements with even atomic numbers usually lie above the curved line, and those with odd atomic numbers usually lie below it.
3 In general, as the atomic numbers increase from 6 to 44 (carbon to ruthenium), the abundances decrease rapidly from 10^7 to 1.
4 The abundances of the elements with atomic numbers greater than 44 decrease, but slowly with increasing atomic number.
5 The abundances of Li, Be, and B are astonishingly low (lie far below the curved line).
6 The abundances of the "iron group" of elements (chromium, manganese, iron, cobalt, nickel, copper, and zinc), with atomic numbers 24 to 30, are much greater (lie above the curved line) than their immediate neighbors.

Table 22.1 summarizes the abundances of elements by groups.

Any theory designed to account for the origin and evolution of the elements must take into account the characteristics of the data presented in Fig. 22.1 and Table 22.1. Scientific theories must always grow out of facts.

22.3 Synthesis of Elements by "Hydrogen Burning"—Hydrogen Fusion

It is hypothesized that the "big bang" initiating the evolution of the universe dispersed the hydrogen far and wide in space, but with the passage of eons of time the gravitational attraction between atoms of hydrogen caused them to agglomerate into huge clouds. As the hydrogen atoms fell toward one another,

TABLE 22.1 *Abundances of Various Elements in the Universe*

Element	Atomic number	Percent of total number of atoms
Hydrogen	1	93
Helium	2	A little less than 7
Carbon to sulfur	6–16	A little less than 0.1
Iron group (10,000 times as abundant as nearest neighbors)	24–30	About 0.001
Heavy elements	44–83	About 0.000,000,1

these clouds collapsed, tremendous pressures developed, and the hydrogen became very dense. As more atoms fell faster and faster into these dense masses, very high temperatures were produced. (Remember the relation between temperature and motion of particles, Sec. 6.2.) When a mass of 10^{33} g hydrogen collects, the density rises to 100 g/cm³ and the temperature to 10 million K (degrees Kelvin). When this happens, a star is born.

At 10 million K the electrons are stripped from the nuclei. When we bring hydrogen nuclei (protons) together with kinetic energies corresponding to this temperature in our manmade atomic accelerators, we find that they unite to form nuclei of the element helium and release a tremendous amount of energy (as shown in the explosion of a hydrogen bomb).

The formation of the stable isotope of helium takes place in steps:

$$\text{}^1_1\text{H} + \text{}^1_1\text{H} \longrightarrow \text{}^2_1\text{H} + \text{}^0_{+1}e + \nu + \text{energy} \tag{22.1}$$

$$\text{}^2_1\text{H} + \text{}^1_1\text{H} \longrightarrow \quad \text{}^3_2\text{He} + \text{energy} \tag{22.2}$$

$$\text{}^3_2\text{He} + \text{}^3_2\text{He} \longrightarrow \text{}^4_2\text{He} + 2\,\text{}^1_1\text{H} + \text{energy} \tag{22.3}$$

In these equations

positron $\text{}^0_{+1}e$ symbolizes a *positron*—a particle with the mass of an electron but with unit positive charge

neutrino ν symbolizes a *neutrino*—a particle with practically zero mass and no electric charge

In each of these reactions, the total mass of the products is less than that of the reactants, and the difference is released as energy in accordance with Einstein's equation: $E = mc^2$ (Sec. 9.5). The positrons produced in Eq. (22.1) unite with electrons, and the two particles disappear with the liberation of energy equivalent to the loss in mass; this "annihilation reaction" may be symbolized as

annihilation energy $$\text{}^0_{+1}e + \text{}^0_{-1}e \longrightarrow \textit{annihilation energy} \tag{22.4}$$

Figure 22.2 shows the process of hydrogen fusion.

Our sun is using up hydrogen by nuclear fusion at the rate of 800,000,000 tons/sec. But we need not worry about running out of solar energy, since the sun has enough hydrogen to continue fusing it for about 5 billion years!

As a star uses up its hydrogen to produce energy, it continues to collapse, and the temperature of the interior rises slowly to 100,000,000 K and the density increases to 100,000 g/cm³. At this temperature, helium fusion begins.

22.4 Synthesis of Elements by "Helium Burning"—Helium Fusion

When two helium nuclei are smashed together in an atomic accelerator, they combine to form a beryllium nucleus. This is very short-lived and tends to

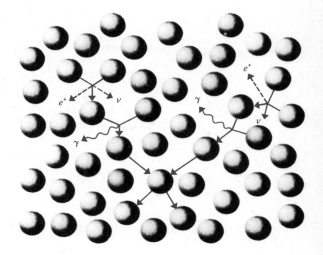

FIGURE 22.2 Hydrogen fusion

split immediately into two He nuclei. We believe that an equilibrium is established in a star between He and Be nuclei. In the presence of plenty of He nuclei, the short-lived Be nucleus can combine with another He to form an "excited" carbon nucleus (a nucleus with abnormally high energy). This radioactive nucleus can emit energy in the form of a gamma ray and become a stable C nucleus. The equations for these nuclear reactions are

$$\begin{array}{ll} {}^{4}_{2}\text{He} + {}^{4}_{2}\text{He} \rightleftarrows & {}^{8}_{4}\text{Be} & (22.5) \\ {}^{8}_{4}\text{Be} + {}^{4}_{2}\text{He} \longrightarrow & {}^{12}_{6}\text{C (excited)} & (22.6) \\ {}^{12}_{6}\text{C (excited)} \longrightarrow & {}^{12}_{6}\text{C (normal)} + \text{energy} & (22.7) \end{array}$$

When three He nuclei are put together to form one C nucleus, there is a loss of mass that is converted to energy. The synthesis of C nuclei by helium fusion thus increases the internal temperature of the star. Figure 22.3 shows this process.

The helium fusion process represented by Eqs. (22.5) to (22.7) skips over the production of lithium, Li, beryllium, Be, and boron, B. We believe that the fusion process is the reason for the very low abundances of these elements noted in Fig. 22.1. The probable source of these elements is discussed in Sec. 22.7.

Helium fusion synthesizes oxygen, neon, and magnesium as follows:

$$\begin{array}{ll} {}^{12}_{6}\text{C} + {}^{4}_{2}\text{He} \longrightarrow {}^{16}_{8}\text{O} + \text{energy} & (22.8) \\ {}^{16}_{8}\text{O} + {}^{4}_{2}\text{He} \longrightarrow {}^{20}_{10}\text{Ne} + \text{energy} & (22.9) \\ {}^{20}_{10}\text{Ne} + {}^{4}_{2}\text{He} \longrightarrow {}^{24}_{12}\text{Mg} + \text{energy} & (22.10) \end{array}$$

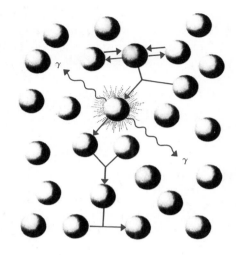

FIGURE 22.3 Helium fusion

22.5 Synthesis of Elements from Nuclei Heavier Than Helium

As a star uses up its helium to produce energy, it collapses further and the temperature rises further. At 600,000,000 K,

$$^{12}_{6}C + ^{12}_{6}C \longrightarrow ^{23}_{11}Na + ^{1}_{1}H + \text{energy} \qquad (22.11)$$

$$^{12}_{6}C + ^{12}_{6}C \longrightarrow ^{24}_{12}Mg + \text{energy} \qquad (22.12)$$

At 800,000,000 K,

$$^{20}_{10}Ne + ^{20}_{10}Ne \longrightarrow ^{24}_{12}Mg + ^{16}_{8}O + \text{energy} \qquad (22.13)$$

At 1.5 billion K, collisions of the atomic nuclei of oxygen and magnesium produce the elements Al, Si, S, P, Cl, Ar, K, and Ca. In each reaction, the mass of the products is less than the mass of the reactants, because energy has been produced. At 2 billion K, various collisions of the nuclei of these elements (Al to Ca) form the nuclei of the "iron group" of elements. Again, in each reaction, mass is converted into energy. By this time the density of the core of the star has risen to 3 million g/cm³!

In all the preceding reactions, the total mass of the product nuclei is *less* than the total mass of the reactant nuclei, since energy has been *liberated*. The processes thus sustain themselves by the continuous production of heat. That is to say, the kinetic energy of the atomic nuclei continues to increase until they are smashing into one another with sufficient speeds to fuse into new nuclei. During the entire sequence of processes, huge amounts of energy are being radiated from the star, but this loss is not sufficient to keep the temperature from rising.

At this point, however, the synthesis of new elements stops. When nuclei unite to form elements heavier than the iron group, energy is *absorbed* by the nuclear reactions taking place. The total mass of the products of these reactions is *greater* than the total mass of the reactants. Because the iron group of elements is at the "end of the line" of the series of processes by which atomic nuclei are synthesized and mass is converted into energy, these elements are much more abundant than those immediately preceding or following the group. The theory of nuclear synthesis is thus in accord with the observed abundances of the elements considered so far.

22.6   The End of a First-generation Star

By the time a star reaches the stage of synthesizing the iron group, it has a layered structure like that suggested in Fig. 22.4, though the boundaries between the layers are more diffuse than those indicated in the drawing. As the iron group forms, the release of energy slows down, and the core contracts very rapidly. When the outer layers of the star collapse into the core, huge amounts of nuclear fuel (the nuclei of the lighter elements) are suddenly made available at very high temperatures, and the star explodes, spewing out into interstellar space the elements it has formed. This spectacular explosion is called a *supernova*, a star that suddenly flares into extraordinary brilliance (as much as 100 million times as bright as our sun) and then quickly dies down to a much less luminous and more dispersed body. A supernova has

supernova

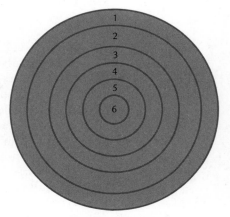

FIGURE 22.4   A layered star

1 Hydrogen fusion; 5 million K
2 Helium fusion; 100 million K
3 Synthesis of C and Ne; 500 million K
4 Synthesis of O, Mg, Na; 800 million K
5 Synthesis of Al to Ca; 1.5 billion K
6 Iron group; 2 to 3 billion K

been observed about once every 400 years during the period when man has been observing and recording special events in the heavens. The Crab Nebula of today, a faintly luminous cloud, is seen in the location where Chinese astronomers recorded the appearance of a supernova in 1054.

22.7   Synthesis of Elements in a Second-generation Star

With the passage of eons of time, gravitational attraction collects the matter thrown out into interstellar space from a first-generation star into a new cosmic cloud. As this collapses, the temperature again rises to millions of degrees. A new kind of hydrogen and helium fusion now occurs. Carbon nuclei combine with hydrogen to form nitrogen nuclei. These in turn combine with helium to form fluorine nuclei.

Collision of helium nuclei with nuclei of light elements liberates large numbers of neutrons along with a great deal of energy. These neutrons are captured by the nuclei of elements in the iron group to build all the heavier elements up through bismuth. Still heavier elements are also formed, but their nuclei are so unstable that they are radioactive and do not accumulate in significant quantities unless they have very long half-lives (like uranium with a half-life of a billion years). Elements heavier than thorium readily undergo fission; this, too, reduces their abundance as soon as they are formed by neutron capture. Though technetium (with an atomic number of 43) has an atomic weight of only 99, its half-life is but 500,000 years, so it did not persist as an element on earth.

spallation

The nuclei of some heavy elements throw off particles larger than α (alpha) particles when bombarded with high-speed particles like protons, α particles, and neutrons. The production of these larger fragments from heavy nuclei is called *spallation*. Nuclear chemists believe that the relatively small amounts of the elements Li, Be, and B (which were skipped over in hydrogen and helium fusion) in the universe were probably produced by spallation from heavy nuclei.

22.8   The Formation of the Earth

The brightness of clusters of stars in our galaxy leads astronomers to believe that our galaxy is about 13 to 15 billion years old. From the ratio of the amounts of $^{235}_{92}U$ to $^{238}_{92}U$ in nature and the half-lives of these isotopes, nuclear chemists conclude that the two were synthesized in approximately equal amounts 6 to 7 billion years ago. Some nuclear physicists believe that synthesis of the elements we have on earth ended about 5 billion years ago. From the ratios of radioactive isotopes in meteorites, geochemists conclude that the solar system formed about 4.5 billion years ago. The oldest rocks on earth are of about this age, confirming the belief that it is also about the age of the earth.

Though cosmologists do not agree completely on the processes that led to the formation of the earth, one hypothesis is that a vast cloud of cold interstellar dust and gas [containing the elements and many compounds of them produced in interstellar space (Sec. 22.9)] was slowly compressed by starlight. As the mass was compressed, gravitational attraction hastened its contraction, and one large mass became the sun. Around this mass turbulent eddies of the remaining gas and dust swirled and contracted until they became the sun's planets and the moons around the planets. Recent studies of the rocks from our moon indicate that it is about the same age as the earth, thus strengthening this hypothesis.

When the gravitational attraction of the earth had consolidated most of the matter that was originally present in the eddy from which it grew, the mass of the earth became essentially constant. Though meteors continue to be attracted from space to the earth, the mass they add is negligible compared to the mass of the earth. At some time in this evolution, heat from radioactivity, and the shrinkage of the gaseous and dusty matter into the solid earth raised the temperature enough to melt the entire earth. Iron and nickel trickled down through the molten mass, and the less dense molten silicates and other compounds, which are the chief components of rocks that have once been melted, floated as an outside mantle on the core. When the system stopped collapsing, the only process producing heat was radioactivity. This was not enough to maintain the high temperature of the earth, and it cooled off by radiating energy into space. While the earth was consolidating out of one of the eddies of dust and gas around the large central mass of the system, this mass was contracting to form the fairly typical star which is our sun.

As the earth cooled, its temperature came into approximate equilibrium; the rate at which it gained heat by radiation from the sun and by internal radioactivity became about equal to the rate at which it lost heat by radiation into space. Today the earth consists of a molten iron-nickel core with a diameter of about 4400 mi, surrounded by a mantle of cooled, solidified rock about 1800 mi thick. On top of the mantle is the crust, which is but a few miles thick.

The crust of the primitive earth was ravaged by crashing electric storms, howling winds, and shattering deluges of rain, which gradually eroded the rock into soil. This was washed into the seas and settled to the bottom, eventually being compacted into sedimentary rock. The ocean floor subsided from the weight of material accumulated on it. Eventually parts of the crust were thrust upward by this subsidence, forming mountains. The collision of great plates of rock beneath the oceans and the continents also thrust large masses of land up into mountains. During the first 1 to 2 billion years after the earth had stopped growing in mass and had reached approximate temperature equilibrium, its atmosphere contained no elemental oxygen, but was largely composed of CO_2, H_2O, N_2, CH_4, and NH_3. There was no life on earth.

22.9 The Origin of Life

Life as we know it today depends upon preexistent life. Cells grow from cells, enzymes come from enzymes. How did the first living cells arise, then, when there were no living cells to make them? How did the ordered sequences of amino acids in the proteins characteristic of living tissues arise when no such tissues existed for patterns? How did the first enzymes (biological catalysts) arise when there were no enzymes to catalyze the formation of these first enzymes? Only within the last 20 years have answers to these fundamental questions been coming from scientific laboratories. Harold Urey, who won the Nobel Prize for his discovery of the isotope of hydrogen with a mass number of 2 (deuterium), became interested in the origin of life. He wondered how amino acids—the most basic stuff of life—could have been formed. Urey postulated that the atmosphere of the earth before life began consisted of a mixture of NH_3, H_2, CH_4, and H_2O, and that the first amino acids were formed by discharges of lightning through this mixture. In 1953 Stanley Miller, a graduate student working for his Ph.D. degree under Urey, prepared a mixture of these gases and passed an electric discharge through it for many hours. Amino acids were formed! Since amino acids are the first step toward the proteins which have unique functions as structures, catalysts, and membranes in the living cell, Miller's results aroused great interest among biochemists. In 1960 the National Aeronautics and Space Administration began liberally subsidizing research on the origin of life, and in the following decade scientists gathered data which helped them devise hypotheses about the origin of life.

Here are the bare bones of a well-supported hypothesis of how life began on our planet. Many simple compounds of carbon, hydrogen, oxygen, nitrogen, and sulfur (and fragments of such compounds) have been detected in interstellar space, in the stars called "red giants," in the atmosphere of the sun, in the atmospheres of other planets in our solar system, and in comets. Some of these are as follows: CH, CH_2, CH_4, C_2H, C_2H_2, CO, CO_2, CN, NH, NH_3, NO, N_2O, H_2O, OH, HS, H_2S, CS, COS, $HCHO$, $HCOOH$, CH_3OH, $HCONH_2$, CH_3CN, and HC_2CN.

With this knowledge in mind, researchers have simulated different kinds of "primitive" atmospheres (mixtures of gases that might have been present in the earth's atmosphere in the early stages of its evolution) and have exposed them to various kinds of energy. Atmospheres containing CH_4, CO, CO_2, NH_3, N_2, H_2S, and H_2O have been exposed to ultraviolet light, electric discharges, radiation from radioactive substances, and heat as simulations of solar radiation, lightning in the atmosphere, radioactivity in earth's rocks, and solar or volcanic heat. As a result of these extensive studies, many amino acids have been synthesized. If hydrogen cyanide, HCN, is present in these atmospheres, purines and pyrimidines (organic compounds whose molecules contain nitrogen and carbon atoms joined in rings) are also formed. The purines and pyrimidines are bases that are incorporated in the genetic units of heredity, deoxyribonucleic acid (DNA) and ribonucleic acid (RNA). They are of key importance in the synthesis of proteins from amino acids in the living cell.

The production of amino acids by energizing primitive atmospheres is accompanied by the formation of simple sugars (Sec. 20.11) and fatty acids (Sec. 20.14). These can be transformed into more complex *macromolecules* (molecules with very large molecular weights) by further exposure to energy in the absence of water. These chemical changes are summarized in Fig. 22.5.

macromolecules

If mixtures of dry solid amino acids are gently heated, proteinlike substances with molecular weights of many thousands are produced; these are called *proteinoids*. Some proteinoids show enzymelike activities. When put into water, proteinoids tend to form physical structures like those observed in living cells. Figure 22.6 shows how the very tiny spheres of proteinoids in water look under a microscope. Some microspheres of proteinoids show fine structures which are so similar to some bacteria that experts are hard put to tell them apart. These microspheres also propagate by budding and fission as do living cells. And they grow and move about when placed in nutrient media. It seems likely that such proteinoids were produced on the surface of the primitive earth and evolved into the simplest kinds of living cells.

proteinoid

FIGURE 22.5 Synthesis of complex organic molecules from simple gases, using various sources of energy

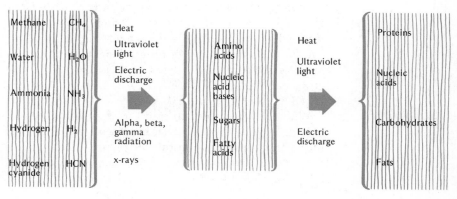

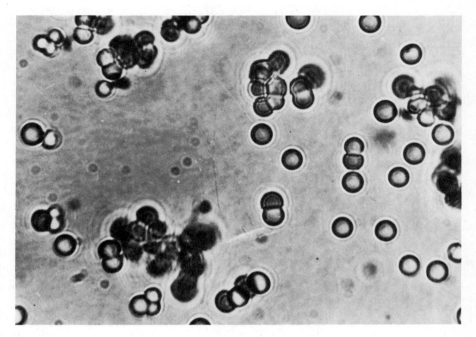

FIGURE 22.6 Photomicrographs of proteinoid microspheres, showing compartmentalization like that in living cells

22.10 The Evolution of Life

Since free (elemental) oxygen quickly destroys the organic compounds produced from the interaction of gases postulated as present in the primitive atmosphere, there must not have been any oxygen or ozone in the primitive atmosphere. The cells which were able to grow and reproduce in the primordial soup of proteinoids may be thought of as the first living organisms. These cells must have used foods from the soup to produce energy for their life processes without using oxygen for respiration as do modern plants and animals. Even today some microorganisms can live without oxygen (Sec. 4.4). Yeasts can ferment glucose to ethanol in the absence of oxygen with the production of CO_2 and the energy they need for living:

$$C_6H_{12}O_6 \longrightarrow 2CO_2 + 2C_2H_5OH + \text{energy}$$

Bacteria ferment the sugar lactose to lactic acid, $CH_3CHOHCOOH$. Alcohol, lactic acid, and CO_2 are waste products from the metabolism of these organisms and must be removed from the environment or the organisms die off from absorbing their own wastes. Evidently some processes took place in the primitive soup which used up such waste products. It is probable that much of the CO_2 introduced into the atmosphere was then dissolved in the oceans and some of it precipitated as $CaCO_3(s)$. Today many anaerobic organisms live in the soil and quickly decompose buried waste products from the metabolism of plants and animals.

Eventually, in the evolution of life, some organisms developed the ability to use the visible and infrared radiation in sunlight to remove CO_2 from the atmosphere by photosynthesis (Sec. 15.2):

$$6CO_2 + 6H_2O \xrightarrow{\text{sunlight}} C_6H_{12}O_6 + 6O_2$$

This process produces only O_2 as a waste product, which easily escapes into the atmosphere. As the concentration of oxygen in the atmosphere built up over millions of years, a layer of ozone, O_3, developed about 20 mi above the earth's surface. This layer absorbs the ultraviolet light in sunshine and protects living organisms from destruction by this lethal radiation. While the ozone layer was developing, all living creatures must have been protected from ultraviolet light by living underwater or in the soil. Once the oxygen concentration in the atmosphere had risen to its present value of about 20 percent, organisms could live aboveground or in surface waters. Some water molecules are decomposed into H_2 and O_2 molecules in the upper atmosphere, where the sun's radiation is very intense. The H_2 probably escaped from the earth, and the heavier O_2 was held in the atmosphere.

Once primitive organisms were established on the earth and in the oceans, the processes of biological evolution led to the proliferation of life with which we are familiar today. The developments sketched in barest outline in the two previous sections are summarized in Table 22.2.

TABLE 22.2 *Chronology of Planet Earth and Its Occupants*

Millions of years ago	Event
4500	Earth solidified
3000	Amino acids existed
2700	Simple organisms like algae first developed
600	Algae, fungi, and soft-bodied marine organisms flourished
500	Clams, starfish, and corals developed
400	First primitive land animals and plants evolved
200	Dinosaurs flourished
100	Small primitive mammals and complex trees developed
1	Abundant, large mammals present; ancestors of man evolved

From this chronology we see that once life was established, it grew in complexity as time went on. About 250,000 molecules are organized into a single-celled organism. One-celled creatures clump together to form many-membered colonies. Special groups within colonies form organs to carry on special processes. A central nervous system develops to control the interaction of organs. The development of circulating warm blood at constant temperature reduces the dependence of organisms on temperature changes in the environment. The rate of reproduction falls off markedly; fish and frogs spawn thousands of offspring per generation, higher mammals and man but a few. Higher animals take care of their offspring for longer portions of their lifetimes. As Julian Huxley, a philosopher-biologist, has pointed out, every one of these additional complexities confers upon the individual or the race increasing power of control over the environment, increasing internal harmony and self-regulative capacity, increasing independence of the outer world, increasing intensity and harmony of mental life.

22.11 Man's Increasing Exploitation of the Environment

We see the same drive toward complexity when we take an overview of human development. About 100,000 years ago our ancestors had evolved into creatures that we dignify with the name "Homo sapiens" (wise man). They lived in groups for protection and to hunt for food. Out of these primitive beginnings developed nomadic tribes that took their food where they found it but made little effort to cultivate it. Then came the herdsmen and farmers who assisted in nature's production of food. With agriculture came life in fixed villages and the increasing complexity of regulations needed when many people live together. As handicrafts and agriculture flourished, trade encouraged the growth of cities and then of nations; always the drive was from simpler to more complex ways of living. With the coming of the industrial age all the peoples of the world have become knitted together into one fantastically complex web—and no man, nor any group of men, is an island.

22.12 The First and Second Laws of Thermodynamics

Thermodynamics is the study of the changes in energy that accompany any kind of transformation of matter. Phenomena as diverse as the falling of a stone, the turning of a wheel, the impingement of light, the burning of fuels, and the growth and reactions of the complex mechanisms of a living cell involve the absorption, emission, redistribution, and transformation of energy, which may be present as heat, light, electricity, mechanical energy of motion, or the chemical energy within all substances. Several laws of nature have emerged from studies of thermodynamics. The first law of thermodynamics states: During any process, energy is conserved; energy is neither created nor destroyed (Sec. 9.4). The second law states: During any process, any system tends to become less ordered (less organized, more chaotic, more mixed up).

22.13 Some Speculation about the Second Law and Increasing Complexity

Is this second law of thermodynamics a contradiction of the increasing complexity (greater organization) which we have observed on earth and in the world of life? There is no contradiction. For the second law deals with the universe as a whole, and we have been concerned with only tiny parts of this universe. The organization of protons, electrons, and neutrons into atoms, of atoms into molecules, of molecules into simple organisms, of simple organisms into complex ones—all these increases in order and organization have been accompanied by increasing disorder and disorganization in those parts of the universe that have supplied us with the energy necessary to carry on organizational processes. Disorder in the universe as a whole has increased with time. When one studies a living system and its nonliving surroundings, one finds that the increase in disorder of the surroundings is always greater than the decrease in disorder in the living portion.

Summary

In the atomic world, protons are fundamental units that can overcome their natural repulsions and form complex atomic nuclei by sharing neutrons. In the molecular world, atoms are the fundamental units that can be bound together to form complex molecules by sharing electrons. In the living world, cells are the fundamental units that are bound together by the sharing of physiological functions. In the human world, persons are the fundamental units that can be bound together by sharing social functions. About 5 billion atoms are organized into one cell, and about 5 billion cells in one man. One is tempted to ask, When we have 5 billion people on earth, will we have achieved "one world"?

When we look at the long sweep of time as we know it, and at what has happened in this span of time, we sense the emergence of a great long-term trend:

As time passes, things tend to increase in complexity

or

Time's arrow points from the simple to the complex.

Life has learned to cope with this increase in complexity and to prosper from it. Faith that the human race, armed with modern science and technology, can learn to establish a dynamic equilibrium with our environment gives us courage to attack the problems of our one world and optimism to support our belief that we can solve them.

New Terms and Concepts

ANNIHILATION ENERGY: The energy released when an electron and a positron encounter one another and disappear; the energy is equivalent to the sum of their masses.

MACROMOLECULES: Molecules of very large molecular weight, usually composed of more than 100 repeated units or building blocks, e.g., proteins and polysaccharides.

NEUTRINO: A subatomic particle of matter having no charge and a mass much less than that of an electron.

POSITRON: A subatomic particle of matter having unit positive charge and a mass equal to that of an electron.

PROTEINOID: A polypeptide having a molecular weight of several thousand.

SPALLATION: A nuclear reaction in which many particles are ejected from an atomic nucleus when it is struck by a particle of sufficiently high energy.

SUPERNOVA: An extremely bright, short-lived celestial object that arises from the explosion of most of the material in a star with the emission of vast amounts of energy.

Testing Yourself

22.1   How do cosmologists account for the fact that hydrogen and helium are the most abundant elements in the universe today?

22.2   What is meant by "hydrogen burning" in a star like our sun?

22.3   What elements are formed by "helium burning"?

22.4   Why do nuclear fusion reactions not take place with elements having atomic weights greater than those of the "iron group"?

22.5   How were the atomic nuclei of atoms heavier than those of the iron group synthesized?

22.6   What are some of the differences between first- and second-generation stars?

22.7   How did the chemicals basic to life probably originate?

22.8   What leads us to believe that the chemical precursors of life originated when there was no oxygen in the atmosphere?

22.9   What is the probable source of the oxygen in the atmosphere today?

22.10 In what ways has man's impact on his environment increased during the development of civilization?

APPENDICES

APPENDIX A   Atomic Weights of the Elements

Element	Symbol	Atomic number	Atomic weight*	Element	Symbol	Atomic number	Atomic weight*
Actinium	Ac	89	(227)	Mendelevium	Md	101	(256)
Aluminum	Al	13	26.98	Mercury	Hg	80	200.59
Americium	Am	95	(243)	Molybdenum	Mo	42	95.94
Antimony	Sb	51	121.75	Neodymium	Nd	60	144.24
Argon	Ar	18	39.948	Neon	Ne	10	20.183
Arsenic	As	33	74.92	Neptunium	Np	93	(237)
Astatine	At	85	(210)	Nickel	Ni	28	58.71
Barium	Ba	56	137.34	Niobium	Nb	41	92.91
Berkelium	Bk	97	(249)	Nitrogen	N	7	14.007
Beryllium	Be	4	9.012	Nobelium	No	102	(253)
Bismuth	Bi	83	208.98	Osmium	Os	76	190.2
Boron	B	5	10.81	Oxygen	O	8	15.9994
Bromine	Br	35	79.909	Palladium	Pd	46	106.4
Cadmium	Cd	48	112.40	Phosphorus	P	15	30.974
Calcium	Ca	20	40.08	Platinum	Pt	78	195.09
Californium	Cf	98	(251)	Plutonium	Pu	94	(242)
Carbon	C	6	12.011	Polonium	Po	84	(210)
Cerium	Ce	58	104.12	Potassium	K	19	39.102
Cesium	Cs	55	132.91	Praseodymium	Pr	59	140.91
Chlorine	Cl	17	35.453	Promethium	Pm	61	(147)
Chromium	Cr	24	52.00	Protactinium	Pa	91	(231)
Cobalt	Co	27	58.93	Radium	Ra	88	(226)
Copper	Cu	29	63.54	Radon	Rn	86	(222)
Curium	Cm	96	(247)	Rhenium	Re	75	186.23
Dysprosium	Dy	66	162.50	Rhodium	Rh	45	102.91
Einsteinium	Es	99	(254)	Rubidium	Rb	37	85.47
Erbium	Er	68	167.26	Ruthenium	Ru	44	101.1
Europium	Eu	63	151.96	Rutherfordium	Rf	104	(259)
Fermium	Fm	100	(253)	Samarium	Sm	62	150.35
Fluorine	F	9	19.00	Scandium	Sc	21	44.96
Francium	Fr	87	(223)	Selenium	Se	34	78.96
Gadolinium	Gd	64	157.25	Silicon	Si	14	28.09
Gallium	Ga	31	69.72	Silver	Ag	47	107.870
Germanium	Ge	32	72.59	Sodium	Na	11	22.9898
Gold	Au	79	196.97	Strontium	Sr	38	87.62
Hafnium	Hf	72	178.49	Sulfur	S	16	32.064
Hahnium	Ha	105	(260)	Tantalum	Ta	73	180.95
Helium	He	2	4.003	Technetium	Tc	43	(99)
Holmium	Ho	67	164.93	Tellurium	Te	52	127.60
Hydrogen	H	1	1.0080	Terbium	Tb	65	158.92
Indium	In	49	114.82	Thallium	Ti	81	204.37
Iodine	I	53	126.90	Thorium	Th	90	232.04
Iridium	Ir	77	192.2	Thulium	Tm	69	168.93
Iron	Fe	26	55.85	Tin	Sn	50	118.69
Krypton	Kr	36	83.80	Titanium	Ti	22	47.90
Kurchatovium	Ku	104	(259)	Tungsten	W	74	183.85
Lanthanum	La	57	138.91	Uranium	U	92	238.03
Lawrencium	Lw	103	(257)	Vanadium	V	23	50.94
Lead	Pb	82	207.19	Xenon	Xe	54	131.30
Lithium	Li	3	6.939	Ytterbium	Yb	70	173.04
Lutetium	Lu	71	174.97	Yttrium	Y	39	88.91
Magnesium	Mg	12	24.312	Zinc	Zn	30	65.37
Manganese	Mn	25	54.94	Zirconium	Zr	40	91.22

*Values in parentheses represent the most stable known isotope.

APPENDIX B Exponential Numbers and Their Use

Very large or very small numbers are conveniently expressed as powers of 10.

1 billion	1,000,000,000	$1 \times 10 \times 10 \times 10 \times 10 \times 10 \times 10 \times 10 \times 10 \times 10$	1×10^9
1 million	1,000,000	$1 \times 10 \times 10 \times 10 \times 10 \times 10 \times 10$	1×10^6
1 thousand	1000	$1 \times 10 \times 10 \times 10$	1×10^3
1 hundred	100	$1 \times 10 \times 10$	1×10^2
1 ten	10	1×10	1×10^1
1 unit	1	1×1	1×10^0
1 tenth	0.1	$1 \times \dfrac{1}{10}$	1×10^{-1}
1 hundredth	0.01	$1 \times \dfrac{1}{10 \times 10}$	1×10^{-2}
1 thousandth	0.001	$1 \times \dfrac{1}{10 \times 10 \times 10}$	1×10^{-3}
1 millionth	0.000,001	$1 \times \dfrac{1}{10 \times 10 \times 10 \times 10 \times 10 \times 10}$	1×10^{-6}
1 billionth	0.000,000,001	$1 \times \dfrac{1}{10 \times 10 \times 10 \times 10 \times 10 \times 10 \times 10 \times 10 \times 10}$	1×10^{-9}

From the above we see that any positive power of ten, 10^n, means 1 multiplied by 10 n times.

Example: $10^4 = 1 \times 10 \times 10 \times 10 \times 10 = 1 \times 10,000$

Thus, 10^4 is easily written as 1 followed by four zeros; 10^n is easily written as 1 followed by n zeros.

Avogadro's number $= 6.02 \times 10^{23}$ or $6.02 \times 100,000,000,000,000,000,000,000$
$= 602,000,000,000,000,000,000,000$

We see also that any negative power of ten, 10^{-n}, means 1 multiplied by $\frac{1}{10}$ n times.

Example: $10^{-4} = 1 \times \dfrac{1}{10 \times 10 \times 10 \times 10} = 1 \times 0.000,1$

Thus, 10^{-4} is easily written as 1 in the fourth decimal place; 10^{-n} is easily written as 1 in the nth decimal place.

The atomic mass unit, amu, is 1.66×10^{-24} g
or $1.66 \times 0.000,000,000,000,000,000,000,001$ g
or $0.000,000,000,000,000,000,000,001,66$ g

APPENDIX C   Comparison of Temperatures on Three Scales

	Fahrenheit	Celsius	Kelvin
Lead metal melts	621	327	600
Temperature for baking bread	400	204	477
Water boils	212	100	373
Normal body temperature	98.6	37	310
Comfortable room temperature	70	21	294
Water freezes	32	0	273
Bitter cold weather, mercury freezes	−40	−40	233
Carbon dioxide (Dry Ice) sublimes	−108	−78	195
Liquid nitrogen boils	−321	−196	77
Absolute zero	−460	−273	0

For converting Fahrenheit to Celsius: $°C = \dfrac{5}{9}(°F - 32)$

For converting Celsius to Fahrenheit: $°F = \dfrac{9}{5}(°C) + 32$

For converting Celsius to Kelvin: $K = °C + 273$

APPENDIX D SI Units

The International System of Units—Le Système International d'Unités

The International Bureau of Weights and Measures, now supported by 40 nations, has established an International System (Système International, SI) of basic units for measuring mass, length, and time and derived units for measuring pressure, amount of heat, electric charge, energy, etc. The names of these units may be modified by appropriate prefixes to give units which are convenient for measuring quantities much larger or much smaller than the defined unit. Some of the prefixes for SI units and their abbreviations are:

micro	μ	one-millionth, 0.000,001, or $\frac{1}{1.000.000}$
milli	m	one-thousandth, 0.001, or $\frac{1}{1000}$
centi	c	one-hundredth, 0.01, or $\frac{1}{100}$
deci	d	one-tenth, 0.1, or $\frac{1}{10}$
deka	dk	ten, 10
hecto	h	one hundred, 100
kilo	k	one thousand, 1000
mega	M	one million, 1,000,000

The abbreviation for the meter (often spelled *metre* in British English) is "m." A micrometer (μm) is one-millionth of a meter, a centimeter (cm) is one-hundredth of a meter, a kilometer (km) is 1000 meters, etc. Some common British units of measurement and their SI equivalents are:

	British Unit	*SI Equivalent*
Length	1 inch (in)	2.54 centimeters (cm)
	1 yard (yd)	0.914 meter (m)
	1 mile (mi)	1.609 kilometers (km)
Area	1 square yard (yd^2)	0.8361 square meter (m^2)
	1 acre	0.4047 square hectometer (hm^2)
	1 square mile (mi^2)	2.59 square kilometers (km^2)
Volume	1 ounce (oz)	29.6 milliliters (ml)
		(1 ml = 1 cm^3)
	1 quart (qt)	0.946 liter (l)
		(1 liter = 1000 cm^3)
	1 gallon (gal)	3.79 liters (l)
Mass (weight)	1 ounce (oz)	28.35 grams (g)
		(1 g = 0.001 kg)
	1 pound (lb)	453.6 grams (g)
	1 ton	907.2 kilograms (kg)

APPENDIX E Substances and Their Formulas

Acetaldehyde, CH_3CHO
Acetic acid, CH_3COOH
Acetone, CH_3COCH_3
Acetylene, C_2H_2
Alanine, $C_2H_4(NH_2)COOH$
Aluminum hydroxide, $Al(OH)_3$
Aluminum oxide, Al_2O_3
Aluminum sulfate, $Al_2(SO_4)_3$
Aminoacetic acid, $CH_2(NH_2)COOH$
Aminomethane, CH_3NH_2
Aminopropanoic acid,
  $C_2H_4(NH_2)COOH$
Ammonia, NH_3
Ammonium chloride, NH_4Cl
Ammonium nitrate, NH_4NO_3
Ammonium sulfate, $(NH_4)_2SO_4$
Arsenic acid, $HAsO_3$
Arsine, AsH_3

Barium bromide, $BaBr_2$
Barium hydroxide, $Ba(OH)_2$
Barium oxide, BaO
Barium sulfate, $BaSO_4$
Benzene, C_6H_6
Beryllium chloride, $BeCl_2$
Beryllium oxide, BeO
Blood sugar, $C_6H_{12}O_6$
Butane, C_4H_{10}
Butanoic acid, C_3H_7COOH
Butanol, C_4H_9OH
Butyric acid, C_3H_7COOH

Calcium carbonate, $CaCO_3$
Calcium fluoride, CaF_2
Calcium hydroxide, $Ca(OH)_2$
Calcium oxide, CaO
Calcium phosphate, $Ca_3(PO_4)_2$
Calcium silicate, $CaSiO_3$
Calcium stearate, $Ca(C_{17}H_{35}COO)_2$
Calcium sulfate, $CaSO_4$
Calcium sulfite, $CaSO_3$
Camphor, $C_{10}H_{16}O$
Carbon dioxide, CO_2
Carbon disulfide, CS_2

Carbon monoxide, CO
Carbon tetrabromide, CBr_4
Carbon tetrachloride, CCl_4
Carbonic acid, H_2CO_3
Cesium fluoride, CsF
Cholesterol, $C_{27}H_{45}OH$
Cinnabar, HgS
Citric acid, $C_3H_4(OH)(COOH)_3$
Cobalt oxide, CoO
Copper sulfate, $CuSO_4$
Corn sugar, $C_6H_{12}O_6$

Decane, $C_{10}H_{22}$
Dextrose, $C_6H_{12}O_6$
Dimethyl mercury, $(CH_3)_2Hg$
Disodium monohydrogen phosphate,
  Na_2HPO_4

Ethane, C_2H_6
Ethanol, C_2H_5OH
Ethyl alcohol, C_2H_5OH
Ethyl butyrate, $C_3H_7COOC_2H_5$
Ethyl formate, $HCOOC_2H_5$
Ethyl iodide, C_2H_5I
Ethylene, C_2H_4
Ethylene chloride, $C_2H_4Cl_2$
Ethylene dibromide, $C_2H_4Br_2$
Ethylene dichloride, $C_2H_4Cl_2$
Ethylene glycol, $C_2H_4(OH)_2$

Formaldehyde, $HCHO$
Formic acid, $HCOOH$
Fructose, $C_6H_{12}O_6$
Fruit sugar, $C_6H_{12}O_6$

Galactose, $C_6H_{12}O_6$
Glucose, $C_6H_{12}O_6$
Glycerine, $C_3H_5(OH)_3$
Glycerol, $C_3H_5(OH)_3$
Glyceryl linoleate,
  $(C_{17}H_{31}COO)_3C_3H_5$
Glyceryl linolenate,
  $(C_{17}H_{29}COO)_3C_3H_5$
Glyceryl oleate, $(C_{17}H_{33}COO)_3C_3H_5$

Glyceryl palmitate,
$(C_{15}H_{31}COO)_3C_3H_5$
Glyceryl palmitoleate,
$(C_{15}H_{27}COO)_3C_3H_5$
Glyceryl stearate,
$(C_{17}H_{35}COO)_3C_3H_5$
Glycine, $CH_2(NH_2)COOH$
Glycol, $C_2H_4(OH)_2$
Glycylalanine,
$H_2NCHOCNHCHCH_3COOH$
Grape sugar, $C_6H_{12}O_6$

Heptane, C_7H_{16}
Hexane, C_8H_{18}
Hydrobromic acid, HBr
Hydrochloric acid, HCl
Hydrogen bromide, HBr
Hydrogen chloride, HCl
Hydrogen cyanide, HCN
Hydrogen fluoride, HF
Hydrogen iodide, HI
Hydrogen peroxide H_2O_2
Hydrogen selenide, H_2Se
Hydrogen sulfide, H_2S
Hydrogen telluride, H_2Te

Iron chloride, $FeCl_3$
Iron hydroxide, $Fe(OH)_3$
Iron (ferrous) oxide, FeO
Iron (ferric) oxide, Fe_2O_3
Iron pyrite, FeS_2
Iron sulfate, $FeSO_4$
Iron sulfide, FeS
Isobutane, C_4H_{10}
Isopentyl acetate, $CH_3COOC_5H_{11}$
Isopropyl alcohol, C_3H_7OH

Lactic acid, $CH_3CHOHCOOH$
Lactose, $C_{12}H_{22}O_{11}$
Lead bromide-chloride, PbBrCl
Lead chromate, $PbCrO_4$
Lead iodide, PbI_2
Lead sulfide, PbS
Levulose, $C_6H_{12}O_6$
Linoleic acid, $C_{17}H_{31}COOH$

Linolenic acid, $C_{17}H_{29}COOH$
Lithium fluoride, LiF
Lithium hydride, LiH
Lithium hydroxide, LiOH

Magnesium chloride, $MgCl_2$
Magnesium hydroxide, $Mg(OH)_2$
Magnesium nitride, Mg_3N_2
Magnesium oxide, MgO
Magnesium sulfate, $MgSO_4$
Magnesium sulfite, $MgSO_3$
Magnesium stearate,
$Mg(C_{17}H_{35}COO)_2$
Maltose, $C_{12}H_{22}O_{11}$
Mercury oxide, HgO
Mercury sulfide, HgS
Methane, CH_4
Methanoic acid, HCOOH
Methanol, CH_3OH
Methyl acetate, CH_3COOCH_3
Methyl alcohol, CH_3OH
Methyl amine, CH_3NH_2
Methylammonium chloride,
CH_3NH_3Cl
Methyl butyrate, $C_3H_7COOCH_3$
Methyl chloride, CH_3Cl
Milk sugar, $C_{12}H_{22}O_{11}$
Monosodium dihydrogen phosphate,
NaH_2PO_4

Nickel oxide, NiO
Nitric acid, HNO_3
Nitric oxide, NO
Nitrogen dioxide, NO_2
Nitrous oxide, N_2O
Nonane, C_9H_{20}
Normal butane, C_4H_{10}
Normal pentyl acetate,
$CH_3COOC_5H_{11}$

Octane, C_8H_{18}
Octyl acetate, $CH_3COOC_8H_{17}$
Oleic acid, $C_{17}H_{33}COOH$
Oxalic acid, $(COOH)_2$
Ozone, O_3

Palmitic acid, $C_{15}H_{31}COOH$
Palmitolic acid, $C_{15}H_{27}COOH$
Pentane, C_5H_{12}
Phosphine, PH_3
Phosphoric acid, HPO_3
Potassium bromide, KBr
Potassium carbonate, K_2CO_3
Potassium nitrate, KNO_3
Potassium sulfate, K_2SO_4
Potassium sulfite, K_2SO_3
Propane, C_3H_8
Propanoic acid, C_2H_5COOH
Propanol, C_3H_7OH

Radium chloride, $RaCl_2$
Radium hydroxide, $Ra(OH)_2$
Rubidium iodide, RbI

Salt, $NaCl$
Selenium dioxide, SeO_2
Silica, SiO_2
Silicon dioxide, SiO_2
Silver chloride, $AgCl$
Sodium acetate, $NaC_2H_3O_2$
Sodium bicarbonate, $NaHCO_3$
Sodium bisulfate, $NaHSO_4$
Sodium carbonate, Na_2CO_3
Sodium chromate, Na_2CrO_4
Sodium hydride, NaH
Sodium hydrogen carbonate,
  $NaHCO_3$

Sodium hydrogen sulfate, $NaHSO_4$
Sodium hydroxide, $NaOH$
Sodium iodide, NaI
Sodium oxide, Na_2O
Sodium stearate, $NaC_{17}H_{35}COO$
Sodium sulfate, Na_2SO_4
Sodium sulfide, Na_2S
Sodium sulfite, Na_2SO_3
Starch sugar, $C_6H_{12}O_6$
Stearic acid, $C_{17}H_{35}COOH$
Stibine, SbH_3
Strontium oxide, SrO
Sucrose, $C_{12}H_{22}O_{11}$
Sugar (sucrose), $C_{12}H_{22}O_{11}$
Sulfur dioxide, SO_2
Sulfuric acid, H_2SO_4
Sulfurous acid, H_2SO_3

Tellurium dioxide, TeO_2
Tetraethyl lead, $Pb(C_2H_5)_4$
Tetramethyl lead, $Pb(CH_3)_4$
Trisodium phosphate, Na_3PO_4

Urea, NH_2CONH_2

Vanadium oxide, VO

Water, H_2O

Zinc oxide, ZnO
Zinc sulfide, ZnS

INDEX

Page numbers in italic refer to pages where definitions are given under New Terms and Concepts.

Brønsted Johannes Nicolaus (1879–
1947), 244–247
Brønsted-Lowry theory, 244–246, 247
Bunsen, Robert Wilhelm von (1811–
1899), 322
Butane, 400–404
Butanoic acid, 414
Butanol, 409
Butyric acid, 414–415

Calcium
in bones, 179, 268, 439
reactions of, 188, 200–201, 359, 362
Calcium carbonate
in biomass, 300–302, 305, 468
in treatment of stack gases, 217
Calcium fluoride, 439
Calcium hydrogen carbonate, 300–302
Calcium hydroxide, 201
in sewage treatment, 271
in treatment of mine wastes, 261–262
in wet scrubber, 217
Calcium iodide, 189
Calcium oxide
in sewage treatment, 271
in treatment of iron ores, 199–200
Calcium phosphate, 271, 439
Calcium silicate, 200
Calcium stearate, 419
Calcium sulfate, 217, 262
Calcium sulfite, 200, 217
Calorie, 79, 92, 93, 430
See also Kilocalorie
Calorimeter, 79–82, 93, 171–172,
429–431
Canal rays, 149, 159
Cannizzaro, Stanislao (1826–1910), 187,
189
Carbohydrate, 307, 412–414
calorific values of, 436
combustion of, 299, 430–432, 435, 439
in diet, 428, 437
metabolism of, 430–432
photosynthesis of, 298–299
See also Cellulose; Starch; Sugar
Carbon
cycle, 297–302
radioactive, 167–169, 180, 300
reactions of, 39–40, 122–124, 202, 218
Carbonaceous material, 219, 230

Carbonate, 298, 300
Carbon dioxide
in atmosphere, 121–123, 133–135, 167,
298–301, 468
in biosphere, 299–303
formation of, 63, 123, 197, 217–218,
220, 255
from metabolism, 430–432, 467
in photosynthesis, 293, 297–299, 302,
305, 443, 468
reactions of, 200, 444–445
from respiration, 234, 299–300, 305
in water, 198, 258, 261, 267, 282–283,
285, 301
Carbon disulfide, 23–24, 43
Carbonic acid, 198, 246, 282, 300
Carbon monoxide
in atmosphere, 121
formation of, 42–43, 46–47, 63, 218,
220
pollution from, 115–117, 124–125,
220–222
reactions of, 218
Carbon tetrabromide, 86
Carbon tetrachloride, 86, 238, 393–395
Carbon-to-carbon bond, 400–424
Carboxyl group, 414–415, 419–420, 423
Carboxyhemoglobin, 124
Carboxylic acid, 415–416, 422, 423
Carver, John Alfred, Jr. (1918–   ), 213
Catalysis, 407, 423
Catalyst, 407, 422, 423
Cathode, 143–149, 159
Cathode corpuscle, 146–148, 159
See also Electron
Cathode ray, 144–149, 159, 323–324
Cathode ray tube, 314, 323–324
Cellulose, 67–68, 72, 413–414
Celsius, Anders (1701–1744), 77
Celsius (centigrade) scale of
temperature, 84, 92, 93, 476
Centimeter, 34, 53, 477
Cesium, 179, 200, 356–359
Cesium fluoride, 188
Chadwick, Sir James C. (1891–   ),
156–157
Characteristic group, 400, 423
Charge cloud, 372, 396
See also Electron cloud
Charge of the electron, 147–149, 159

Mixture, 21–24, 29, *30*, 235
    heterogeneous, 20–21, *30*, 235
    homogeneous, 20–21, *30*, 234–235
Molal boiling point elevation, 237, 247, *248*
Molal freezing point lowering, 237–239, 247, *248*
Molality, 237, *248*
Molar volume, 107–109, *110*, 193
Mole, 50–52, *54*, 237
Molecular structure, 371–396, 400–422
Molecular weight, 50, *54*
Molecule, 28, *30*, 42–44
    diatomic, 106–107, *110*, 375, 377
    and ions, 244–248
    macromolecule, 466, *471*
    motion of, 98–110, 237–238, 278–279
    See also Kinetic-molecular theory
Monatomic ion, 356, 366, 367
Monosaccharide, 413, *424*
Monosodium dihydrogen phosphate, 245

Natural gas, 218-220, 230, 300, 405
    liquefied, 220-221, *230*
Natural law, 38, *54*
Natural science, 12–13, *15*
Negative electric charge
    on electron, 142–148
    on ions, 240
    produced by rubbing, 140–142
    and stream of water, 381–382
Neon, 194
Neptunium, 175–176
Net ionic equation, 243–244
Neutral solution, 199, *207*
Neutralization, 198–199, 206, *207*, 242–244, 419
Neutrino, 459, *471*
Neutron, *159*
    bombardment with, 170, 174–176
    discovery of, 156–158
    in nuclear reactions, 167, 170, 174–176
Neutron moderator, 175, 177, *181*
Newton, Sir Isaac (1642–1727), 315, 318–319
Nickel oxide, 220
Nitric acid, 202, 304
Nitric oxide, 115, 121, 126–129, 304. *See also* Nitrogen oxide

Nitrogen
    in air, 122, 126, 153, 191
    atom, 390–392
    cycle, 303–306, 308
    family, 202, 392
    fixed, 303–306, *308*, 309
    metabolism of, 431
    in nuclear reactions, 156, 167, 173, 463
    in polluted water, 257–259
    properties of, 202
    reaction with hydrogen, 104–107, 286–288
    reaction with oxygen, 126
    in sewage treatment, 272
    See also Denitrification; Haber process
Nitrogen dioxide, 20, 113, 115, 121–122, 126–129, 304
Nitrogen oxide (NO), 304, 465
Nitrogen oxides, 115–116, 121, 126, 135–136, 219–222, 307
Nitrous oxide, 121, 122, 306, 465
Noble gas, 206, *207*
    discovery of, 191–192, 194
    stable octet in, 358–359, 366–367
    See also Rare gas; Zero group
Nonane, 400
Nonelectrolyte, 236–239, 247, *248*, 394. *See also* Electrolyte
Nonmetal, 25, *30*, 188–189, 197–198, 201–202, 363–367. *See also* Metal
Nonmetallicity, 201–202, *207*. *See also* Metallicity
Nonvolatile solute, 237–238, 247, *248*
Normal boiling point, 101, *110*
Normal butane, 403, 404
Normal pentyl acetate, 415
Nuclear energy, 163–181
Nuclear fission, 170–175, 179–180, *181*, 463
Nuclear fusion, 170–173, *181*, 459–463
Nuclear power plant, *see* Atomic power plant
Nuclear reaction, 156, *159*, 163–181, 459–463
Nuclear reactor, 174–180, *181*, 229. *See also* Breeder reactor
Nucleus, *see* Atomic nucleus
Nutrient, 62, 73, 257–259, 297

Photo Credits

From American Museum of Natural History, chapter opening photo, chapter 1
From Nicholas Sapieha/Stock, Boston, p. 12
Courtesy of Walter C. McCrone Associates, Inc., p. 18
From Mortimer Abramowitz, p. 32
From Brown Brothers, p. 41
Courtesy of St. Regis Paper Company, Inc., p. 56
From United Press International, p. 60
From Peter Menzel/Stock, Boston, p. 66 *top left*
Courtesy of American Iron & Steel Institute, p. 66 *top right*
Courtesy of Glass Containers Manufacturers Institute, p. 71
Courtesy of Los Alamos Scientific Laboratory, p. 74
Courtesy of Taylor Instrument Company, p. 89 *top*
From Culver Pictures, Inc., p. 92
From Brown Brothers, p. 105
Courtesy of General Motors Corporation, Research Laboratories, p. 112
From Rapho Guillumette Pictures, p. 114
Courtesy of Los Angeles Air Pollution Control District, p. 127
Courtesy of Lawrence Berkeley Laboratory, p. 138
From Culver Pictures, Inc., p. 150
Courtesy of General Electric Company, Space Division, Energy Systems Programs, p. 162
From Brown Brothers, p. 164
From Fritz Goro, Time Life Picture Agency, © Time, Inc., p. 184
From Phillip G. Trager, photo previously published by *Scientific American*, p. 210
Courtesy of Pacific Gas & Electric Company, p. 223
From Dr. Lewis R. Wolberg, pp. 232, 250
Courtesy of U.S. Department of the Interior, Fish & Wildlife Service, p. 258
From Dr. Lewis R. Wolberg, p. 276
From John Urban/Stock, Boston, p. 292
Courtesy of NASA, p. 312
From Brown Brothers, p. 315
Courtesy of High Altitude Observatory, Division of National Center for Atmospheric Research, National Science Foundation, p. 330
From John Urban, p. 352
From American Museum of Natural History, p. 370
From Dr. Lewis R. Wolberg, p. 398
Courtesy of Merck, Sharp and Dohme Research Laboratories, Division of Merck & Co., Inc., p. 426
Courtesy of NASA, p. 432 *(Figure 21.3)*
Courtesy of U.S. Army Institute of Environmental Medicine, p. 433 *(Figure 21.4)*
Courtesy of University of Arizona News Service, p. 444 *(Figure 21.5)*
Courtesy of NASA, p. 454
Courtesy of Dr. Sidney W. Fox, Institute for Molecular and Cellular Evolution, University of Miami, Florida, p. 467 *(Figure 22.6)*

Illustrations drawn by B. J. and F. W. Taylor.